# Surface Modification Technologies

## AN ENGINEER'S GUIDE

*edited by*

## T. S. SUDARSHAN
Materials Modification Inc.
Falls Church, Virginia

MARCEL DEKKER, INC.      New York and Basel

Library of Congress Cataloging-in-Publication Data

Surface modification technologies: an engineer's guide / edited by
T.S. Sudarshan.
    p.   cm.
  Includes bibliographies and index.
  ISBN 0-8247-8009-4 (alk. paper)
  1. Surfaces (Technology)  2. Coatings.  I. Sudarshan, T. S.
TA418.7.S915   1989
620'.44--dc20                                              89-7761
                                                            CIP

This book is printed on acid-free paper.

MARCEL DEKKER, INC.
270 Madison Avenue, New York, New York 10016

Current printing (last digit):
10  9  8  7  6  5  4  3  2  1

PRINTED IN THE UNITED STATES OF AMERICA

# Preface

The increasing need for advanced-generation materials that have excellent friction, wear, corrosion, and optical properties has led to a rapid expansion in the field of surface modification technologies. Some of the innovative applications were previously either unexplored or thought to be infeasible.

Although some surface technologies such as plating have been successfully applied as industrial processes for a good part of this century, several concepts have been extended or expanded with the general advance in science and technology. This has led to a new generation of composite surface coatings that can almost satisfy the requirements of several demanding applications. The range and scope of modifying surfaces have also been extended significantly by developments in the fields of plasma, lasers, and electron beams, all of which are now increasingly being applied in the field of surface technologies. Several innovative applications have therefore seen the light of day in the fields of microelectronics and communication, power plants, transportation, and heavy-duty machinery used in manufacturing.

In spite of these advances in applications, a comprehensive book outlining several of these technologies for use by engineers and/or shop floor personnel has been noticeably missing. It is hoped that this book will fill such a requirement. The book has been designed for the average reader but could also serve as a useful text for a senior-level course in materials processing, focusing for a full semester on surface modification technologies. Principles associated with a number of individual processes have been outlined and the physical properties associated with the surfaces highlighted. Some new applications associated with the coating or alloying of surfaces are described, providing a view of the vast scope of use of some of these techniques.

Finally, I wish to thank all the contributors to this book for devoting their precious time to enhancing a general awareness of their fields and sharing some of their valuable experiences.

T. S. Sudarshan

# Contents

# Contributors

**Deepak G. Bhat** New Coating Development Department, GTE Valenite Corporation, Troy, Michigan

**Ruth Chatterjee-Fischer** Institut für Harterei Technik, Bremen, Federal Republic of Germany

**Sidney Dressler** SECO/WARWICK Corporation, Meadville, Pennsylvania

**Peter Frach** Technical Physics Department, Manfred von Ardenne Research Institute, Dresden, German Democratic Republic

**Ullrich Heisig** Technical Physics Department, Manfred von Ardenne Research Institute, Dresden, German Democratic Republic

**John Keem** Research and Development Department, Ovonic Synthetic Materials Company, Troy, Michigan

**Hillary Solnick-Legg** Ionic Atlantic, Inc., Atlanta, Georgia

**Keith O. Legg** Ionic Atlantic, Inc., Atlanta, Georgia

**P. A. Molian** Department of Mechanical Engineering, Iowa State University of Science and Technology, Ames, Iowa

**Indira Rajagopal** National Aeronautical Laboratory, Bangalore, India

**Siegfried Schiller** Technological Research and Development Department, Manfred von Ardenne Research Institute, Dresden, German Democratic Republic

# 1

# Composite Coatings

INDIRA RAJAGOPAL  *National Aeronautical Laboratory,*
*Bangalore, India*

## 1  INTRODUCTION

For engineering applications a material is often chosen to satisfy
the strength requirement. That material may not possess satisfactory
surface properties such as wear resistance, corrosion resistance,
abrasion resistance, electrical conductivity, and solderability. To
impart specific surface properties, coatings are used. Though there
are many methods of coatings, electroplating and electroless plating
are extensively used because they are economical, versatile with
respect to types of coatings, and easy to reproduce.

Electrodeposition of metals and alloys is well known. Recently,
composite materials have gained importance in engineering. Coating
composite material is useful in high technologies, for example, aerospace and nuclear engineering.

An ideal material having the above set of properties is very
desirable in high technology applications in aeroengines, modern
gas turbine engines, automobile industries, and so on, where the
material problems are acute. It is essential to use a suitable protective coating to minimize wear, erosion, high-temperature oxidation,
and corrosion. Electroplating can produce a combination of desired
surface properties with suitable reinforcements.

It is well known that dispersion-strengthened composite materials—
metals or alloys containing a fine dispersion of hard, inert particles—
exhibit higher mechanical properties and better microstructural
stability at higher temperatures. A composite material contains at
least two components and each component imparts its assets to the

newly synthesized material while suppressing their shortcomings. As a result, the composite material is more useful than any of its components.

Electrodeposited composites, as stronger materials, could be used at ordinary and elevated temperatures as longer-wearing, corrosion-resisting materials and as cutting tools. The advantage of electroforming composites is that any complex shape can be produced by this technique without elaborate equipment. Even a conventional plating shop could undertake the job of electroforming composites without extra expenditure. Electrodeposition of composites will soon become an industrial reality.

The electrolytic codeposition process has the following features:

1.  The ability to coat to close tolerances, thus often eliminating the cost of postmachining.
2.  The bond strength of the coatings is higher than 7000 MPa even on high chromium steels.
3.  The process does not require applied pressure of heating, hence the metallurgical properties of the substrate metal/alloy remain unaffected. There is no risk of distortion of the finished components.
4.  The coating can be obtained very uniformly even on complex contours by suitable design of anodes, jigs, plating parameters, and so on. For specialized applications, this process seems to be the only viable technique—for example, coating small bores with a wear-resistant material, thus permitting the adoption of designs not possible before.
5.  The process is highly economical, and a suitable hard coating can be applied to soft substrate metals.
6.  The process is capable of giving a wide range of metal-particle compositions, thus making it possible to tailor the coating for the required surface characteristics.
7.  The process lends itself to automation.

Electrocomposites can be classified into five categories.

1.  Particle-dispersed metal composites (PDMC)
2.  Fiber-reinforced metal composites (FRMC)
3.  Electroless metal composites
4.  Layered and laminated composites
5.  Optical composites

## 1.1 Particle-Dispersed Metal Composites

Particle-dispersed metal composite coatings can be obtained as a surface coating or as an electroformed material by suitable methods.

The coatings are obtained when insoluble materials in fine powder form are added to an electroplating bath, and electrodeposition is continued in the conventional manner. During electrolysis the particles are incorporated into the deposit, and a composite deposit is formed. For preparation of electrocomposites for practical application, the particles are held in suspension. Air agitation, mechanical agitation, magnetic stirring, and fluidizing with a flowing electrolyte are among the methods used.

Composites can be plated on all substrates over which electroplating can be done. The thickness of the coatings depends upon the size of the particles, the nature of the particles, and the nature of the metal deposited. Metals which can be plated with particles are cobalt, copper, gold, chromium, iron, cadmium, lead, nickel, zinc, and their alloys.

The particle additives include the following oxides of Al, Zr, Ti.

Carbides of Ti, Ta, Si, W, Cr, Zr, B, Ni
Nitrides of B, Si
Borides of Ti, Zr, Ni
Sulfides of Mo, W
Graphite, mica, stearates, PTFE, and diamond

The conditions of electroplating are to be chosen in such a manner as to secure the formation of a composite with suitable volume fraction of dispersed phase.

It is possible to increase the volume fraction of the dispersed phase in the composite by

a. Increasing the volume of the particles in solution
b. Increasing the current efficiency of the bath
c. Adding certain surface active agents to promote codeposition
d. Adding certain monovalent cations such as Tl, Rb, or Cs

PDMCs are used to

1. Increase the wear resistance, abrasion resistance, and creep resistance of metals or alloys ($Ni + SiC$, $Pb + TiO_2$)
2. Increase the corrosion resistance, for example, by using micro discontinuous chromium plating on composite nickel-plated steel ($Ni + Al_2O_3$)
3. Impart a dry self-lubricating coating ($MoS_2$ with Ni or Cu)
4. Increase the strength at elevated temperatures ($Ni + Al_2O_3$)
5. Offer a heat-treatable alloy (Ni + Cr powder)
6. Provide coatings useful in nuclear applications ($Ni + Pu$, $Ni + UO_2$)

The success of PDMC materials is dictated by the following factors:

1. The volume of particles in suspension determine the volume of codeposited particles.
2. Critical lower limit of particle size.
3. Uniformity of suspension.
4. Position and geometry of the cathode.
5. Conditioning and pretreatment of the particles.

Important variables in codeposition are the metal plate, particle size and volume, bath throwing power, and pH, CD, current form, and stirring efficiency (where CD = current density). Thus by using a variety of suspended particles and a variety of metal-plating processes, a wide range of composite electroplates can be obtained with special properties. The advantages of using composite coatings are many. The process is simple and economical, and the coatings with desired properties can be applied on any metal or alloy to alter the surface characteristics. The coatings are required by designers, engineers, and technologists in the chemical, marine, mechanical, agricultural, aerospace, electrical, electronics, and computer industries, among many others.

## 1.2  Preparation of Electrocomposites

In the preparation of PDMC materials, the particles in question are to be kept in suspension by suitable agitation in the electro-plating baths. Two techniques are commercially used: the *plate pumping process* and the *liquid/air process*.

### Plate Pumping Process

The plate pumping process uses the principle that when close-fitted reciprocating motion is given to a perforated plate in the plating bath, there will be good turbulence in the bath in a controlled manner. The plate pumping process seems to be quite popular for the production of composite coatings in Britain (Kedward and Wright, 1978) and in Germany (Celis and Roos, 1982). The main idea is to produce the liquid jets through the perforations, which prevents the powder from settling down at the bottom of the tank. This agitation can be used with or without conventional air agitation. A schematic diagram of a plate pumper unit is given in Fig. 1.

### Liquid/Air System

In the liquid/air system, the process operates by drawing liquid from the top of the solution through a pump. The liquid is recircu-

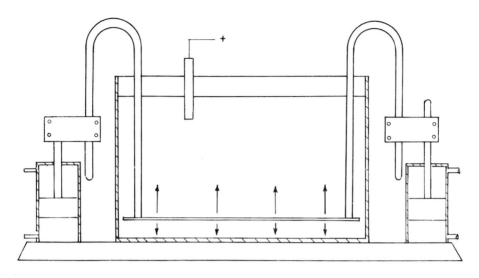

FIGURE 1   Schematic layout of the plate pumper ECC forming plant.
(From Kedward, 1975.)

lated through the inlet of conical bottom of the cell. This is a closed-
loop circulation. This agitation can be supplemented by air agitation.
A schematic diagram for the liquid/air electrocomposite forming
process is given in Fig. 2. The overall effect is violent agitation
of the solution and uniform dispersion of the particles.

## 1.3   Mechanism

Dispersion composite coatings containing the metal-oxide particles
in a metal matrix have been prepared by electroplating and electroless plating methods. However, the mechanism by which these particles can become incorporated into a growing deposit is not fully
established.

   The mechanism of particle entrapment in a metal matrix has
been proposed by many workers as follows:

1.   Mechanical collision theory of the particles, or mechanical entrap-
     ment.
2.   Electrostatic interaction of the particles to the metal electrode,
     or electrophoresis.
3.   Codeposition of particles by the chemical bonding to the electrode,
     or two-stage adsorption.

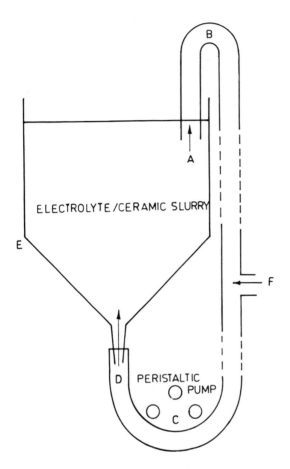

FIGURE 2   Liquid/air ECC forming process. (From Kedward, 1975.)


A good review of existing theories of particle entrapment on
the codeposition of metals is given in Raj Narayan and Narayanan
(1981) and Celis and Roos (1982).

It is postulated (Martin, 1965) that the cathode efficiency is
important in determining whether codeposition is possible. If the
metal deposition is very rapid, any particle that is delayed at the
cathode will be engulfed in the deposit. Martin (1965) pointed out
that the fields involved in the electrodeposition (0.1-0.3 V/cm)
were too low to invoke a mechanism based on electrophoresis, and
he concluded that the agitation of the bath caused the suspended
particles to impinge and adhere momentarily to the cathode surface,
with the consequent entrapment in the growing crystals.

Brandes and Goldthorpe (1967) showed that $Al_2O_3$ could codeposit satisfactorily with nickel from acid baths but not with copper from acid baths (though the cathode efficiency in both cases was nearly 100%). They concluded that particle entrapment in both cases was nearly 100% and that it depended on the microthrowing power of the plating bath rather than just the metal deposition rate. Baths known to have good microthrowing power, such as the acid copper-plating bath, may not be suitable for entraping the particles, since the particles will be displaced from the cathode surface when the deposit grows from beneath. This explains why it is possible to codeposit particles from an alkaline cyanide copper bath and not from an acid bath. This hypothesis, however, does not explain all the observed facts. Sautter (1963) found that the $Al_2O_3$ content in nickel deposited from a Watt bath containing various concentrations of suspended $Al_2O_3$ was affected by pH, CD, and temperature. He proposed that the factors affecting the entrapment of $Al_2O_3$ particles into Ni deposits seemed to be the adsorption of particles and the electrostatic interaction of the particles to the cathode.

Tomaszewski et al. (1969) found that the pronounced increase in the entrapment of $Al_2O_3$ was observed when $Ti^+$, $Cs^+$, or $Rb^+$ ion was introduced in the acid $CuSO_4$ bath, and they proposed that the surface charge of the $Al_2O_3$ is most significant in promoting the codeposition of the particles into copper deposits in the $CuSO_4$ bath.

Snaith and Groves (1972), from their experimental results, suggested that the hydrodynamic transport to the vicinity of the cathode and subsequent mechanical entrapment of the particles were the most important factors in the electrodeposition of composite coatings.

Sykes and Alner (1974) also studied the dispersion coatings of the $Cu$-$Al_2O_3$ system and proposed that the adsorption of $Al_2O_3$ on the cathode surface seemed to be the most important factor in the codeposition reaction.

In 1972 Guglielmi (1972) discussed the kinetics of deposition of inert particles. To explain the peculiarities shown by the codeposition of inert particles from electrolytic baths, he proposed a mathematical model based on two successive adsorption steps (Fig. 3). In the first step the inert particles are loosely adsorbed on the cathode and they are in equilibrium with the particles in suspension being still surmounted by adsorbed ions and solvent molecules. This first step is postulated to be substantially physical in character. The second adsorption step is thought to be field assisted and therefore substantially electrochemical in character; it produces a strong adsorption of the particles to the cathode. The strongly adsorbed particles are then progressively submerged by the growing metal. Guglielmi arrived at the following equation:

$$\frac{C}{\alpha} = \frac{Wi_0}{NF \ dv_0} \ e^{(A-B)_\eta} \left( \frac{1}{K} + C \right)$$

where

C = concentration of suspended particles
$\alpha$ = volume fraction of particles in the deposit
W = atomic weight of the electrodeposited metal
d = density of the same
$\eta$ = valence
F = Faraday constant

The parameters $i_0$ and A are related to the metal deposition and are the constants in the Tafel equation (which gives a relation between CD i and the overpotential) $\eta$:

$$i = i_0 e^{A\eta}$$

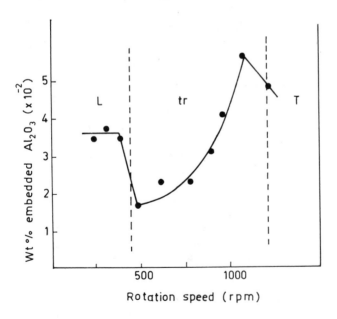

FIGURE 3 Weight percent of embedded alumina in copper from acid copper plating bath versus the rotation speed of rotating disk electrode. (From Celis and Roos, 1982.)

The parameters $V_0$ and B play a symmetrical role to the parameters $i_0$ and A but are related to the inert particle deposition. K is derived from the Langmuir adsorption isotherm and depends essentially on the interaction between particles and cathode.

Guglielmi verified the validity of his model for the codeposition of titania and silicon carbide with nickel from a nickel sulfamate bath.

Based on results of streaming potential and adsorption studies on alumina in nickel and copper electrolytes, Foster and Kariapper (1973) showed that the nature of the acquired surface charge of the alumina was an important factor in its ability to be codeposited. They concluded that

1. Codeposition occurs in nickel plating baths due to strong adsorption of nickel ions on the particle surface.
2. Codeposition of $\gamma$-alumina does not occur in acid copper plating baths because cation adsorption on the particle surface is very small.
3. Addition of thallium and rubidium ions to the acid copper bath produces a large positive charge on the alumina surface, thereby promoting codeposition.

Celis and Roos (1977) from their results on the codeposition of $\alpha$-$Al_2O_3$ in copper plating baths with and without the addition of $Tl^+$ ions concluded that the second adsorption step is rate-determining. This explains why the particles are codeposited easily when $\eta$ is increased.

The relative importance of the various factors that could contribute to the intensity of the interaction between particles and cathode surface has been reported by Foster and White (1981). Using cathode vibration and short flash coating, they could make a distinction between the instantaneous bonds and their physical and electrostatic components and the fine dependent bond created by the deposition of metal around the particle. The formation of a significant solid bond in a nickel solution at pH 4.5 is so rapid that it can practically be considered as instantaneous. This is not the case in a copper solution at pH 0.5, where alumina particles can rise on the advancing cathode surface before ultimately becoming engulfed. Limited published information is available on the mechanism of electroless composite plating. Honma et al. (1980) showed experimental data agreeing with an adapted Guglielmi model based on mixed potential theory.

Guglielmi's model based on two adsorption steps is a valid representation of the process taking place during codeposition.

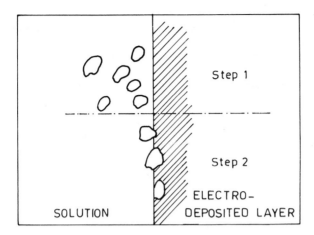

FIGURE 4   Schematic view of the two adsorption steps as in
Guglielmi's model. (From Guglielmi, 1972.)


    Further work is necessary to clarify the reactions occurring
in the hydrodynamic and electrochemical double layers at the
cathode/solution interface and to clarify the relative effects of the
different codeposition parameters (Fig. 4).
    From our experience we postulated the following mechanism
for codeposition (Rajagopalan, 1982):

*Stage 1*

The transport of suspended particles from the bulk of the plating
solution to the cathode surface. During this stage the mode and
intensity of agitation and the cathode configuration are the most
important factors affecting the formation of a uniform particle
suspension at the interface between the cathode and the plating
solution. The particles are kept suspended by agitation.

*Stage 2*

The particles of suspended solid moving within the liquid can hit
the cathode. Some of them will bounce. A few can come to rest
at the cathode surface because their kinetic energy is just equal
to the energy absorbed by the metal on impact. Such particles
begin to fall because of gravity. Because of roughness on the metal
surface and on the particle, the particle can remain in contact with
the cathode for a short time.

*Stage 3*

In this time the ions adsorbed onto the particle near the particle/cathode
interface may get reduced, thereby bonding the particle to the metal

surface. Herein comes the choice of the cathode configuration, position, CE, and the $\phi$ potential and the ion bridges. The volume percentage of the particles in the composite coating steadily increases when the cathode is altered from the vertical position to the horizontal mode. Even as little as 5° inclination of the cathode greatly enhances the entrapment of the particles. The conditions should be favorable for the particles at the cathode surface to remain there for a critical time till the metal ions are discharged at the particle/cathode interface. Herein comes the choice of the mode of agitation, size of the particles, nature of the metal deposits, and position of the cathode. Thereafter the particles get progressively coated and are finally "embedded." Nearly 90% of the particles could be codeposited in a horizontal mode (Fig. 5).

## 1.4 Properties of Powder-Dispersed Metal Composites

The status of the electrocomposites and the properties of powder-dispersed metal composites are reviewed by Raj Narayan and Narayanan (1981), and Celis and Roos (1982). An excellent survey of the principal techniques for production of wear-resistant coatings on metals from the vapor, liquid, and solid states is summarized by Celis and Roos (1984). In general, it is known that conductive particles give much higher codeposition than nonconductive ones

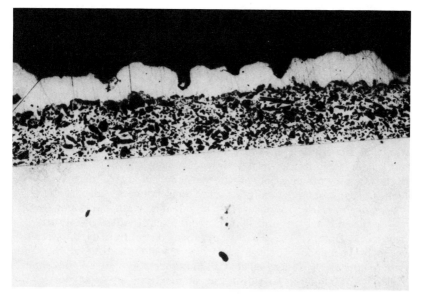

FIGURE 5   WC-Ni composite coating (horizontal mode): WC = 90%. (From Indira Rajagopal, 1985.)

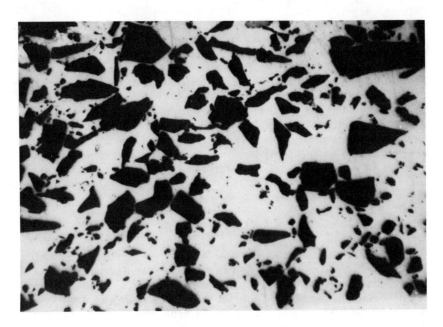

FIGURE 6   SiC-Ni composite coating: size of SiC = 1 mm or more.
(From Indira Rajagopal, 1985.)

and that large particles give perembedded particle a much higher
volume percentage than small particles.

   In order to get a large volume fraction of the dispersed particles
or codepositing large particles, horizontal electrodes can be used
in a suitable plating bath. Figure 5 shows the 90% v/v tungsten
carbide particle in a nickel matrix, and Figs. 6 and 7 show the
codeposited coarse silicon carbide particles greater than 1 mm (which
cannot be kept in suspension) and SiC < 0.05 mm in a nickel matrix
by this plating technique. In sediment codeposition or occlusion
plating, sedimentation on the upper side of the horizontal cathode
takes place. It is evident that in such cases the bath agitation
and the shape and position of the cathode mainly determine the
volume percent of codeposited particles (Ghouse et al. 1980).

   For engineering purposes the physical and chemical natures
of the particles are of utmost importance along with the nature
of the metal matrix. For wear resistance, very fine hard particles
have to be incorporated; for antistick or cold starting or dry lubrication problems, soft agglomerated particles have to be incorporated.
For many engineering purposes, a codeposition of 30-40 vol.% of
particles (such as alumina and silicon carbide) is necessary, and
this can be obtained on horizontal surfaces.

The soft barium sulfate particles codeposited with the copper provide antistick and good slip properties to the copper surface. The hard aluminum oxide particles when very fine and deposited in high volume percent provide wear resistance. Since the particles do not form alloys with the copper, the electrical and heat conductivities are practically unchanged.

Interest in cermets stems from the useful engineering properties they exhibit. The dispersion of fine nonmetallic particles in the metal matrix brings about an improvement in the wear properties, strength, hardness, and corrosion behavior.

The formation of dispersion-hardened alloys with improved mechanical properties by electroforming technique was reported by Brown and Gow (1972). Dispersion-strengthened alloys of copper, nickel, and aluminum have also been developed (British Patent, 1972) by utilizing whiskers or short fibers of $Al_2O_3$ or quartz.

## 1.5 Lubricating Coatings

In lubricating coatings the second phase is chosen for its ability to confer lubricating properties. Solid lubricants are usually used in extremes of temperatures or under low pressure where conventional lubricants such as grease or oil cannot be used. They are used for both rolling and sliding contacts.

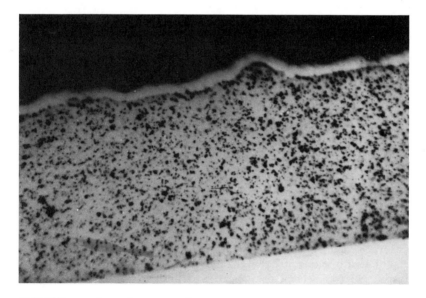

FIGURE 7   Ni + SiC composite coating × 1000: size of SiC = 0.01 mm. (From Indira Rajagopal, 1985.)

To minimize friction or wear, solid film lubricants should

1. Provide low shear strength and have low coefficient of friction
2. Form an integral bond with the substrate and adhere under loading.
3. Provide uniform coverage on the substrate and a clean surface
4. Be free from impurities
5. Be stable under working
6. Have uniform particle size
7. Not contaminate the resultant material after wear

Metal matrix solid film lubricants are capable of working satisfactorily under very high temperatures or cryogenic temperatures or under higher pressures than conventional organic-based lubricants. The lubricant materials recommended for composite coatings are graphite, polytetrafluoroethylene (PTFE), $MoS_2$, $WS_2$, mica-graphite fluoride, and so on, and the matrix materials include copper, nickel, or zinc (Young, 1974, 1975).

Mostly nickel is employed as a matrix. However, there are some papers describing the use of copper as a matrix for codeposition of graphite or $MoS_2$. Vest and Bazzarre (1967) have deposited Ni-$MoS_2$ composite coatings from a nickel sulfamate bath, the details of which are given below.

| | |
|---|---|
| Nickel sulfamate | 43 oz/gal |
| Boric acid | 4.5 oz/gal |
| Wt.% $MoS_2$ | 20 (60 v% $MoS_2$) |
| | 0.25 (25 v% $MoS_2$) |
| Temperature | 120°F |
| Current density | 25 A/ft$^2$ |
| pH 2 | 60 v% $MoS_2$ |
| pH 5 | 25 v% $MoS_2$ |

The composite containing the Ni-$MoS_2$ coating is used in the orbiting plane experiment package (OPEP) on the Eccentric Geophysical Observatory (ECO) spacecraft, which is an unconventional device. This is a scanning mechanism which was designed to convert rotational motion into oscillatory motion. The requirement is that the lubricating coating should not outgas any product. The coatings containing 25 v% $MoS_2$ in a nickel matrix had very low coefficient of friction (0.05-0.18) in the range of other $MoS_2$-containing films. However, the load-carrying capability of this composite film on a soft substrate was not so high and they were suitable only for low loads. Ni-$MoS_2$ composite coatings exhibit a textured surface which is considerably softer than pure nickel.

Solid film lubricants or dry film lubricants provide a lubricating film free of impurities and also seem to provide a coating that combats corrosion. Electrolytic coedeposition of $MoS_2$ with metals seems to offer a versatile and convenient method of making bonded solid lubricants on metal components which are meant to have minimum wear (Yuko Tsuya, 1974).

Lubricating properties of some codeposited composites of metal-graphite fluoride deposits have been studied. Copper was a better matrix, in this case, than nickel. The wear resistance of the composites were improved only when there was sufficient quantity of the graphite fluoride (GF) and GF was found to be a better lubricant in water or oil than in air.

From the experimental results of copper- and nickel-base self-lubricating composite plating, the following conclusions were drawn (Yuko Tsuya, 1974).

1.  The addition of as little as 1% CF in v/v 1% copper- and nickel-base platings reduces the coefficient of friction and wear rate of the mating surface.
2.  Significant improvement in wear resistance of the composite coating is obtained only if the solid lubricant concentration in the bath exceeds a certain volume percent (3% in the case of Cu-GF plating).
3.  Copper is a better matrix than nickel in the self-lubricating codeposited composite platings with respect to both the wear rate and load-carrying capacity under various conditions.

Tomoszewski et al. reported that graphite was codeposited with copper from acid baths even in the absence of promoters. Cu-BaSO$_4$ composite coatings, which can be used for sliding contact applications because of the antisticking properties of the BaSO$_4$ particles, could be prepared from acid sulfate baths in the presence of promoters.

Tl$^+$ enhances the codeposition of $MoS_2$ with silver from a cyanide bath or a cyanoferrate bath. Saifullin (1973) had reported the codeposition of graphite with silver from iodide baths.

Steel fasteners are cadmium plated in the automobile industry to overcome stick-slip problems. The coefficient of friction and the integrity of the coating play a significant role in this stick-slip phenomenon. Cd plating has been used for its antigalling properties. The increasing cost of cadmium and effluent treatment problems necessitated finding substitutes for Cd plating. Though zinc coatings are considered as alternatives, the galling nature of the deposits can produce stick-slip problems. However, when graphite is codeposited with zinc from a neutral chloride bath, the coatings compare very favorably with cadmium coatings for antigalling applications (Donalowski and Morgan, 1983).

## 1.6  Coatings for Improving Wear Resistance

Composite coatings with the particulate ceramic-dispersed phase
are known to have improved hardness and wear resistance. The
increased wear resistance of the composite coating is largely due
to the wear properties of the hard particles. It is assumed that
when a composite coating is brought into contact with a sliding
counterface the hard particles bear the wear load. Hence various
attempts are made to codeposit $Al_2O_3$, $TiO_2$, and carbides of Si,
W, Cr, Ti, diamond, and others, in a wide range of matrix metals
such as Ni, Cr, Co, Cu, and Fe.

Kedward (1969) reported that a certain amount of metal-to-metal
contact also occurs in practice due to nonuniformity of the applied
load, and pointed out the importance of selecting the matrix metal
and the dispersed phase. He recommended a hard matrix and a
hard dispersoid (e.g., Co and Cr carbide).

### Nickel-Based Electrocomposites

Ni-based electrocomposites with alumina, titania, refractory carbides
and nitrides, graphite, $MoS_2$, $WS_2$, mica, quartz, diamond, and
so on, are prepared from standard nickel sulfate and nickel sulfamate
baths.

$Al_2O_3$ (submicron-sized $\alpha$ or $\gamma$ form) have been codeposited
with nickel. In all cases the volume fraction of the codeposited
particles was proportional to the volume percentage of the dispersed
particles in the solution up to 200 g/L. There was an increase in
the alumina content in the deposit with increase in CD, and the
codeposition was not satisfactory at pH less than 2.

The microhardness of $NiAl_2O_3$-composite coatings was higher
than that of a pure nickel coating, and there was an increase in
hardness with increased $Al_2O_3$ content (300-400 VPN Ni, 450-700
VPN Ni-$Al_2O_3$ composite), where VPN is the Vickers pyramid number).

The inclusion of $Al_2O_3$ particles in the nickel matrix increased
the ultimate tensile strength (UTS) and yield strength (YS). Sautter
(1963) observed an increase of YS from 8 to 35 kg/mm$^2$ by dispersing
3.5 to 6 vol.% of $Al_2O_3$. The oxidation resistance of the composite
coating was higher than Ni, but the corrosion resistance was poor.

The structure of electrodeposited nickel alumina composites
at ambient and elevated temperatures was studied in detail by Gilliam,
McVie, and Phillips (1966). They concluded that the enhanced strength
of nickel composites at higher temperatures was due to the blocking
of glide dislocations by the particles. Sinha et al. (1973) recommended
a heat treatment of nickel alumina composites in vacuum at 650°C
to minimize brittleness and enhance ductility.

For Ni-$TiO_2$ composites, properties such as hardness, UTS,
and YS increased due to incorporation of $TiO_2$. However, the

mechanical properties of Ni + TiO$_2$ are inferior to Ni-Al$_2$O$_3$ composites. The composites had lower porosity, increased corrosion resistance, and better retention of mechanical properties at higher temperatures.

The hardening produced in the composite is not caused entirely by the dispersed particles. Refinement of the grains of the matrix nickel (in the presence of the particles) also contributes.

Figure 8 shows a significant increase in hardness of Ni-Al$_2$O$_3$ and Ni-TiO$_2$ alloys as compared with nickel and the retention of hardness at elevated temperature (Greco and Baldauf, 1968). The retention of yield strengths of the dispersed alloys after heat treatment is very significant.

Kingery (1953) has reported that fine ZrO$_2$ particles have less interfacial energy than Al$_2$O$_3$ in a nickel matrix and hence ZrO$_2$ particles may have better stability in a nickel matrix. Joshi and Totlani (1983) have shown that the codeposition of ZrO$_2$ particles with nickel increases the microhardness, yield strength, ultimate tensile strength, and abrasive wear resistance and decreases ductility. They have shown that annealing the deposits results in a decrease in their mechanical properties, but even in the annealed state Ni-

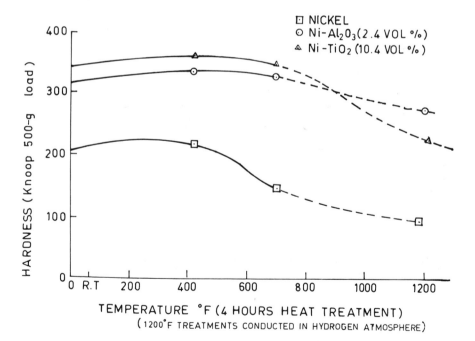

FIGURE 8  Variation of microhardness versus heat treatment for nickel, Ni-Al$_2$O$_3$, and Ni-TiO$_2$. (From Greco and Baldauf, 1968.)

$ZrO_2$ composite deposits retain much higher mechanical properties than do pure nickel deposits. There is an improvement in the abrasive wear characteristics of $Ni+ZrO_2$ deposits on annealing.

Both yield strength and ultimate tensile strength of the composite electroplates are higher compared with those of pure nickel electrodeposits. The increase of yield strength and ultimate tensile strength of 6.5% $ZrO_2$ nickel composite over pure nickel plate is around 40% and 25%, respectively. A composite electroplate containing 4% and 6.5% $ZrO_2$ retained tensile strength of 363 and 417 $MN/m^2$, respectively after vacuum annealing at 1000°C for 1 h. For electroformed pure nickel the UTS was found to be less than 98 $MN/m^2$ under similar heat treatment conditions. However, the percentage elongation of nickel was decreased from 28 to 2-8.5% for $Ni-ZrO_2$ composites containing 2.1-6.5% $ZrO_2$. (See Figs. 9 and 10.)

Abrasive wear resistance of electrodeposited nickel and $Ni + ZrO_2$ composite coatings in the as-plated and heat-treated states showed that there is a substantial improvement in the wear resistance of the composite coating after annealing. For high-temperature wear applications preferably between 800-1000°C, this composite is recom-

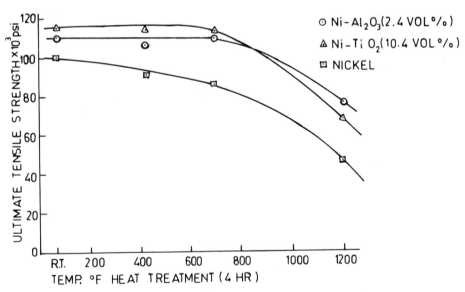

(1200°F Treatment conducted in hydrogen atmosphere)

FIGURE 9   Variation between UTS and heat treatment for nickel, $Ni-Al_2O_3$, and $Ni-TiO_2$. (From Greco and Baldauf, 1968.)

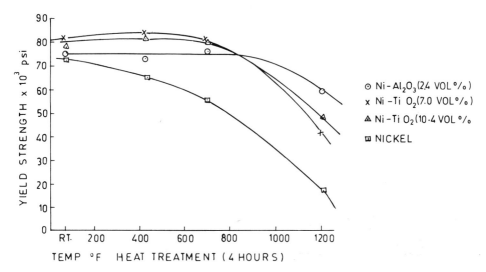

FIGURE 10  Variation between yield strength and heat treatment for Ni, Ni-Al$_2$O$_3$, and Ni-TiO$_2$. (From Greco and Baldauf, 1968.)

mended by the authors since Ni + ZrO$_2$ composites are more stable at higher temperatures. This is because ZrO$_2$ particles are stable in the nickel matrix at higher temperature. These coatings are known to have high-temperature oxidation resistance also.

The effect of anions on the formation of electrodeposited composite coatings was studied in detail by Tomaszewski (1976) and had shown that the amount of BaSO$_4$ codeposited with Ni is maximum with a NiSO$_4$ bath.

The structure and morphology of electrodeposited composites were reported to be highly dependent on the electrical conductivity of the codeposited particle (Zahavi and Kerbel, 1982). Uniform coatings were obtained with nonconductive particles, but porous and nonuniform deposits were obtained with highly conductive particles. The oxides of metals were codeposited with nickel and the composites were smooth, whereas that of carbides of chromium resulted in rough composites which can be made uniform in the presence of suitable addition agents.

Nickel and electroless dispersion coatings with the incorporation of silicon carbide were developed for contact surfaces in internal combustion engines. These coatings were found to be quite successful for improved wear resistance. PDMC coatings are also used to protect plastics, compression molds, casting molds, as well as pumps, spindles, and other parts subject to friction. For combating wear at high temperatures Co-chromium carbide composite coatings are recommended (e.g., for aircraft engines). Recently Co-$Cr_2O_3$ composite coatings are shown to have superior wear properties at elevated temperatures.

Nickel-SiC composites under the trade name Nikasil was developed in Germany in 1963 for the lining of their range of reciprocating engine cylinders. These coatings have been applied to all kinds of reciprocating engines operating under severe conditions. They have excellent mechanical properties, such as good abrasion resistance and slip properties. They are used for the production of a large variety of Al cylinders.

The oxidation characteristics of electrodeposited nickel composites containing up to 13% silicon carbide particles at high temperature was studied extensively by Stot and Ashby (1978). They found that the introduction of 12 vol.% SiC particles caused a significant decrease in the oxidation rate, due largely to internal oxide-derived particles acting as barriers to $Ni^{2+}$ diffusing through the NiO scale. Until now, industrially developed applications were mainly restricted to the Ni-SiC system. The major breakthrough of composite plating is actually taking place in the automobile industry with the use of electrolytic Ni-SiC coatings in a new generation of motor engines made of weight-saving Al alloys. Some of these Ni-SiC coatings have been introduced under trademarks such as Nikasil and Elnisil. A few European and Japanese automobile companies have already introduced this technology in their most recently developed models. One application of Ni-SiC composite coating is to protect Ti helicopter rotor axial engine blades against erosive wear (Fig. 11). Electrodeposition of nickel-silicon nitride composites was studied in detail by Joshi and Totlani (1981) as an improvement over oxide-dispersed composites.

The composites of Ni-$Al_2O_3$ and Ni-$TiO_2$ do not retain the enhanced mechanical properties of the composites once they are vacuum annealed at temperatures above 750°C. Apart from the softening of the matrix, this result is reported to be due to particle coarsening, which leads to nonuniform distribution of dispersoid in the nickel matrix. This particle coarsening causes nonuniform dislocation stress fields around the coarsened particles. Silicon matrix $Si_3N_4$ is a very strong hard material and can retain its strength up to 1200°C. It has good oxidation resistance and good dimensional stability. Hence $Si_3N_4$ particles will coarsen only slowly and will retain their enhanced mechanical properties to much higher temperatures. They have reported that the $Si_3N_4$ content in the deposit increases with the percentage volume of $Si_3N_4$ in the bath linearly (1 vol.% in the bath, 10 vol.% in the deposit at 2A/$dm^2$). They found that the microhardness of the composite increased with

FIGURE 11  Ni-SiC coated Ti-6A1-4V helicopter rotor axial engine blades. (From Indira Rajagopal, 1982.)

an increase in $Si_3N_4$ content in the deposit, and there was no change in microhardness after vacuum annealing at 350°C for 1 h. There was no blistering or cracking of the electrodeposits even when the deposit was heated at 900°C. There was a sharp decrease in microhardness for deposits annealed at 500-900°C.

A composite of 4.46 vol.% $Si_3N_4$ retained a tensile strength of 42 kg/mm$^2$ after vacuum annealing at 900°C. In comparison, pure nickel plate electroformed from the same bath under identical plating conditions retained less than 10 kg/mm$^2$ after annealing at 900°C.

However, the composites containing more than 9 vol.% were brittle in the as-plated condition with percentage elongation in the range of 2.5-8.5%, compared with 28% for pure nickel electroplates.

The volume percent of $Si_3N_4$ embedded in the composite versus the volume of the particles in the bath is shown in Fig. 12, and the effect of CD on percentage particle inclusion is given in Fig. 13. The hardness and UTS of the composite remain practically constant when $Si_3N_4$ in the deposit exceeds 4% (Figs. 14 and 15).

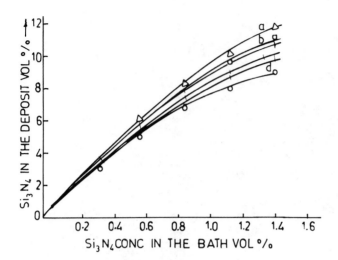

FIGURE 12    Effect of $Si_3N_4$ concentration in the bath on the $Si_3N_4$ content of the electrodeposits: a. CD = 2 A/dm$^2$; b. CD = 2.5 A/dm$^2$; c. CD = 3.0 A/dm$^2$; d. CD = 3.5 A/dm$^2$; e. CD = 4 A/dm$^2$; f. CD = 4.5 A/dm$^2$. (From Joshi and Totlani, 1981.)

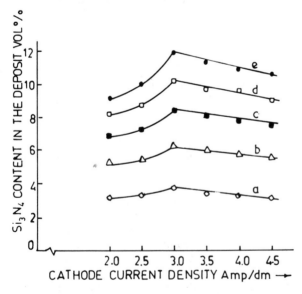

FIGURE 13    Variation in the $Si_3N_4$ content of the deposits with cathode CD $Si_3N_4$ conc. in bath: a. 0.31 vol.% (v/v); b. 0.56 v/v; c. 0.84 v/v; d. 1.12 v%; e. 1.34 v%. (From Joshi and Totlani, 1981.)

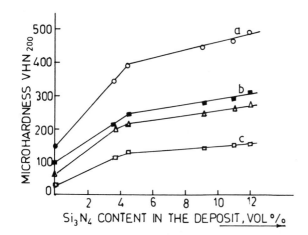

FIGURE 14   Variation of H (VHN 200) of the deposits with $Si_3N_4$ content of the deposit: a. as-plated; b. vac. annealed at 500°C 1 hr; c. vac. annealed at 750°C 1 hr; d. vac. annealed at 900°C 1 hr. (From Joshi and Totlani, 1981.)

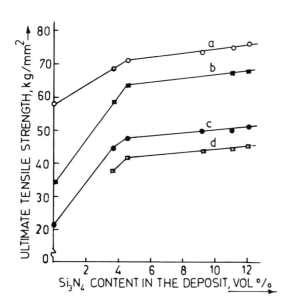

FIGURE 15   Variation of UTS of the deposit with $Si_3N_4$ content. Other conditions are the same as for Fig. 14. (From Joshi and Totlani, 1981.)

By incorporating a high volume of tungsten carbide in a nickel matrix, Kedward (1969) obtained improved mechanical strength. A composite having high inclusion content (over 60% v/v) of medium particle size (10-20 µm) were used on components subject to wear at ambient temperatures under nonimpacting conditions. On the other hand, for applications at elevated temperatures, tantalum carbide particles were recommended. Much better high-temperature performance was obtained by use of cobalt-zirconium boride. Cobalt-zirconium diboride, which has an inclusion content of less than 15% v/v of large particles (15-35 µm) could be used successfully at elevated temperature and high loads.

A coating of Ni-TiC on stainless steel (s.s.) screws helps to combat seizure problems (Kedward, 1967). For combating fretting wear, Ni-B nitride composite is recommended. Table 1 lists the nickel-based coatings with their important properties. The micro-structures of the coatings consist of a metal matrix in which the ceramic particles are randomly dispersed. Optical studies indicate that ceramic is present as individual particles or agglomerates, depending upon their nature. The ceramic contents of the composite coatings range from 30 to 50% by volume.

Photo micrographs Figs. 16-21 show a variety of particle sizes and that the percentage incorporation in a nickel matrix can be achieved by varying the bath composition and operating conditions.

## Cobalt-Based Composites

*Cobalt-chromium carbide composite.* For dry sliding wear, nickel cannot be used since it adheres to most other metals and alloys. The nickel composite, though it performs better than nickel coatings, still is not the right type of material for certain high-temperature applications for combating wear. Cobalt, on the other hand, because of its cph structure possess good wear and frictional properties. Hence the cobalt composite coatings with hard particle reinforcement exhibit good wear properties.

Cobalt-chromium carbide composite coatings have better wear characteristics up to 800°C (Kedward and Wright, 1978). However, they deteriorate at temperatures above 800°C due to carbide breakdown and coarsening of the particles. The oxidation resistance of electro-deposited cobalt-chromium carbide coatings can be significantly improved by carrying out a heat treatment to form a homogeneous alloy. The industrial interest for the cobalt-chromium carbide coatings is solely related to R & D work of a British company which promotes this type of composite coating for its favorable high-temperature behavior on aircraft parts. It is shown that an electro-deposited composite coating of a cobalt matrix with 30 vol.% chromium carbide powder (2-4 µm dia) is capable of controlling certain forms

TABLE 1 Electrodeposited Composite Coating Produced Based on Nickel and Cobalt

| Matrix | Additives Type | Additive concentration Particle size (μm) | Additive concentration Electrolyte (g/L) | Deposit (v/v) | Deposit Hardness (HV) | Surface finish (min CLA) | Preparation parameters Current density (A/dm²) | Preparation parameters Plating rate (μm/hr) |
|---|---|---|---|---|---|---|---|---|
| Nickel | Aluminum oxide ($Al_2O_3$) | 1–2 | 500 | 35 | 551 | 25–35 | 3.0 | 35 |
| Nickel | Boron carbide ($B_4C$) | 0–8 | 500 | 30 | – | 15–20 | 3.0 | 35 |
| Nickel | Boron nitride (BN) | 1–5 | 33 (ball milled) | 12 | 435 | 30–40 | 10.0 | 80 |
| Nickel | Chromium oxide ($Cr_2O_3$) | 0–1 | 500 | 30 | 533 | 10–15 | 6.0 | 50 |
| Nickel | Nickel oxide (NiO) | 5–10 | 500 | 15 | – | 20–30 | 3.0 | 30 |
| Nickel | Chromium carbide ($Cr_3C_2$) | 2–5 | 500 | 30 | 401 | 15–20 | 3.0 | 35 |
| Nickel | Silicon nitride ($Si_3N_4$) | 1–2 | 500 | 15 | – | 80–90 | 3.0 | 30 |
| Nickel | Titanium carbide (TiC) | 3–5 | 500 | 20 | 500 | 12–15 | 3.0 | 35 |
| Nickel | Tungsten carbide (WC) | 0–1 | 333 | 20 | 480 | 15–30 | 12.0 | 140 |
| Nickel | Tungsten disilicide ($WSi_2$) | 2–3 | 500 | 25 | 470 | 40–70 | 3.0 | 35 |
| Nickel | Zirconium diboride ($ZrB_2$) | 3–5 | 500 | 25 | 515 | 30–35 | 3.0 | 35 |
| Nickel | Zirconium carbide (ZrC) | 3–5 | 500 | 20 | 533 | 30–40 | 3.0 | 35 |
| Nickel | Zirconium oxide (ZrO) | 0.5–1 | 500 | 20 | 453 | 23–35 | 3.5 | 40 |
| Cobalt | Boron nitride (BN) | 1–5 | 33 | 15 | – | 25–35 | 6.0 | 70 |
| Cobalt | Chromium carbide ($Cr_3C_2$) | 3–5 | 500 | 20 | 430 | 10–15 | 3.5 | 40 |
| Cobalt | Tantalum carbide (TaC) | 3–8 | 500 | 45 | 470 | 15–25 | 3.5 | 45 |
| Cobalt | Tungsten carbide (WC) | 0–1 | 330 | 25 | 400 | 15–25 | 3.5 | 40 |
| Cobalt | Silicon carbide (SiC) | 1–2 | 500 | 25 | 400 | 10–15 | 3.5 | 40 |
| Cobalt | Zirconium diboride ($ZrB_2$) | 3–5 | 500 | 30 | 420 | 15–25 | 3.5 | 40 |

FIGURE 16  Ni + Al$_2$O$_3$ nickel composite × 500. (From Indira
Rajagopal, 1985.)

of wear on aircraft engines at temperatures up to 800°C. The wear
resistance of electrodeposited cobalt-chromium carbide is much
superior to flame-sprayed composite coatings (Kedward and Wright,
1978).

Cobalt-chromium oxide composite coating for high-temperature
wear resistance was studied by Martin Thoma (1984). The Cr$_2$O$_3$
was chosen for the following reasons:

1.  It has a low density; hence it was easy to maintain the particles
    as a suspension.
2.  Cr$_2$O$_3$ was quite stable in the bath, and it cannot be oxidized
    or reduced in the bath.

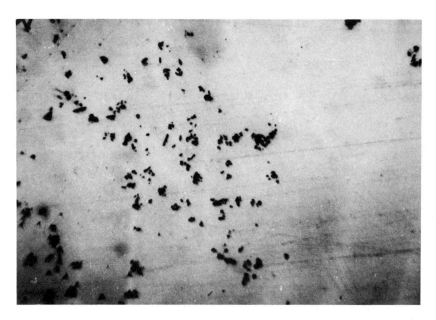

FIGURE 17   Ni + $MoS_2$ composite × 500. (From Indira Rajagopal, 1985.)

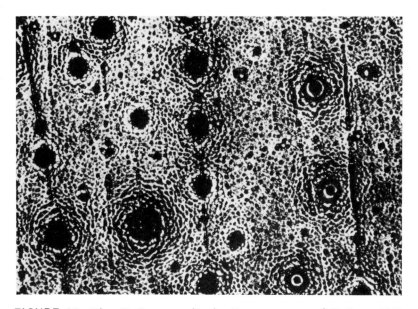

FIGURE 18   Ni + $B_4C$ composite in the presence of U.S. × 1000. (From Indira Rajagopal, 1985.)

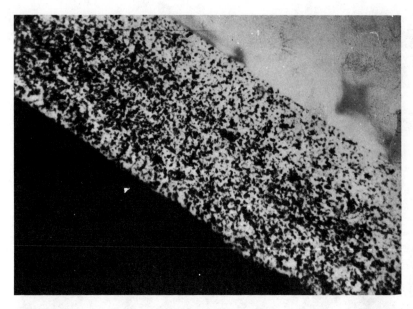

FIGURE 19   Ni + TiC composite × 500. (From Indira Rajagopal, 1985.)

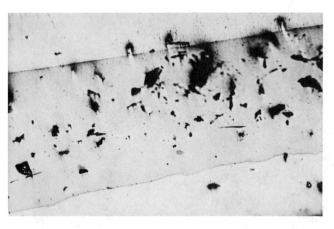

FIGURE 20   Ni-Co + diamond composite × 500. (From Indira Rajagopal, 1985.)

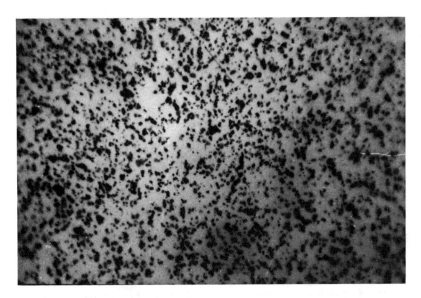

FIGURE 21 Ni + mica composite × 500. (From Indira Rajagopal, 1985.)

3. $Cr_2O_3$ did not react with the cobalt matrix at high temperatures and was thermally stable, unlike chromium carbide. The temperature-related reactions which depend on the stability of the included particles and the reactivity of the matrix were studied by Thoma. It was shown that Ni-SiC composites exhibit exothermic peaks at 400° and 600°C (probably due to the formation of $Ni_3Si$). Chromium carbide is shown to react with the cobalt matrix at 700°C. It was shown that incorporation of 20-25% $Cr_2O_3$ particles enhanced the wear resistance of cobalt composites even at 400-700°C. From 400 to 700°C there was no change of wear because of the formation of $Co_3O_4$. The $Cr_2O_3$ content in the bath should be about 200-250 g/L.

*Applications.* The $Cr_2O_3$-Co composite coating on metals have many desirable properties which are required for applications at elevated temperatures. The adhesive strength of the Co-$Cr_2O_3$ dispersion coating on many metals is 200-250 N/mm$^2$. The composite coating exhibits outstanding wear resistance from 300° to 700°C. This coating is quite suitable where good performance of components of high temperature with wear resistance is required. The potential applications are for compressor turbines or even combustion chamber components in aircraft engines.

Typical applications (Kedward and Wright, 1978) for the cobalt-
chromium carbide composite which resulted in improved wear performance include

1.  Compressor stator and shroud assembly
2.  Engine mounts
3.  Interstage seal located in nickel alloy turbine disks
4.  Brake piston (salvaging of worn-out Al alloy pistons)
5.  Steering body for aircraft nose wheel

The use of Tribomet in the aeroengine industry is discussed at
length by Kedward (1974).

The electrodeposited composite coatings produced based on
nickel and cobalt are summarized in Table 1 (Kedward et al., 1976).
The $Co-Cr_3C_2$ coatings have many commercial applications where
the prime requirement is for counterface compatibility. The coating
has been shown to be capable of combating adhesive wear against
many metals, alloys, and ceramics. However, due to its relatively
low bulk hardness in comparison to other wear-resistant materials,
its use for combating abrasive wear is limited. The composites of
$CoW-Cr_3C_2$ have much superior abrasive wear resistance than $Co-$
$Cr_3C_2$ do, and hence its application in aerospace industries has
increased in recent times.

The actual wear performance of a plated composite depends
on the size, shape, and volume content of the particles and their
intrinsic wear properties. The crystal structure, mutual solubility,
with the contacting materials, hardness, and surface energy all
contribute to the wear behavior of the composites. Table 2 shows
the important physical properties that affect wear for a number
of metals and ceramics which have been successfully codeposited
(Kedward, 1969).

*Codeposition of Tin Alloys*

ITRI developed composite coatings of tin with some nonmetallic
inclusions such as silicon carbide and $MoS_2$. Silicon carbide composite
coatings should have improved abrasion and wear resistance with
better frictional characteristics, while soft $MoS_2$ could produce coatings with good antifriction characteristics. The composite coatings
of tin were prepared by a conventional codeposition technique from
both acid and alkaline electroplating baths. ITRI found that the
optimum concentration of SiC in solution for maximum inclusion in
the deposit was 400 $g/dm^3$. It increased to a maximum and then
decreased with increasing current density. SiC is deposited more
readily from alkaline than from acid electrolytes. Tin $MoS_2$ deposits
from the acid stannous sulfate bath were crystalline, shiny, and

TABLE 2 Physical Properties of Certain Materials Which Affect Wear Frictional Behavior

| Material | Structure | Density | Melting point (°C) | Young's modulus ($10^6$ lb/in$^2$) | Surface energy (erg/cm$^2$) | Hardness (kg/mm$^2$) | $\bar{p}$ ($10^{-8}$ cm) |
|---|---|---|---|---|---|---|---|
| Zirconium dioxide | Cubic | 5.7 | 2,700 | 35.0 | 530 | 1,150 | 0.46 |
| Alumina | Hexagonal | 3.98 | 2,000 | 60.0 | 740 | 2,190 | 0.34 |
| Titanium carbide | Cubic | 4.92 | 3,030–3,200 | 65.0 | 900 | 2,400 | 0.38 |
| Zirconium carbide | Cubic | 6.8 | 3,030–3,400 | 60.0 | 600 | 2,100 | 0.28 |
| Magnesium oxide | Cubic | 3.58 | 2,850 | 4.4 | 670 | 500 | 1.3 |
| Vanadium carbide | Cubic | 5.48 | 2,730 | 63.0 | 1,250 | 2,500 | 0.5 |
| Nickel | Cubic | 8.19 | 1,455 | 30.0 | 1,700 | 210 | 81.0 |
| Cobalt | Hexagonal | 8.7 | 1,495 | 30.0 | 1,530 | 125 | 12.0 |
| Gold | Cubic | 19.3 | 1,063 | 1.2 | 1,120 | 58 | 19.0 |
| Silver | Cubic | 10.5 | 961 | 11.0 | 920 | 80 | 11.0 |
| Tin | Tetragonal | 7.31 | 232 | 6.0 | 570 | 5.3 | 11.0 |
| Copper | Cubic | 8.43 | 1,083 | 11.0 | 1,100 | 80 | 14.0 |

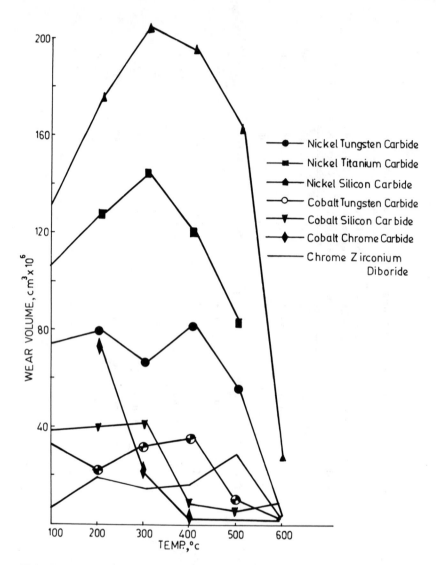

FIGURE 22   Wear resistance of composite coatings. (From Kedward, 1969.)

smooth, and might have potential applications for producing surfaces
for applications in bearings.

## Electrodeposition of Lead Composites

The strengthening of electrodeposited lead and lead alloys was studied
by Wiesner et al. (1970) by codepositing with $Al_2O_3$, $TiO_2$, or $BaSO_4$
particles. They also found that the lead fluoborate baths with lignin
sulfonic acid and coumarin gave whisker-free thick electroforms
of lead, and hence this bath was chosen for dispersion hardening
with particle codeposition. They found that as high as 150 g/L
$BaSO_4$ was required for incorporating an appreciable quantity of
particles in the deposit. Though $BaSO_4$ was dispersed in the deposit,
they had reported that there is no dispersion hardening on the
electroformed lead or its alloy. $TiO_2$ (0.03-1 μm) could be codeposited
with lead when the particle concentration in the bath was as high
as 150 g/L. They had reported that the mechanical strengths of
electroformed lead and its alloys was 0.5%. $TiO_2$ had the highest
creep resistance, and the dispersion of $TiO_2$ in electroformed lead
alloys increases the strength of the deposit.

Figure 23 shows the creep curves for electroformed lead and
some electroformed lead alloys at 23°C. They had also observed
that the bath containing 0.23 g/L each of peptone and resorcinol
invariably yielded lead deposits with increased hardness, and this
was attributed to the carbon codeposition.

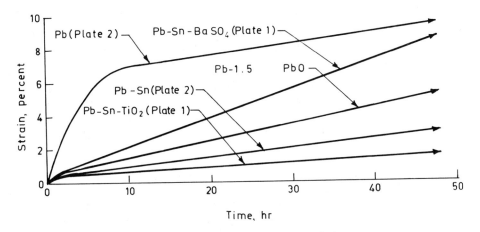

FIGURE 23  Creep curves for electroformed lead and some electroformed lead alloys. (From Wiesner et al., 1970.)

## Copper

With acid copper sulfate plating baths many of the fine powders
such as barium sulfate aluminum oxide and silicon carbide, which
very readily codeposit with nickel plating baths, do not codeposit
from the standard copper sulfate plating baths. However, it was
found (Tomasgewski et al., 1969) that bath-soluble salts of monovalent cations when present in the standard acidic copper plating
baths of either low or high concentration of copper ions help the
codeposition of dispersed fine particles (such as barium sulfate,
strontium sulfate, aluminum oxide, titanium oxide, zirconium oxide,
SiC, $B_4C$).

Of the monovalent cations, thallium, cesium, and rubidium were
outstanding in the promotion of the codeposition of the fine bath-
insoluble inorganic powders. As little as 0.05 g/L of $Tl^+$ was effective
in promoting the deposition of the bath-dispersed fine powders.
The salts of monovalent cations may be carbonates, sulfates, acetates,
hydroxides, fluorides, etc. Amino acids such as alanine, EDTA,
and other amino sequestering agents, especially those with more
than one amino group, give excellent results in promoting codeposition
in acid copper sulfate plating baths.

When codeposition from the acid copper fluoborate bath was
studied, it was found that unlike the acid copper sulfate bath,
many of the bath-insoluble nonconducting powders codeposited with
the copper from the plain bath. The aliphatic polyamine compounds,
amino acids, and the thallous cesium and rubidium help to maximize
the codeposition on vertical cathodes.

*Treatment of the powders.* The commercial grades of the powders
are sieved to get particles with size ranging from 1, 3, 5, 10 μm
and 20 μm. The appropriate size particles are washed with alcohol,
trichloroethylene, or acetone to remove any dirt or oily matter.
The powder is treated with 1:1 HCl for 3 h and then washed with
1:1 nitric acid. The powder is washed thoroughly and treated with
10% solution of sodium sulfate for 2 hr. The particles, after thorough
washing, are soaked in a small amount of the plating solution for
24 hr and then blended with solution as a slurry by stirring at
high speeds. The solution is then decanted, and the powder is
thoroughly washed and added to the plating bath in required proportions (Indira Rajagopal et al., 1981). The percentage codeposition
on a vertical cathode of a fine bath-insoluble nonconducting powder
in the metal plate depends on many factors in the various types
of aqueous electroplating baths. Besides pH, temperature, nature
of the metal, anions, cationic additives, neutral additives, particle
size, etc., there are the important factors of concentration of particles dispersed in the bath and the degree of agitation of the bath
at the cathode surface.

*Conclusions.* Bath-insoluble nonconducting fine particles, such as barium sulfate, alumina, and silicon carbide, dispersed in acid copper sulfate plating baths do not codeposit appreciably on vertical cathodes, unlike nickel baths.

Addition to the acid copper plating baths of monovalent cations such as thallium and cesium ions and amino acids promoted extensive and uniform codeposition of the fine particles. The monovalent cations or the amino compounds do not codeposit with the powders to any appreciable extent. The pH of the copper bath has little influence on the amount of codeposition. The amines, especially polyamines and amino acids, also aid codeposition of particles from other plating baths, such as acidic nickel, zinc, and cadmium baths.

Codeposition of anatase titania with copper from an acidic copper sulfate bath was studied in detail by Athavale and Totlani (1982).

The codeposition of copper with inert particles such as $Al_2O_3$, SiC, $ZrO_2$, $Cr_3C_4$, $TiN_4$, etc., have received less attention in electrocomposite technology. This is because the particles cannot be codeposited easily with copper from an acid sulfate bath. It has been reported by Tomaszewski et al. (1969) that addition of certain cationic promoters such as $CS^+$, $Tl^+$, and $Rb^+$ ions to the plating solutions will promote the codeposition of inert particles with copper.

Chen et al. have reported that if the particles are of a particular crystalline form such as $\alpha$-$Al_2O_3$ or rutile $TiO_2$ they can be codeposited with copper in the absence of promoters.

Lakshminarayan et al. (1976) have shown that the chloride ions inhibit the codeposition of $\alpha$- or $\gamma$-$Al_2O_3$ with copper. Totlani et al. have shown that the presence of 10 ppm of chloride ions to the bath of $TiO_2$ containing 0.54 vol.% $TiO_2$ decreased the codeposition of $TiO_2$ from 6.57 to 1.97 v/o at a CD of 2.5 $A/dm^2$, and the presence of 20 ppm $Cl^-$ completely inhibited the codeposition of $TiO_2$ from a standard acid sulfate copper plating bath. They found that the $TiO_2$ codeposited in the absence of Cl ions was a function of $TiO_2$ concentration in the bath and the CD. They reported that the addition of monovalent ions such as Tl, Rb, or Cs had a beneficial effect on the codeposition of $TiO_2$ with copper, and chloride ions have no inhibiting effect up to a concentration of 20 ppm, which is not otherwise the case.

It was also pointed out that the electrical conductivity of Cu-$TiO_2$ deposits decreases as the amount of $TiO_2$ in the deposits increases by about 4-10% for deposits containing 2.17 to 4.76 vol.% $TiO_2$.

## Gold Composites

Because of its very high tarnish resistance, its high electrical conductivity, and low contact resistance, gold is extensively used
in the electronics industry in electrodeposited form as a contact
surface. For most sliding contact applications (as in the case of
fingers of printed circuit boards) pure gold is too soft and has
an inherent tendency to cold-weld on the mating surface, thus
wear of electrodeposited gold. Hence, it is customary to use a gold-
nickel or gold-cobalt (0.1% Ni or Co) alloy coating to improve the
wear properties and hardness of the deposit. However, by alloying
cobalt or nickel, the mechanical properties are improved, but the
contact resistance and the electrical conductivity, especially at
high temperatures, are badly affected.

Composite plating of gold with dispersed ceramic particles was
considered an alternative way to produce gold deposits with improved
mechanical properties and improved electrical and thermal properties.
The use of gold-based composites for increased strength and creep
resistance has been studied by Sautter (1964). A codeposited gold-
tungsten carbide composite was reported to have better welding and
metal transfer characteristics of reed switch and relay applications
(Peiffer et al., 1969). The sliding wear resistance of gold electrodeposits, while maintaining tarnish resistance and electrical properties,
was studied by Larsen (1976). This was achieved by the use of composite gold electroplates in which conducting tarnish-resistant refractory
carbide particles, such as TiC, were uniformly dispersed in a pure
gold matrix. The gold composite coatings were prepared from acid
or alkaline cyanide gold plating solutions. An increase of hardness
by two times was achieved with 10% V-TiC in gold deposit without
sacrificing other qualities. Buelens, Celis, and Roos (1983) have
shown that, up to 20 g/L of $\alpha$-$Al_2O_3$ of particle size 0.5 $\mu$m, the
morphology of gold deposit from the citrate or phosphate is not
changed at normal CD. It is shown that the addition of small particles
lowers the overpotential required to obtain a given CD. Highest
codeposition occurred at pH 4 and 60°C and at a CD of 0.5-1 $A/dm^2$.

## Chromium-Based Wear-Resistant Coatings

Chromium is attractive as a matrix metal because of its good wear
and oxidation properties. Addison and Kedward (1977) have produced
chromium composites containing up to 5% alumina from a hexavalent
chromium electrolyte containing 400 g/L $Al_2O_3$ and 1 g/L thallous
sulfate. They have reported that up to a temperature of 200°C
the wear resistance of Cr-$Al_2O_3$ composite coatings was higher than
that of pure chromium.

Greco and Baldauf (1968) successfully codeposited $TiO_2$ with
chromium from hexavalent baths and the hardness of the chromium

composite with 1% $TiO_2$ was reported to be two times higher than that of hard chromium. Similar results with improved hardness of $Cr-Al_2O_3$ was reported by Raj Narayan and Chattopadhyay (1982). Kamat (1979) reported that cesium oxide up to 5% could be codeposited with chromium from a hexavalent bath in the presence of tetraethylene pentamine and promoters such as $Ti^+$, $Li^+$, $NH_4^+$, and so on, and larger amounts of the promoters over 5 g/L gave deposits with over 5% inclusions. The consumption of thallium nitrate was about 1 g/50 A·hr/L of bath.

The details regarding the codeposition of refractory oxides, nitrides, carbides, and so on, were given in detail by Young (1974), and the importance of the addition of promoters such as $Tl^+$ was discussed. Addison and Kedward (1977) codeposited $Cr_3C_2$ particles with chromium from a trivalent bath. The wear resistance of the composite coating is lower than that of electrodeposited chromium coatings over a temperature up to 400°C. Young (1975) had reported that, next to diamond, $B_4C$ was most effective in increasing the wear resistance of the coating. The chromium $ZrB_2$ coatings exhibited good wear characteristics between 300° and 400°C. He had studied (1974) the codeposition of diamond particles from the hexavalent chromium baths and found that the inclusion of diamond particles increases with the $Tl^+$ concentrate in the bath and if the diamond particles are plated with electroless nickel; it decreased with increase in CD. The wear resistance of the Cr-diamond composite coatings was very good. The feasibility of codeposition of WC with Cr from hexavalent chromium plating bath was reported by Raj Narayan and Singh (1983).

## 1.7 Composite Diamond Coatings

By the electroplating technique it is possible to incorporate diamond powder into nickel and thus electroform metal-bonded abrasives in any desired shape. In the past two decades, electroforming of diamond tools (Fig. 24) has become an accepted industrial production technique. Nickel was selected as the coating metal because of its higher strength. In the resin-bonded wheel, three types of wear occur: (1) loss of abrasive grains by impacting and frictional forces, (2) chipping or fracture of the grains, and (3) resin failure due to overheating. These problems arise due to a poor chemical bond between grain surface and resin. Electroplated nickel-diamond coatings have very good bonding characteristics, and since nickel acts as a heat sink also, metal-bonded diamond tools last longer, even in dry grinding operations.

The important considerations in producing diamond electroplated surfaces, apart from pure electrochemical factors, are the diamond grit size the quantity of diamond in the electrolyte, and the method

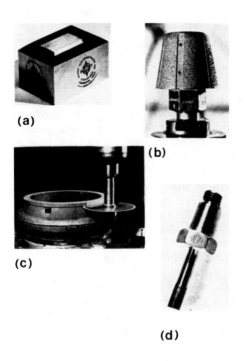

**(a)**

**(b)**

**(c)**

**(d)**

FIGURE 24   Metal-bonded abrasive tools. (From Spiro, 1967.)

of deposition. Since diamond particles have high hardness, low
coefficient of friction, and high thermal conductivity, the metal-
diamond composite coating has many desirable properties. The baths
used for plating are of the sulfamate type (Zahavi, 1983). A cross
section of a typical diamond composite is given in Fig. 25 (Indira
Rajagopal, 1985). The experimental procedure the correlation of
wear-resistant properties of the diamond composite versus particle
size, particle density, coating uniformity, diamond shape and type,
particle size distribution, and matrix properties are discussed by
Roshan (1978). The manufacture of the electrodeposited diamond
wheel and its grinding action is discussed by Sato and Suzuki (1982)
in their excellent article. The applications of diamond-dispersed
composite coating include hand and machine files, hones, cutoff
disks, dressing tools, and grinding wheels mainly for form grinding.
In all applications the process is a multipoint cutting action carried
out by the diamond particles in the nickel coat (Patents, 1981)
with significantly reduced wear of tools. The excellent performance
of shock-synthesized diamond in the finishing of watch jewels,
natural and synthetic gems, precision metal parts, precision electronic

FIGURE 25    Cross section of diamond-impregnated composite × 500 coating. (From Indira Rajagopal, 1985.)

components, and tungsten carbide tools and dies is well known (Sharp, 1975).

In recent times, the most challenging problem is to find a suitable material to machine the high-temperature reusable surface insulation (HRSI) tiles. They are used in a variety of shapes and sizes in each space shuttle, and the site material is produced from silica fiber, which inflicts heavy wear during machining with conventional tools. Electroplated diamond tools could machine these tiles (Daniel, 1981) to the required specifications.

## 1.8    Coatings for Improving the Corrosion Resistance

Considerable improvements in the corrosion resistance of bright nickel-chromium plate can be obtained by depositing the chromium with a controlled microporous structure. Microporous or cracked chromium can be obtained by (1) plating chromium from special electrolytes containing selenic acid as additives, (2) plating chromium over stressed nickel, and (3) use of codeposited nonconducting materials during nickel plating, which provide tiny insulated areas on which subsequently applied chromium does not deposit. Certain decorative nickel deposits produced by codepositing various very fine bath-insoluble nonmetallic particles in the plate have been found to lead to improved corrosion resistance when they are used as the top plate in multiple layer systems before the final chromium is applied.

The improved corrosion resistance can be understood in the following way. The large area of nickel exposed in the pores or cracks results in a very low current density in each pit or crack due to the uniform distribution of the corrosion currents. Under such favorable circumstances the corroding current is wholly or substantially directed to the nickel, and the chromium deposit is relatively uncorroded and remains cathodic to the nickel. With the highly porous chromium overlay, electrochemical corrosion of the underlying nickel is uniformly distributed instead of being highly concentrated over a smaller number of pits as in ordinary chromium plate. In other words, the corrosion current density is higher in ordinary chromium-plated nickel, whereas with microcracked chromium plated nickel the corrosion current density is smaller.

Microcracked chromium can be plated from chromium baths containing selenates (Safraneck, 1960). These deposits have shown excellent corrosion resistance in accelerated and outdoor tests. Its disadvantages are that covering power is (1) poor, (2) the deposits should be thick enough to produce cracks and large locked-in residual stresses often cracks the underlying nickel plate, and (3) a blue coloration develops with thick deposits. A dual chromium plating process, which includes deposition from warm conventional plating solution followed by plating from chromium plating baths containing selenates, has been used to overcome the problems encountered with microcracked chromium. Though the corrosion resistance of dual chromium deposit is good, the higher thickness and cost and greater difficulties in maintaining the bath have discouraged its use in commercial practice.

An entirely new approach was made to get a microcracked chromium using the conventional chromium plating process. To accomplish this a 1/2- to 3-min post-nickel strike (PNS) of highly stressed nickel is plated on bright nickel followed by conventional chromium plating. This process needs an additional nickel plating step.

A new, simple, indirect method of obtaining consistently uniform microporous decorative chromium plate has been developed (Brown and Tomaszewski, 1960; Clauss and Klein, 1968). In this process, dense nonmetallic powders such as certain insoluble sulfates, oxides, and oxalates, are codeposited with nickel. When such nickel deposits are followed by about 0.01 mil of chromium, a high degree of corrosion protection is obtained because a fine, favorable porosity pattern is induced in the chromium. Chromium plate of conventional thickness (0.1 mil) was reported (Brown, 1965). Dense microporosity is induced in the thin final chromium plate when it is plated on top of decorative bright nickel plate containing multitudinous, fine nonmetallic particles

codeposited with the nickel and embedded in the surface. The microporous chromium increases the anodic area and thereby decreases
the rate of corrosion pitting penetration into the nickel plate. When
this bright nickel plate containing the codeposited discrete particles
is used as the final bright nickel plate of a multiple-layer nickel
composite plate, exceptionally good corrosion protection is afforded
to the underlying metal especially in recessed areas. The beneficial
effects are particularly significant and striking for complex-shaped
articles. The types of bath-insoluble powder employed include sulfates, oxides, silicates, fluorides, carbides, and silicides. The
particle size of the powders recommended is .02-.05 μm. They were
dispersed in the plating bath by air agitation and codeposited with
the semibright and fully bright nickel. The concentration of the
powder in the solution is around 50 g/L, and the pH of the nickel
plating bath is 2-3.5 at 50°-60°C. The presence of brighteners
helps increase the rate of codeposition of powders, which in turn
results in maximum corrosion resistance when the chromium plate
is applied. The principles and practical details regarding the production of satin nickel by codeposition of finely dispersed solids are
given by Tomaszewski et al. (1963), and improvements in corrosion
protection with decorative cermet Ni-Cr plate is dealt with by Brown
and Tomaszewski (1966).

The nickel seal process is a proprietary process that utilizes
a composite nickel deposit for producing microporous chromium.
In this process one, two, or three layers of bright nickel are followed
by a nickel flash containing nonmetallic particles. This forms microcracked layers that adhere well on chromium plating.

## 1.9   Heat-Treatable Alloys/Metal Coatings

Bazzard and Boden (1972) have shown that chromium powders can
be codeposited with nickel, and the composite can be treated to
850°-1050°C to produce Ni-Cr alloy coatings. This method seems
to offer a new way to prepare alloys. This is an interesting development in the area of electrocomposite coatings. In their process,
Bazzard and Boden (1972) used a high-speed nickel sulfamate bath
to codeposit chromium particles to give high rates of electrodeposition
required for forming. By controlling the size and concentration
of particles as well as by using rotating cathode with sponge-covered
scrapers, they obtained Ni-Cr composite with low porosity. They
have shown that heat treatment at 1050°C for 12 hr caused diffusion
of chromium particles in an electrodeposited nickel matrix and thereby
formed a nickel-chromium alloy which cannot be obtained as an alloy
plate from conventional plating methods.

Heat-treated electrocomposites of palladium and silver particles
with nickel have been reported to increase the corrosion resistance

of composite coatings in nitric acid solutions by three times over pure
nickel coatings (Saifullin and Nadeeva, 1970). It has been shown that
cupro nickel, stellite, and ball-bearing steels similar to EN 31 can be
produced by this promising technique (Kedward, 1969). It is reported
that the codeposition of tungsten particles with copper resulted in electroforms of copper with improved erosion resistance and antiarcing
properties. This type of electroform is particularly useful as an
anode material in electrochemical machining or in spark erosion
machining applications (Kedward, 1977).

Dust chromium and tungsten powders have been codeposited
with nickel. Such mixtures of metals can be heat treated to diffuse
the metals. Alloys which are difficult to plate from alloy baths can
be prepared this way. Chromium powder can be codeposited in
an iron nickel bath followed by heat treatment to diffuse the chromium
into the matrix. Metal powders and ceramics codeposited simultaneously with chromium and cobalt-molybdenum alloys are being developed
as high-temperature corrosion and oxidation-resistant coatings.
EpiK (1975) codeposited boron particles with nickel and diffusion
bonded the deposits to obtain composite boride coatings. Samsonov
(1977) reported that simultaneous codeposition of chromium and
boron particles with nickel from a chloride bath resulted in a composite with excellent wear characteristics after suitable heat treatment
at 900°-1000°C. Similar work with iron as the matrix and with boron
as the dispersoid has also been reported by Fedorchenko (1975)
and Guslienko (1976). Iron boride coating with improved wear
resistance was obtained by heat treatment at 1150°C, which resulted
in a uniform distribution of boron in iron and formed a solid solution.
Multicomponent alloy coatings could be produced by particle codeposition (Foster, Cameron, and Caren, 1985). In their process, alloy
coatings of nickel or cobalt with chromium, aluminum, and yttrium
(MCrAlY) have been produced by codepositing a prealloyed powder
with nickel or cobalt from a conventional plating solution followed
by suitable heat treatment. By using a rotating barrel and suitable
agitation, they showed that alloy coatings can be produced to specification over a wide CD range. This has been demonstrated on
turbine blade geometries and deposits containing up to 40 vol.%
of prealloyed powder, which after heat treatment produced alloys
typically used as overlay coatings for gas turbine blades. This
process is simple, elegant, and economical for coating MCrAlY on
turbine blades. This process is not yet commercially feasible.

## 1.10 Coatings for Specialized Applications

Composites are also used in the nuclear energy field. Materials
such as $UO_2$ and Pu have been codeposited with nickel to electroform

material which can be used as an ion detector or as a fuel element. Recently, a electrolytic codeposition process has been put to a novel use (Macmillan et al., 1980). Particles of simulated defense and civilian radioactive waste calcines have been incorporated into copper. This method of making metal matrix composite forms that are comparatively stronger and less porous (at the same time highly leach resistant) than metal matrix composite forms made by pressing and sintering has considerable advantages over the casting method because of its use of ambient temperatures and pressures. The radioactive wastes can be easily codeposited with metals of the iron group or with copper. The precipitated heavy metal oxides and hydroxides of thorium, uranium, and neptunium can be codeposited with nickel (Petit et al., 1977). Neutron-absorbing materials such as boron and its compounds codeposited with nickel could be used as a reactor control material. The details regarding production of copper-boron carbide composites used in nuclear industries are covered in the patents of Kennecott Corporation (1981).

## 1.11  Summary and Conclusions

A composite material is a heterogeneous mixture of two or more homogeneous phases bonded together, and it usually has its own distinctive properties unobtainable in either phase individually. The great success of electrocomposites is due not only to their superior mechanical properties, lightweight features, etc., but they also offer the designer an ability to tailor a material to his design requirement.

Metals as materials are potentially superior to both polymers and ceramics in providing composites with good overall mechanical properties, being particularly suitable in situations where good transverse properties are important. They also offer distinct advantages in the temperature range 400°-1000°C where oxidation problems in metals are not yet severe and polymers are forbidden because of their inferior thermal stability.

The use of appropriate coatings with required surface properties on a bulk material chosen for its engineering properties and ease of manufacture not only allows conservation of scarce and expensive materials, but it also reduces component cost and extends its life. Although much progress has been made in the development of new materials, there exists at present no single material which can withstand all the extreme operating conditions in modern technology. Protection of materials from hostile environments has, therefore, become a technical and economic necessity. An ideal material suitable for applications over a wide range of operating conditions should have at least the following properties:

1.  Good mechanical properties over a wide temperature range
2.  Resistance to thermal and mechanical cycling
3.  Resistance to wear, erosion, and heat
4.  Chemically inert so as to withstand corrosion and high-temperature oxidation

Composite plating is indeed a reproducible process, provided important electrolysis conditions such as local current density, bath composition, and fluid flow along the cathode are constant. The actual industrial success of Ni-SiC composites plating has been proved beyond doubt.

By a variety of suspended ceramics and several metal plating processes, a wide range of composite electroplates can be produced, each with its own special properties. The electrocomposite coatings offer the engineer and designer a new means of combating wear, abrasion, and lubrication problems.

Composite plating with copper, nickel, and cobalt as the metal matrix is already being done on an industrial scale. The future of composite plating appears to be in the production of materials with new properties, by means of which urgent technological problems may be solved. By way of illustration, one may mention the production of phosphorescent metallic coatings as well as the production of anode coatings that have a special catalytic effect.

## References

Addison, C. A. and Kedward, E. C. (1977). *Trans. Inst. Met. Finish.*, *55*:41.

Athavale, S. N. and Totlani, M. K. (1982). *J. Electrochem. Soc.*, *31*:119.

Bazzard, R. and Boden, P. J. (1972). *Trans. Inst. Met. Finish.*, *50*:63, 207.

Brandes, E. A. and Goldthorpe, D. (1967). *Metallurgia, 76*:195.

British Patent (1972). 1300, 642.

Brown, D. S. R. and Gow, K. V. (1972). *Plating, 59*:45.

Brown, H. and Silman, H. (1965). *Trans. Inst. Met. Finish.*

Brown, H. and Tomaszewski, T. (1966). Proc. of Surface 66, p. 88.

Bullens, C., Celis, J. P., and Roos, J. R. (1983). *J. App. Electrochem., 13*:541.

Bullens, C., Celis, J. P., and Roos, J. R. (1985). *Trans. Inst. Met. Finish., 63*:6.

Bussen, M. and Perinand, M. (1981). Proc. Colleque. Innovations Technologiques dans les Traitments de Surfaces, Paris, p. 9.

Cameron, B. P., Foster, J., and Carew, J. A. (1979). *Trans. Inst. Met. Finish., 57*:113.

Celis, J. P. and Roos, J. R. (1977). *J. Electrochem. Soc.*, *124*:1508.

Celis, J. P. and Roos, J. R. (1982). *Reviews in Coatings and Corrosion*, *5*:1.

Celis, J. P. and Roos, J. R. (1984). *Surface Engineering* (Ram Kossowsky and S. C. Surghal, ed.) Martinus Nighoff, Boston, p. 614.

Chen, E. S., et al. (1971). *Met. Trans.*, *2*:937.

Clauss, R. J. and Klein, R. W. (1968). Proc. Int. Metal Finishing Conference.

Daniel, P. (1967). *Industrial Diamond Review*, *27*:466.

Daniel, P. (1981). *Industrial Diamond Review*.

Donalowski, W. A. and Morgan, J. K. (1983). *Plating and Surface Finishing*, *70*:48-51.

Doskar, J. and Gabriel, J. (1967). *Met. Finish.*, *65*:71.

Economes, J. and Kingery, W. D. (1953). *J. Amer. Ceramics Soc.*, *36*:403.

Edwards, J. A. (1959). *Products Finishing*, *12* (1).

Epik, A. P., Guslienko, Yu. A., and Sverdlok, N. A. (1975). *Chem. Abs.*, *83*:17692.

Fedorchenko, I. M., et al. (1975). *Chem. Abs.*, *83*:17585 f.

Foster, J. and Cameron, B. (1976). *Trans. Inst. Met. Finish.*, *54*:178.

Foster, J., Cameron, B. P., and Carew, J. A. (1985). *Trans. Inst. Met. Finish.*, *63*:115.

Foster, J. and Kariapper, A. M. J. (1973) (1974). *Trans. Inst. Met. Finish.*, *51*:27, :87.

Foster, J. and White, C. (1981). *Trans. Inst. Met. Finish.*, *59*:8.

Fry, H. and Morris, F. G. (1959). *Electroplating and Metal Finishing*.

Ghouse, M., Viswanathan, M., and Ramachandran, E. G. (1980). *Met. Finish.*, *78* (4):44.

Gilliam, E., McVie, K. M., and Phillips, M. (1966). *J. Inst. of Met.*, *94*:228.

Greco, V. P. and Baldauf, W. (1968). *Plating*, *55*:250.

Guglielmi, N. (1972). *J. Electrochem. Soc.*, *119*:1009.

Guslienko, Yu. A. (1976). *Chem. Abs.*, *84*:186530 k.

Honma, H., Ohtaka, and Mitsin, H. (1980). Proc. 10th Int. Cong. Met. Finish., p. 241.

Indira Rajagopal (1984). *Products Finishing*, Jan:12.

Indira Rajagopal (1985). Unpublished work.

Indira Rajagopal et al. (1981). Proc. 2nd Int. Symp. on Industrial and Oriented Electrochemistry, SAEST, India, 6.29.1.

Joshi, M. N. and Totlani, M. K. (1981). Proc. 2nd Int. Conf. on Electroplating and Metal Finishing, SAEST, Bombay, p. 89.

Joshi, M. N. and Totlani, M. K. (1983). *Know Your Product*, Nov:41.

Kamat, G. R. (1979). *Plating and Surface Finishing, 66*:56.
Kedward, E. C. (1967). *Engineering*, Feb:1.
Kedward, E. C. (1969). *Metallurgia, 79*:225.
Kedward, E. C. (1977). *Cobalt, 3*:53.
Kedward, E. C. and Wright, K. W. (1978). *Plating and Surface Finishing, 65* (8):38
Kedward, E. C., Addison, C. A., and Tennett, A. A. B. (1976). *Trans. Inst. Met. Finish., 54*:8.
Kedward, E. C., Wright, K. W., and Tennett, A. A. B. (1974). *Tribology International*, 7:221.
Kennecott Corp. (1981). U.S. Patent, 4,249,998; 4,235,917.
Kilgore, C. R. (1962). *Products Finishing, 27.*
Kingery, W. D. (1953). *J. Amer. Ceramics Soc., 36*  :362.
Lakshminarayan, G. R., Chen, E. S., and Sautter, F. K. (1976). *Plating and Surface Finishing, 63* (4).
Larson, C. (1976). *Electroplating and Metal Finishing*, 12.
Linderbeck, D. A. and Macalonan, C. G. (1974). *Industrial Diamond Review*, Mar:84.
Macmillan, N. H., Ray, P., and Shaffer, P. T. B. (1980). *Nuclear Technology, 51* (11):97.
Martin, P. W. (1965). *Metal Finishing Journal*, 11:399.
Martin Thoma, M. T. (1984). *Plating and Surface Finish., 71*:51.
Nuoko, V. P. and Shrier, L. L. (1973). *J. App. Electrochem.,* 3:137.
Peiffer, et al. (1969). Proc. 17th Annual National Relay Conf., p. 1.
Petit, G. S., Wright, R. R., Kwanoski, T., and Weber, C. W. (1977). *Plating and Surface Finish., 64*:46.
Rajagopalan, S. R. (1982). Materials processing in space: A workshop (Ramaseshan, S. and Tiwari, M. K., eds.) Indian Academy of Sciences, p. 115.
Raj Narayan and Chattopadhyay. (1982). *Surface Technology,* 16:227.
Raj Narayan and Narayanan, B. H. (1981). *Reviews on Coatings and Corrosion, 4* (2):113.
Raj Narayan and Surjit Singh. (1983). *Metal Finishing, 81*:45.
Ramaswamy, A. and Totlani, M. K. (1977). *J. Electrochem. Soc.* (India), 26.
Roos, J. R., Celis, J. P., Buelens, C., and Goris, D. (1984). Process Metallurgy (Warren, I. H., ed.), Elsivier, Amsterdam, p. 197.
Roos, J. R., Celis, J. P., and Helsen, J. A. (1977). *Trans. Inst. Met. Finish., 55*:113.
Roshon, D. D. (1978). *IBM J. Res. and Dev., 22* (6):681.
Safranek, W. J. et al. (1960). *Plating, 47*:1027.
Saifullin, R. S. and Nadeeva, F. I. (1970). *Chem. Abs., 73*:126374a.

Saifullin, R. S. et al. (1973). *Prikl. Electrokhemi.*, *1-2*:50.

Samsonov, G. V., Zhunkovskii, G. L., Luchka, M. V., and
Kindrachuk, M. V. (1977). *Chem. Abs.*, *86*:1627089.

Sato, K. and Suzuki, K. (1982). *J. Metal Fin. Soc. of Japan, 33*:285.

Sautter, F. K. (1963). *J. Electrochem. Soc.*, *110*:557.

Sautter, F. K. (1964). *Metall, 18*   :596.

Seyle, E. J., et al. (1957). *Proc. Amer. Electroplater's Soc.*, *44*:29.

Sharp, W. F. (1975). *Wear, 32*:315.

Sinha, P. K., Dhananjayan, N., and Chakrabortti, H. K. (1973).
*Plating, 60*:55.

Snaith, D. W. and Groves, P. D. (1972). *Trans. Inst. Met. Finish.*,
*50*:95.

Stott, F. H. and Ashby, D. J. (1978). *Corrosion Science, 18*:183.

Stott, F. H., Lin, D. S., and Wood, G. C. (1973). *Corrosion
Science, 13*:449.

Sykes, J. M. and Alner, D. J. (1974). *Trans. Inst. Met. Finish.*,
*52*:28.

Tomaszewskii, T. W. (1976). *Trans. Inst. Met. Finish.*, *54*:45.

Tomaszewskii, T. W., Clauss, R. J., and Brown, H. (1963). *Proc.
Amer. Electroplater's Soc.*, 169.

Tomaszewskii, T. W., Tomaszewski, L. C., and Brown, H. (1969).
*Plating, 56*:1234.

U.S. Patent (1981). 4275528.

U.S. Patent (1981). 4302300.

Vest, C. E. and Bazzare, D. F. (1967). *Met. Finish.*, *65*(11):52.

Wapler, H., Spoonen, T. A., and Balfour, A. M. (1980). *Tribology
International*, 21.

Ward, J. J. B. and Christie, I. R. A. (1974). *Trans. Inst. Met.
Finish.*, *52*:87.

White, C. and Foster, J. (1978). *Trans. Inst. Met. Finish.*, *56*:92.

White, C. and Foster, J. (1981). *Trans. Inst. Met. Finish.*, *59*:8.

Wiesner, H. J., Fray, W. P., Vandervoort, R. R., and Raymond,
E. L. (1970). *Plating, 57*:358.

Williams, R. V. (1966). *Electroplating and Metal Finishing, 19*:92.

Williams, R. V. and Martin, P. W. (1964). *Trans. Inst. Met. Finish.*,
*42*:182.

Wilson, K. S. and Malak, T. P. (1970). 6th S.D.C.E. Conf. Paper
No. 43.

Withers, J. C. (1962). *Products Finishing, 26*:62.

Young, J. P. (1974). *NBSIR, 74*:614.

Young, J. P. (1975). *Plating and Surface Finishing, 62*:348.

Yuko Tsuya, Hiroshi Vemura, Voichiro Gkamoto, and Shigehiko-
Kurosaki. (1974). *ASLE Trans.*, *17* (3):229-235.

Zahavi, J. and Hazan, J. (1983). *Plating and Surface Finishing,
70*:57.

Zahavi, J. and Hava Kerbel (1982). *Plating and Surface Finishing,
69*:76.

## 2   FIBER-REINFORCED METAL COMPOSITES (FRMC)

Examples of composite materials are paper, foam material (e.g.,
cork), carbide cutting tools, pneumatic tires, and fiber-reinforced
polymers and metals. Fiber reinforcement of polymers produces
strong, stiff material with low density. A much more attractive
feature of this class of composites is that it offers the designer
the ability to tailor a material to his design requirements. The
recent design innovations in the aircraft industry, such as flexible
wing, mission-adaptive wing, and forward-swept wing, have come
about because graphite and boron-fiber-reinforced polymer composites
could be tailored directionally in the fabrication of the wing to
control the load distribution.

However, fiber-reinforced polymer composites (i) cannot be
used at higher temperatures, (ii) have a tendency to delaminate,
and (iii) have poor resistance to erosion. To overcome these problems,
fiber reinforcement in a metallic matrix has been investigated. Such
composites may not solve these other problems, but they can certainly
stand much higher temperatures.

Several techniques, such as liquid metal infiltration, compaction
and sintering of powders, vapor deposition, spraying, and electroplating, can be used for fabricating fiber-reinforced metal matrix
composites. Electroforming these composites is attractive because
the method, in general, produces the required component without
additional processing, and because it is carried out at lower temperatures (<100°C), so there is no risk of damaging the fiber.

Two methods of fabrication have been used for fiber-reinforced
polymer composites. One is filament winding, which consists of
winding over a mandrel, a resin-coated fiber, at an angle of 0°
to 90° (or much larger) to the axis of rotation, with subsequent
curing and extraction to realize a composite structure. Another
process is to compact, by applying pressure and temperature, tapes
of fibers held by a partially cured resin matrix. Both processes
have been studied for making fiber-reinforced metal composites
by electroforming. These investigations and the properties of electroformed FRMC are discussed below.

### 2.1   The Mechanics of Fiber Reinforcement

The aim of FRMC is to combine the ductility of metal with the strength
and stiffness of a ceramic or metal filament. To achieve this, fibers
must be separated from each other and oriented in the direction
of the load. They must be strongly bound to the matrix. When
a load is applied to such a system, the strain is the same for the
matrix and the fiber. Because the fiber strength is much greater

than that of the matrix, the load on the fiber has to be larger than that on the matrix to make the strains on both equal. In other words, in a fiber-reinforced composite the fibers carry the major portion of the load and the matrix metal simply serves as a binder to transfer the load.

The principles of fiber reinforcement have been discussed by Kelly and Davies (1965) in their classic paper and by Kelly (1973). The ultimate strength of such composites, $S_c$, with continuous fibers is expressed by the "mixture rule"

$$S_c = S_f V_f + S_m'(1 - V_f) \qquad 2.1$$

where $S_f$ is the breaking strength of the fibers, $S_m'$ is the stress present in the matrix at the fracture strain of the fibers, and $V_f$ is the volume fraction of the fiber. The volume fraction of the matrix, $V_m$, is equal to $1 - V_f$. This rule of mixtures is valid for mixtures with $V_f$ larger than a critical volume fraction $V_c$ defined by

$$V_c = \frac{S_{u,m} - S_m}{S_f + (S_{um} - S_m)} \qquad 2.2$$

where $S_{u,m}$ is the tensile strength of matrix.

An alternative way of expressing the mixture rule would be

$$\frac{\sigma_f}{\sigma_m} = \frac{E_f}{E_m} \frac{V_f}{V_m} \qquad 2.3$$

where $\sigma_f$ and $\sigma_m$ represent, respectively, the stress in the fiber and that in the matrix.

Figure 26 shows a schematic sketch of the stress-strain diagram for fiber, matrix, and composite. It is readily seen that at any given strain, the stress on the fiber is more than that on the matrix. At A, both the fiber and the matrix are strained elastically. At B, the fiber is still strained elastically, but the matrix is strained plastically. At C, the fiber fractures, rapidly followed by the fracture of the matrix at $C_m$. It would be interesting to make some calculations making use of Eq. 2.3. For a boron fiber in an aluminum matrix the $E_f/E_m$ ratio is about 5.5. At 50% volume fraction of fiber, the fiber will be taking 5.5 times more load than the matrix. With a graphite whisker in an aluminum matrix, the fiber will be taking a load 14 times larger than the matrix for the same volume fraction of fiber because the modulus of elasticity for the graphite whisker is 14 times that of aluminum. In general, the load taken by the fiber increases with an increase in fiber modulus of elasticity and volume fraction.

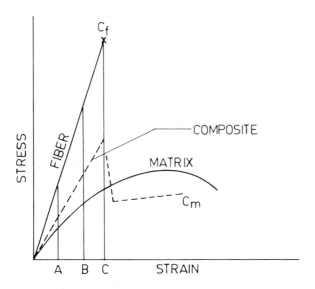

FIGURE 26   Schematic sketch of stress-strain curve for fiber-
reinforced composite.

The modulus of elasticity of fiber-reinforced composite, $E_c$, is

$$E_c = E_f V_f + \left( \frac{d\sigma_m}{de_m} \right)_e (1 - V_f) \qquad\qquad 2.4$$

where $(d\sigma_m/de_m)_e$ is the slope of the stress-strain curve of the
matrix at the strain of the composite.
        The above discussion is somewhat simplified but is reasonably
accurate for practical purposes. However, deviations may be noticed
if the orientation of the fibers is not strictly parallel to the load
direction. The strength may also decrease if the filaments are not
continuous. From the point of view of load transfer, a fiber whose
length is 10 times that of its diameter behaves just like a fiber
of infinite length. If the aspect ratio (i.e., the ratio of length
to diameter) is 9, load transference is 97% of that when the fibers
are infinitely long (Ramaseshan, 1971).
        The strength of a fiber-reinforced composite is lowered if the
fibers are inclined in the direction of the applied tensile stress.
The strength is extremely sensitive to the angle of inclination $\phi$,
as illustrated in Fig. 27. A practical way of reducing this extreme
sensitivity of fibrous composite to orientation is to bond laminated
sheets each containing aligned fibers. The layers are stacked such

that the misorientation alternates (i.e., $+\phi$, $-\phi$, $+\phi$, ...). This arrangement permits a fair amount of strength to be realized even when the angle of misalignment is large (Fig. 28).

Whenever it is not possible to produce a component by laminating sheets of fibrous composite, helical winding at a high angle or at 0° and 90° to the axis of rotation is employed to get strengthening in more than one direction. For a component of complex shape, such a filament winding technique is the obvious choice.

## 2.2 Choice of Fibers and Matrices

An engineer usually wants materials in a structure to be strong, light, and stiff; that is, the specific strength (ultimate tensile strength S divided by specific gravity $\rho$) and the specific stiffness

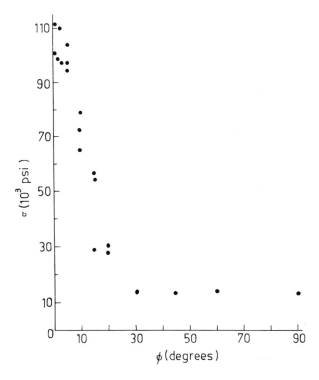

FIGURE 27 Measured variation of tensile strength $\sigma$ with angle $\phi$ between fiber direction and tensile axis. The specimen consisted of $Al-SiO_2$ fiber (50% by volume) composite. (From Kelly, 1973; reproduced with the kind permission of Oxford University Press, London.)

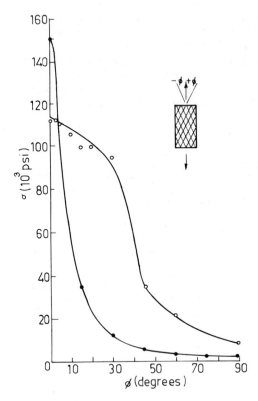

FIGURE 28  Measured variation of tensile strength σ with angle
φ for specimens containing alternate layers of fibers. In each layer
the fibers are continuous and parallel. Alternate layers are at +φ
and -φ to the tensile axis. Open circles = 40% SiO$_2$ fiber in Al matrix.
Full circles = 66% E glass fiber in epoxy resin. (From Fig. 5.23b of
Kelly 1973; reproduced with kind permission of Oxford University
Press, London.)

E/ρ  should be large. Therefore one should choose fibers and
matrices with high strength, large elastic modulus, and low density.

    Metallic composites (i.e., composites of nonmetallic fiber in
metallic matrix or metallic filaments in metal matrix) possess higher
density than polymeric composites. Hence, specific strengths of
metallic composites will be smaller than those of fiber-reinforced
polymers.

    Whiskers, which are single crystalline filaments, can be used
since they are available with diameters from submicron to several

microns and have aspect ratios of 50 to 10,000 (Roos, 1971). Properties of some of the whiskers are given in Table 3. In view of their large aspect ratio, they will behave like a continuous filament as far as strength is concerned, but the major problem with this material is alignment.

Fibers, both metallic and nonmetallic, are therefore the obvious choice for use as reinforcement. The desired characteristics are

1. Relatively high strength and stiffness
2. Easy handling
3. Compatible with the matrix

Properties of some of the common fibers are given in Table 4. Some of the fibers that have been employed for electroforming FRMC are tungsten, steel, boron, graphite, SiC, and glass.

Metals have many valuable properties that make them ideal candidates for matrix material. They have three basic properties, namely, high strength, stiffness, and relative insensitivity to extremes of temperatures. Plastic flow contributes to preventing notch sensitivity of the composite. These properties give metals considerable advantages over polymers or ceramics as matrix material. However, in most pure metals dislocations move rapidly under relatively low applied stresses, so the metallic matrix makes little contribution to the elastic modulus of the composite and is liable to failure by fatigue. They are excellent binders for strong fibers, particularly at elevated temperatures.

The matrix

1. Holds the fibers together
2. Protects the surfaces of the fibers
3. Separates the fibers to prevent cracks from spreading from one fiber to another
4. Distributes the applied forces evenly on the fibers

## 2.3 Electroplating FRMC

Incorporating a continuous filament or discontinuous fiber into an electrodeposit is simple in principle but poses several difficulties in practice. Most of these difficulties concern obtaining a continuous, void-free matrix and securing a large volume fraction of fiber.

Two processes are possible. The first is the filament winding method, and the second is processing an electroformed warp sheet. Both techniques depend upon electroforming. The suitability of various electroplating baths for this purpose has been discussed by Harris et al. (1971). The composition and operating conditions of baths considered by them as suitable are given in Table 5.

TABLE 3  Properties of Whiskers

| Material | Specific gravity ρ | Tensile strength, $S_f$ × $10^{-3}$ ($MP_a$) | Elastic modulus, $E_f$ × $10^{-6}$ ($MP_a$) | Specific strength, $S_f/\rho$ × $10^{-3}$ | Specific stiffness, $E_f/\rho$ × $10^{-6}$ | Diameter (μm) |
|---|---|---|---|---|---|---|
| Sapphire (alumina) | 3.95 | 20.69 | 0.428 | 5.24 | 0.108 | 3–10 |
| BeO | 2.85 | 13.10 | 0.345 | 4.6 | 0.12 | 10–30 |
| $B_4C$ | 2.52 | 13.79 | 0.483 | 5.47 | 0.192 | — |
| SiC | 3.21 | 20.69 | 0.483 | 6.44 | 0.15 | 1–3 |
| SiN | 3.18 | 13.79 | 0.379 | 4.34 | 0.119 | — |

Source: From Roos (1971).

TABLE 4  Properties of Some Common Fibers

| Material | Specific gravity ρ | Tensile strength, $S_f$ × $10^{-3}$ ($MP_a$) | Elastic modulus, $E_f$ × $10^{-6}$ ($MP_a$) | Specific strength, $S_f/\rho$ × $10^{-3}$ | Specific stiffness, $E_f/\rho$ × $10^{-6}$ |
|---|---|---|---|---|---|
| Boron | 2.574 | 2.069–3.448 | 0.379–0.414 | 0.8–1.34 | 0.147–0.161 |
| SiC | 3.459 | 2.069–2.758 | 0.483 | 0.6–0.8 | 0.14 |
| Tungsten | 19.234 | 4.275 | 0.4 | 0.22 | 0.02 |
| Steel | 7.804 | 4.137 | 0.2 | 0.53 | 0.026 |
| E-glass | 2.546 | 3.448 | 0.072 | 1.35 | 0.028 |
| S-glass | 2.49 | 4.827 | 0.085 | 1.94 | 0.034 |
| Graphite | 1.49 | 2.758 | 0.372 | 1.85 | 0.25 |

Source: From Greco, Wallace, and Cesaro (1969).

TABLE 5 Classification and Operating Conditions of Electroplating Solutions

| Deposited metal | Electrodeposition solution[a] | Solution composition and normal operating condition (CD = current density) | Possible application of composite | Type of deposit |
|---|---|---|---|---|
| Nickel | Sulfamate: 1, 1, X | Nickel sulfamate 600 g/L<br>Nickel chloride 10 g/L<br>Boric acid 40 g/L<br>Temperature 60–70°C<br>pH 3.8–4.2<br>CD up to 8600 A/m$^2$ | High-temperature structural | Continuous |
| Nickel/cobalt alloy (60/40) | Sulfamate: X | Nickel sulfamate 200 g/L<br>Cobalt sulfamate[b] 24 g/L<br>Boric acid 35 g/L<br>Temperature 45°C<br>pH 3.8–4.2<br>CD 320–540 A/m$^2$ | High-temperature structural | Continuous |
| Copper | Sulfate: X | Copper sulfate 188 g/L<br>Sulfuric acid 74 g/L<br>Temperature 15–50°C<br>CD 320–540 A/m$^2$ | Bearings | Continuous |
| | Fluoborate: X | Copper 120 g/L<br>Fluoboric acid 31 g/L<br>Temperature 27–49°C<br>CD up to 4300 A/m$^2$ | | Continuous |

(continued)

Table 5 (continued)

| Deposited metal | Electrodeposition solution[a] | Solution composition and normal operating condition (CD = current density) | | Possible application of composite | Type of deposit |
|---|---|---|---|---|---|
| [Copper] | Pyrophosphate: b, Y | Copper pyrophosphate<br>Potassium pyrophosphate<br>Ammonia<br>Temperature 50–60°C<br>pH 8.0–8.6<br>CD up to 320 A/m² | 94 g/L<br>300 g/L<br>2 ml/L | | Semicontinuous |
| | Cyanide: a, 2, Y | Copper<br>Free potassium cyanide<br>Temperature 54–66°C<br>pH 11.5<br>CD 215 A/m² | 38 g/L<br>28 g/L | | Discontinuous |
| Bronze (90/10) | Copper cyanide/stannate: a, 3, Y | Copper<br>Tin<br>Sodium hydroxide<br>Free sodium hydroxide<br>Temperature 60–65°C<br>pH 8.5–9.0<br>CD 540 A/m² | 32 g/L<br>16 g/L<br>10 g/L<br>15 g/L | Bearings | Discontinuous |
| | Copper cyanide/pyrophosphate: c, Y | Copper<br>Stannous tin<br>Stannic tin<br>Free potassium cyanide<br>Temperature 60°C | 15 g/L<br>0.5 g/L<br>4 g/L<br>5–10 g/L | | Discontinuous |

| | | | | |
|---|---|---|---|---|
| | | pH 9.5<br>CD 215–430 A/m$^2$ | | |
| Lead | Fluoborate: a, X | Lead 100 g/L<br>Free fluoboric acid 44 g/L<br>g/L<br>Temperature 15°C<br>CD 160–215 A/m$^2$ | Bearings | Continuous |
| Lead/tin alloy (70/30) | Fluoborate: a, X | Lead 62 g/L<br>Tin 28 g/L<br>Free fluoboric acid 40 g/L<br>Temperature 27–38°C<br>CD 320 A/m$^2$ | Bearings | Continuous |
| Silver | Cyanide: Y | Silver cyanide 44 g/L<br>Potassium cyanide 45 g/L<br>Sodium carbonate 15 g/L<br>Potassium hydroxide 4 g/L<br>Temperature 38–47°C<br>CD up to 1180 A/m$^2$ | Bearings | Semicontinuous |
| Aluminum | Ethereal: Z | Aluminum trichloride 16 g/L<br>Lithium aluminum hydride in ether<br>Temperature 23°C<br>CD 160–1180 A/m$^2$ | Structural medium temperature | Continuous |

[a]Manufacturer

a = W. Canning and Co., Ltd (UK); b = Albright and Wilson Mfg., Ltd. (UK); c = The Enquist Chemical Co. Inc. (USA); X = simple salt solution; Y = complex salt solution; Z = nonaqueous solution.

Trade name

1 = Ni-Speed; 2 = Cuprax; 3 = Penybron Pyrobrite Lustralite

[b]Made by dissolving cobaltous carbonate in sulfamic acid.

*Source:* From Harris et al. (1973).

*The Continuous Filament Winding Method*

In this process, winding the filament on the mandrel and plating it
are carried out simultaneously. For winding, a drive mechanism
similar to that of the coil maker is used. The mandrel is immersed
in the plating bath and rotated. The filament is supplied through
a distributor which is connected to a traverse mechanism. This
mechanism is connected to a variable-ratio micrometer drive so that
the spacing between windings can be varied. The traverse mechanism
automatically reverses its direction after traveling a preselected
distance, thereby winding layer after layer over the required length.
The inclination of the distributor is adjusted to vary the angle of
the winding. The speed of mandrel rotation is varied by varying
the rpm of the drive motor. The above mechanism permits winding
the filament over a desired length at any preselected angle and
spacing. A typical setup is shown schematically in Fig. 29 (Harris
et al., 1971). The experimental arrangements employed by others
(Baker, Harris, and Holmes, 1967; Hansen, DuPree, and Lui, 1968;
Withers and Abrams, 1968; and Wallace and Greco, 1970) are similar.

Harris et al. (1971) have investigated in detail the influence
of various parameters on the porosity of the composite made by
this process. Their results are summarized below:

For conducting fibers, porosity increases with an increase in
fiber volume fraction, winding angle, fiber diameter, and number
of fibers wound simultaneously. Consequently, the above method
can only be employed for composites with a small angle of winding.
An example of the same is shown in Fig. 30. In this case, the shape
is simple and hence permits a low winding angle. In a more complex
shape, like the one shown in Fig. 30, one might expect that a small
winding angle would lead to slippage, especially at the conical section,
because of the large cone angle (30°). However, simultaneous winding and plating avoids this difficulty in conducting fibers because
plating bonds the filaments and thereby prevents slippage.

Obviously, pressure vessels with optimum winding angle cannot
be electroformed satisfactorily by this procedure. Harris et al.
(1971) suggest that this may be overcome with a sophisticated polar
winding machine which would wind a number of layers with 0° angles
followed by a number of layers with 90° angles. The apparatus
proposed by them is shown schematically in Fig. 31.

Baker, Harris, and Homes (1967) found for conducting fibers
that an increasing rate of deposition at constant volume fraction
results in higher porosity. Harris et al. (1971) investigated this
aspect and showed that such an increase in porosity is due to
deposition on the conducting filament during its travel from the
surface of the solution to the mandrel. The porosity becomes independent of current density if the distance between the plating

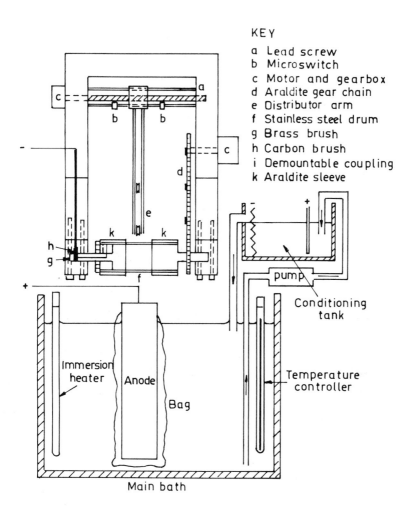

KEY

a Lead screw
b Microswitch
c Motor and gearbox
d Araldite gear chain
e Distributor arm
f Stainless steel drum
g Brass brush
h Carbon brush
i Demountable coupling
k Araldite sleeve

FIGURE 29  Schematic sketch of the winding frame and electroforming setup. The winding frame is shown above the bath with a view to depict its details. (From Harris et al., 1971; reproduced with the kind permission of the Institute of Metal Finishing, London.)

FIGURE 30   Electroformed FRMC produced by filament winding.
(Left) Low-angle wound tube of Ni-W wire composite. (Right) More
complex solid of revolution consisting of copper-nickel wires in
a nickel matrix. Both were electroformed with Ni sulfamate bath
of W by Canning & Co. Ltd., U.K. at 0.32 A/cm$^2$ and 60°C. (From
Harris et al., 1971; reproduced with the kind permission of The
Institute of Metal Finishing, London.)

solution and the top of the mandrel is minimal. In other words,
the extent of coating on the fibers before they are wound is negligible.

The major problem of electroforming FRMC is porosity, due
to formation of voids. Alexander, Withers, and Macklin (1967)
classified the porosity in conducting fiber composites into two types.
They are represented in Fig. 32. Voids of the first type arise because the rate of growth of deposit on the conducting fiber is larger
than that on the substrate. This results in the impingement of growth

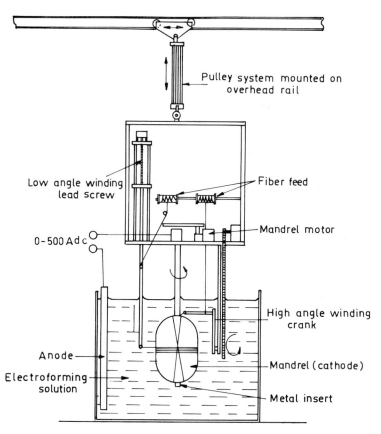

Pulley system mounted on overhead rail

Low angle winding lead screw

Fiber feed

Mandrel motor

0-500 Adc

High angle winding crank

Anode

Mandrel (cathode)

Electroforming solution

Metal insert

FIGURE 31 Schematic of a proposed polar winding machine for electroforming pressure vessels. (After Harris et al., 1971; reproduced with the kind permission of The Institute of Metal Finishing, London.)

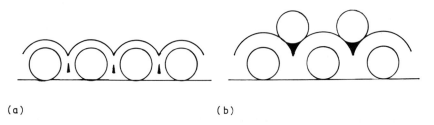

(a)                              (b)

FIGURE 32 Schematic representation of void formation in electroforming FRMC with conducting fiber. (From Alexander, Withers, and Macklin, 1967.)

fronts from neighboring filaments before the growth front of the
substrate reaches the point of intersection. This type of void will
occur irrespective of whether it is a single layer or multilayer.
In contrast, type 2 voids occur only in multilayer windings. As
Fig. 33 shows, the void arises when the filament sits in the valley
between the fibers of the previous layer. Such valley formation
is the result of uniform growth of deposit on all sides of the filament.

The above types of voids are not desirable because they entrap
electrolytes, which causes corrosion in service. They also lower
the strength of the composite. Several types of solutions have been
suggested. One is a geometric solution (Wallace and Greco, 1970).
Referring to Fig. 33, we can see that the distance between the
plane passing through the centers of fibers of a row and the substrate will equal r, the fiber radius, irrespective of the distance
of separation, S, of the fibers. If S < 2r, the diameter of the fiber,
then void formation will certainly occur even if the rates of deposition on the fiber and substrate are the same. However, if S = 2r,
void formation may not occur, if the microthrowing power of the
bath is good. The probability of void formation will rapidly decrease
as S is increased beyond 2r. Larger distance of separation are
to be avoided because they result in lower fiber volume fraction
and, hence, lower composite strength.

From the above discussion, it is clear that the electrolyte used
for electroforming must have good microthrowing power. Thus a
simple salt bath is preferred. An electrolyte with good leveling
power will be even better. One disadvantage is that such leveled
deposits are brittle.

A brittle matrix can affect the strength of the composite.
Donovan and Watson-Adams (1970) got by this problem by using

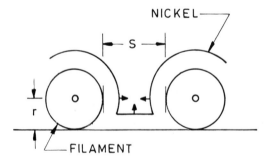

FIGURE 33  Schematic representation of growth fronts over conducting fibers to show the importance of distance of separation of adjacent
fibers. (From Wallace and Greco, 1970.)

cadmium salt as a leveling additive for Watt's bath. Cadmium salt
does not cause brittleness to the same extent as organic leveling
agents do. They showed that type 1 voids do not form when a
Watt's bath containing cadmium is employed to electroform a Ni-steel
wire composite. In addition, they found that even the plane of
weakness, caused by the impingement of growth fronts of neighboring filaments, disappears. The plane of weakness does not affect
strength in the direction of the fiber in strongly oriented unidirec-
tional composites. It will influence strength in the lateral direction.
Hence, in practical composites the voids are likely to affect the
strength and cause an undue sensitivity to fatigue.

Periodic reverse (PR) current plating has also been suggested.
Reports on the benefits from the technique are contradictory.
Donovan and Watson-Adams (1970) found that it did not reduce
the voids for the Ni-steel wire composite. But Withers and Abrams
(1968) found it to be useful for an Al-B fiber composite. One difference between the two systems is that in the former case the fiber
is conducting, whereas in the latter case the filament is nonconducting. Perhaps, the observed difference in respect of PR plating is
attributable to this.

Postplating heat treatment has been used for eliminating the
voids. The results of Harris et al. (1971) show that heat treatment
at 800°C significantly reduces the voids in Ni-copper-nickel wire
(38 μm) composite (Fig. 34). Donovan and Watson-Adams (1970)
found that, by heat treating, the plane of weakness, due to impingement boundaries, can be removed. Healing of voids by heat treatment
is far easier with fine fibers (0.5 μm) according to Broutman and
Krock (1967).

Carbon fibers, though conducting, pose some special problems.
They are generally available only as multifilament tows. If they
are used for winding, the plating does not "penetrate" into the
center of the tow. To get a uniform deposit coverage, the two is
given a preliminary coating of the matrix material while it is in
a relatively open condition. It is then wound on a drum and plating
is carried out simultaneously, as shown in Fig. 35. As the tow
is wound onto the drum, the fibers pack together and plating is
mainly confined to the outer surface. Adjacent tows are then bound
together by further plating. Stripping the electroform from the
drun results in a warp sheet of aligned carbon fibers in the matrix,
which is suitable for hot-pressing.

Indira Rajagopal et al. (1987) showed that deposits can be
"thrown" into the interior or carbon fiber tows by a suitable sur-
factant, which can wet carbon fiber, and by a pulse plating technique. Loose carbon fibers and microstructure of a warp sheet
plated by this technique are shown in Fig. 36. The coating on the
individual fiber is perfect in all regions of the tow.

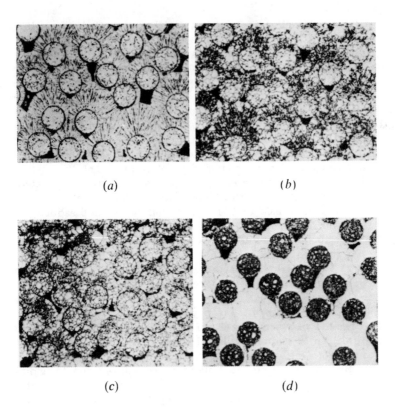

(a)                                                    (b)

(c)                                                    (d)

FIGURE 34   Photomicrographs showing the effect of heat treatment
on the microstructure of electroformed nickel composite containing
35% by volume of 38-μm Cu-Ni wire. (a) As-plated × 135. (b) Heat
treated at 500°C for 3 hr × 135. (c) Heat treated at 575°C for
3 hr × 135. (From Harris et al., 1971; reproduced with the kind
permission of The Institute of Metal Finishing, London.)


There is a difference in the way a deposit grows around a
conducting fiber, in contrast to an insulating fiber. In the latter
case the deposit does not grow on the filament, but only grows
round it and encapsulates it. The neighboring fibers do not act
as "thieves." Hence, voids of the types discussed above should
not occur in composites with insulating fibers. This conclusion is
corroborated by the observations of Withers and Abrams (1968),
Wallace and Greco (1970), and Indira Rajagopal et al. (1980, 1987).
However, the results of Harris et al. (1971) are contrary to this
expectation. They find more voids in insulating fiber composites.

Harris et al. (1971) explain their results in terms of the differences in growth pattern between conducting fiber (Fig. 34) and insulating fiber composites, which were pointed out by Withers and Abrams (1968). The growth pattern for the latter case is shown schematically in Fig. 37. The deposit grows between fibers. The growth front advances more rapidly outward than laterally. Consequently, a hump is formed, leading to valleys which result in voids of type 2. However, at points of crossover, the fiber does not sit in the valleys, thereby leading to a locally higher volume fraction. Such regions of disarray cause the fibers to slip, thereby increasing the voids.

The present authors feel that the voids noticed by Harris et al. (1971) are mainly due to slippage of fibers. In conducting fibers the filaments are bonded, by the deposition that takes place on it, as soon as they contact the mandrel. Such a bonding is clearly not possible in insulating fibers. Because of this difference, slippage is more likely in insulating fibers, especially with continuous winding. Wallace and Greco (1970) and Indira Rajagopal et al. (1980, 1987) used a batch process, described in the next section, which consists of alternately winding and plating. There is less slippage with this method.

Even though nonconducting fibers can be used as such for electroforming composites, some authors have preferred to make

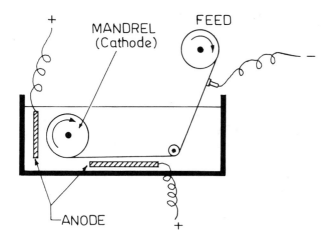

FIGURE 35 Schematic illustrating plating of carbon fiber tone prior to winding to facilitate penetration of the deposit into the center of the tow.

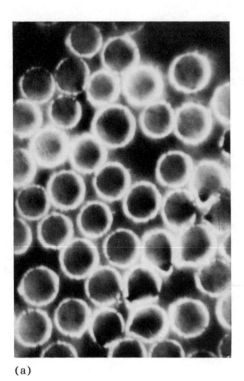

(a)

FIGURE 36  Pulse nickel-plated carbon fiber. (a) Plated carbon
fiber. (b) Scanning electron micrograph of cross section of nickel-
plated carbon fiber.

the fibers conducting by precoating them. For instance, Cooper
(1970) made silica fibers conducting by cracking acetylene on their
hot surface, and then used these coated fibers for composite preparation. A photograph from his paper is reproduced in Fig. 38 since
it clearly shows that the growth pattern of a conducting fiber is
virtually identical to the schematic representation in Fig. 33.

There are instances of R & D groups showing a preference
for nonconducting fibers. For example, Wallace and Greco (1970)
preferred to make boron surfaces nonconducting by coating it with
boron nitride and then use them for making composites.

*Electroforming FRMC by Alternate Winding and Plating*

FRMC components could be conveniently prepared by a modified
electroforming procedure. The fibers glass, tungsten, and carbon

(b)

are wound on a stainless steel mandrel which is electrodeposited
with the metal (chosen as matrix) to a thickness of a few microns.
This is immersed in a plating bath of the matrix metal. A layer
of the metal is deposited by low-current-density electrolysis till
all the fibers are completely covered by the electrodeposit. Then
the next layer is wound, and the plating is continued as before.
Thus layer after layer the composite is built up. The electroplating
conditions are to be so chosen that the deposit will not preferentially
grow outward and will not be laminated with relatively poor adhesion
between laminae. This procedure was used by Wallace and Greco
(1970) and Indira Rajagopal et al. (1980, 1987).

Indira Rajagopal et al. (1980, 1987) electroformed Cu-glass
fiber and Ni-tungsten fiber composites by this method. To facilitate
adhesion and to improve uniformity of distribution of the deposit
thickness, they used pulse plating. Tubes of Cu-glass fiber and
Ni-tungsten wire composites electroformed by this method are shown
in Fig. 39 and Fig. 40.

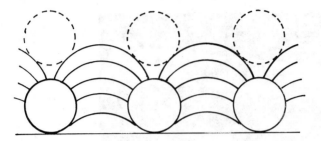

**FIGURE 37** Schematic representation of growth pattern for insulating fiber composite.

## Preparation of Warp Sheet

Good fill without voids can be maintained with filaments up to 50 μm diameter. However, more advanced reinforcements like boron, SiC, and beryllium are available only in larger diameters (100 μm). These are conducting and hence pose problems of void formation. As pointed out earlier, carbon fibers pose difficulties since they are only available as tows.

A convenient method of using them would be to combine electroforming and hot-pressing. The special advantages (minimal damage of fibers and absence of reaction between fiber and matrix) during fabricating of the total electroforming of the composite will be lost.

For hot-pressing, warp sheets are needed. These are electroformed by winding one layer of fibers on a plated cylindrical mandrel and plating until the fibers are covered. The electroformed layer is then slit, separated from the mandrel, flattened, and used as a warp sheet for subsequent processing by hot-pressing.

## Electroforming FRMC with Whiskers as Reinforcement

Whiskers have high strength and are available with aspect ratios of 50 or more. It is pointed out in Section 2.1 that theoretical strength can be realized for discontinuous filament composites if their aspect ratio is greater than 10 and if they are aligned perfectly. Aligning the whiskers is a major problem. General Technologies Corporation (GTC) has electroformed SiC whisker composites with good alignment (Withers and Abrams, 1968). The process employed for alignment is not disclosed.

## 2.4  Properties of FRMC

Nickel-tungsten filament composites have been studied by O'Brien, Martin, and Williams (1966), Withers and Abrams (1968), Wallace

FIGURE 38  Microstructure of partially electroformed copper-silica composite. Silica fibers have been made conducting by cracking acetylene on their hot surfaces.

FIGURE 39  Copper-glass fiber composite electroformed by alternate winding and plating.

FIGURE 40   Ni-W filament composite electroformed by alternate winding and plating.

and Greco (1970), Harris et al. (1971), and Indira Rajagopal et
al. (1980, 1987). Mechanical properties reported by a few authors
are given in Tables 6 and 7 and Fig. 41. It is seen that the rule
of mixtures is obeyed and that volume fractions up to 60% can be
achieved. Wallace and Greco (1970) showed that heat treatment
at 650°C for half an hour "healed" the voids.

Boron fiber-nickel matrix composites have been studied by Withers
and Abrams (1968) and Wallace and Greco (1970). The former group
used the filaments as such and obtained sound composites. Their
data on mechanical properties are shown in Table 6. They found
that composite strength improved after heat treatment at 300°C
(Table 8). They ascribed this improvement to an increased bonding
of the filament and matrix. They also noticed that boron filament
reacts at elevated temperatures with nickel-forming nickel boride.

Wallace and Greco (1970) studied a Ni-B fiber system, making
use of 100-μm boron with or without a boron nitride coating. By
metallographic studies they arrived at the following conclusions.
(1) Unprotected boron fibers reacted with the matrix at temperatures
of 650°C and above. No reaction zone was seen at 500°C. (2) Boron-
nitride-coated boron filaments did not react with the matrix when

TABLE 6   Tensile Strength of Electroformed Composites

| Reinforcing filament | Volume percent filament | Tensile strength × $10^{-3}$ (MPa) | Modulus × $10^{-6}$ (MPa) |
|---|---|---|---|
| None | 0 | 0.621–0.827 | 0.138 |
| Tungsten (2 to 4 mil) | 20 | 1.034 | 0.186 |
| Tungsten | 20 | 1.172 | 0.172 |
| Tungsten | 30 | 1.145 | 0.207 |
| Tungsten | 50 | 1.517 | 0.234 |
| Boron (4 mil) | 15 | 0.793 | 0.193 |
| Boron | 23 | 0.827 | 0.207 |
| Boron | 35 | 1.103 | 0.221 |
| Boron | 42 | 1.296 | 0.221 |
| Silicon carbide (4 mil) | 20 | 0.69 | 0.207 |
| Silicon carbide | 40 | 1.034 | 0.276 |
| Silicon carbide | 50 | 1.276 | 0.310 |

*Source*: From Withers and Abrams, 1968.

TABLE 7   Tensile Strength of Electroformed Fiber Composite

| Material | UTS expt (MPa) | UTS calc (MPa) |
|---|---|---|
| Ni | 690.5 | — |
| Ni+W (5%) | 773.4 | 759.6 |
| Ni+W (10%) | 814.8 | 828.6 |
| Ni+W (20%) | 1035.5 | 1001.2 |
| Ni+W (30%) | 1312 | 1242.4 |
| Ni+W (50%) | 1415 | 1380.7 |

*Source*: From Indira Rajagopal et al., 1987.

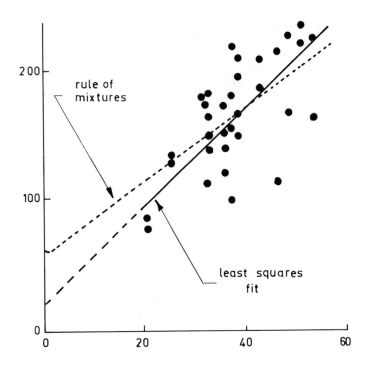

VOLUME PERCENT TUNGSTEN

FIGURE 41   Tensile strength of Ni-W filament composite as a function
of volume fraction of W fibers. (From Wallace and Greco, 1970;
reproduced with the kind permission of American Electroplaters
Society, USA.)

TABLE 8   Effect of Elevated Temperature on Tensile Strength
of Electroformed Ni-B Fiber Composite (25% by volume B fiber)

| Condition | Composite strength $\times 10^{-3}$ (MPa) | Composite modulus $\times 10^{-6}$ (MPa) | Matrix |
|---|---|---|---|
| As deposited | 0.862 | 0.152 | 100 |
| 300°C | 1.0 | 0.193 | 90 |
| 500°C | 0.414 | 0.193 | 64 |
| 700°C | 0.207 | 0.124 | — |

*Source:*   From Withers and Abrams (1968).

TABLE 9  Effect of Heat Treatment on the Bond Strength Between Boron Fiber and Matrix in Ni-B Fiber Composite

| Boron filament | Heat treatment | Average bond strength (MPa) |
|---|---|---|
| BN coated | None | 19.44 |
| Uncoated | None | 12.62 |
| BN coated | 500°C | 36.61 |
| Uncoated | 500°C | 22.11[a] |

[a]Best value obtained for uncoated boron.
*Source*: From Wallace and Greco (1970).

the composite was exposed for a short time to a temperature of 650°C. However, at 900°C, a reaction zone was seen, indicating breakdown of the boron-nitride coating.

By microprobe analysis they found that the reaction zone contained a diffusion layer and reaction products of more than one stoichiometry. They also found an increase in bond strength upon heat treatment irrespective of whether the boron filaments were coated or not. Table 9 shows their results.

Ni-SiC fiber composites were studied by Withers and Abrams (1968). SiC appeared to be more stable than boron. Even after 100 h at 600°C no evidence of a reaction was seen. However, interaction between the fiber and matrix was seen in long-term tests at 900°C.

Cu-W fiber composites have been studied as a model composite system by several people. References to their work can be found in Withers and Abrams (1968). These authors obtained Cu-W fiber composites (40% volume fiber) with a strength of 1323 MPa.

Indira Rajagopal et al. (1987) investigated Cu-glass fiber composites prepared by alternate winding and plating. Their results are shown in Table 10. It is seen that the rule of mixtures is obeyed. They found that composites with volume fractions higher than 35% did not have adequate strength in a direction perpendicular to the fiber, indicating that the matrix is not continuous at such large volume fractions.

Al-B fiber composites have a special attraction because their specific strength will be much higher than the other metal matrix composites. However, electroforming this composite has additional problems. Aluminum can be deposited only from nonaqueous baths operated in a dry nitrogen atmosphere. Withers and Abrams (1968)

TABLE 10   Tensile Strength of Electroformed
Copper-Glass Fiber Composite

| Vol.% glass fibers | UTS expt (MPa) | UTS calc (MPa) |
|---|---|---|
| Cu + 2% | 225.5 | 216.3 |
| Cu + 5% | 236.3 | 230.2 |
| Cu + 10% | 255.4 | 253.2 |
| Cu + 15% | 294.2 | 279.7 |
| Cu + 20% | 299.1 | 304 |
| Cu + 30% | 348.1 | 345.3 |

prepared Al-boron fiber composites by using the plating bath
developed by Couch and Brenner (1952). To improve the quality
and smoothness of the electrodeposit, they used periodic current
reversal. They noticed that boron behaved as a nonconductor in
this bath, whereas in the nickel bath it was a conductor. Therefore,
a single-layer composite could be readily formed, while multilayer
composites were difficult to produce without applying a conducting
coating on the boron fiber. Table 11 shows the strength of multilayer composites.

    To improve Al-B composites, electroformed monolayers were
stacked in a die and hot-pressed. By this procedure, the authors
obtained 40 vol.% composite with a typical tensile strength of 1103
MPa and an elastic modulus of $0.193 \times 10^6$ MPa.

TABLE 11   Tensile Strength of Electroformed Al-B Fiber Composites

| Volume percent filament | Tensile strength $\times 10^3$ (MPa) | Modulus $\times 10^6$ (MPa) |
|---|---|---|
| 12 | 0.276 | 0.103 |
| 15 | 0.207 | 0.124 |
| 20 | 0.283 | 0.124 |
| 25 | 0.428 | 0.152 |

*Source*: From Withers and Abrams (1968).

TABLE 12   Tensile Strength of Electroformed Nickel–SiC Whisker Composites

| Volume percent filament | Tensile strength $\times 10^{-3}$ (MPa) | Strength utilization of incorporated whiskers $\times 10^{-6}$ (MPa) |
|---|---|---|
| 8 | 1.482 | 0.0096 |
| 9 | 1.717 | 0.0117 |
| 10 | 1.558 | 0.009 |
| 10 | 1.986 | 0.0131 |
| 10 | 2.255 | 0.0159 |

*Source*: From Withers and Abrams (1968).

Hanson, DuPree, and Lui (1968) have also discussed electroforming Al–B fiber composites. They used chemically silvered boron filaments and obtained composites with approximately 15 vol.% of fiber. The strength of this composite was 249.6 MPa and had an elastic modulus of $0.097 \times 10^6$ MPa. Comparing these values with aluminum, they concluded that the composite showed a 250% increase in tensile strength and an 87% increase in modulus. These authors predicted that an Al–B fiber composite with 40% reinforcement would compare favorably with the properties of Ti-6Al-4V titanium alloy as far as specific strength and specific modulus were concerned.

Withers and Abrams (1968) showed that sound Ni–SiC whisker composites could be prepared by infiltrating SiC whisker roving by nickel electroforming. The mechanical properties of these composites are shown in Table 12. Comparing these values with Ni–SiC fiber composites, we see that the whisker composites are much stronger than the fiber composites.

Tungsten-wire-reinforced nickel-iron-alloy matrix composites fabricated by electroforming have been investigated by Marsden and Jakubovics (1977). They studied the effect of fiber reinforcement on the magnetic properties of composites. Their main results are summarized below. In composites with negative magnetostriction, the maximum permeability decreased with increasing fiber volume fraction. When the magnetostriction was positive, the maximum permeability increased with the volume fraction of fiber, reaching a peak value at $V_f \simeq 0.1$. This magnetic behavior could be understood in terms of stresses induced in the matrix, during cooling

from the heat treatment temperature. The stresses were caused
by the difference in the thermal expansion between fiber and matrix.
By theory it was shown that the peak coincided approximately with
the volume fraction at which the maximum uniaxial elastic stress
was expected to form in the matrix. Above this value of $V_f$, the
uniaxial and transverse stresses became sufficiently high to cause
plastic deformation in the entire matrix, leading to the observed
fall in the maximum permeability, which, however, remained above
that shown by the unreinforced matrix.

## 2.5  Conclusions

Fiber-reinforced metal matrix composites can be electroformed either
by simultaneously winding and plating or by alternately winding
and plating. By these techniques, sound composites, containing
about 35% by volume of nonconducting fiber or 50% by volume of
conducting fiber, can be realized in practice. The filament winding
technique would be limited to shapes that are solids of revolution.
Whenever a shape cannot be generated by winding or when a higher
volume fraction is needed, a hot-pressed warp sheet (which is nothing
but an electroformed monolayer fiber composite) will have to be
used. Whisker composites have also been electroformed. The major
problem of whisker composite is whisker alignment.

Electroformed FRMC obeys the mixture rule. Among the properties, only the strength and resistance to temperature have been
investigated. At sufficiently high temperatures, the fibers degrade
by reaction with the matrix. The highest temperature at which
FRMC can be used will be limited by fiber-matrix interaction.

## References

Alexander, J. A., Withers, J. C., and Macklin, B. A. (1967).
    NASA, CR-785:9
Baker, A. A., Harris, S. J., and Holmes, E. (1967). *Metals and
    Materials*, 1:211.
Broutman, L. J. and Krock, R. H. (1967). *Modern Composite Mate-
    rials*, Addison-Wesley, p. 244.
Cooper, G. (1970). *Composites*, 1:153.
Couch, D. E. and Brenner, A. (1952). *J. Electrochem. Soc.*, 99:234.
Donovan, P. D. and Watson-Adams, B. R. (1970). PARDE Memo-
    randum, 12/70, Royal Armament Research and Development
    Establishment, UK.
Greco, V. P., Wallace, W. A., and Cesaro, J. N. L. (1969). *Plating*,
    56:262.
Hanson, R. N., Du Pree, D. G., and Lui, K. (1968). *Plating*,
    55:247.

Harris, S. J., Baker, A. A., Hall, A. F., and Bache, R. J. (1971). *Trans. IMF, 49*:205.

Indira Rajagopal, Rajam Jayaraman, William Grips, Vasudevan, N. and Rajagopalan, S. R. (1980). Proc. Second Int. Symp. on Industrial and Oriented Basic Electrochemistry, SAEST, India, 6.29.

Indira Rajagopal, Rajam, K. S., Gripps, V. K. W., and Rajagopalan, S. R. (1987). Proc. INCEF 86, McGraw-Hill (in Press).

Kelly, A. (1973). *Strong Solids*, Clarendon Press, Oxford, pp. 157-220.

Kelly, A. and Davies, G. J. (1965). *Met. Reviews, 10*:1.

Marsden, A. L. and Jakubovics, J. P. (1977). *J. Mater. Sci., 12*:434.

O'Brien, D. J., Martin, P. W., and Williams, R. V. (1966). *Applied Materials Research, 5*:241.

Ramaseshan, S. (1971). Proc. Seminar on Newer Materials for Industrial Application, Institution of Production Engineers, Bangalore Section, p. 7.

Roos, J. H. (1975). *Materials Res. Standards, 11*(5):11.

Wallace, W. A. and Greco, V. P. (1970). *Plating, 57*:342.

Withers, J. C. and Abrams, E. F. (1968). *Plating*, June:605.

## 3   ELECTROLESS COMPOSITES

The idea of combining two materials to yield a composite led to the development of electrocomposites, which were discussed in the previous section. The successful commercial exploitation of Ni-SiC electrodeposited composite for coating Wankel engine cylinders in Germany stimulated research on deposition of composites, in 1965, from electroless baths.

Normally, contamination of an electroless plating bath by solid particles is to be avoided since it can cause autocatalytic decomposition of the bath. But, for the deposition of electroless composite, inert particles have to be added to the bath. This may increase the surface area loading to about 10,000 $cm^2$/L which is around 800 times greater than that generally recommended. In spite of these apparent problems, R&D efforts were directed toward electroless composites in many centers because electroless deposition results in uniform coating even on a complex shape, and because there is no extra buildup on edges and sharp corners.

One of the earliest electroless composites to be plated was deposition of a thin layer of electroless nickel-diamond as an intermediate layer to improve corrosion resistance of nickel-chromium electrodeposits (Odekerken, 1966).

In 1966, Metzger and his colleagues successfully deposited electroless nickel (EN)-$Al_2O_3$ composite (Metzger and Florian, 1976). Development of EN-SiC under the proprietary name Kanisil (Metzger, 1972), EN-diamond by scientists of DuPont Company and a variety of others followed.

### 3.1 Experimental Procedure for the Deposition of Electroless Composite

Figure 42 shows schematically an experimental arrangement for the deposition of electroless composites. A cylindrical vessel with a conical bottom and an outer jacket is used. The electrolyte with dispersed particles at the appropriate temperature is pumped through the bottom. The outer jacket collects the electrolyte that overflows. The substrate to be coated is suspended in the inner compartment. To keep the particles in suspension, the electrolyte is agitated mildly.

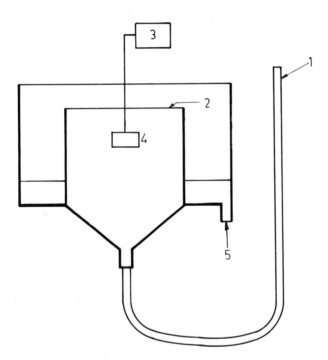

FIGURE 42   Schematic of an experimental setup for deposition of electroless composites. 1. Inlet for thermostated electroless composite bath. 2. Overflow level. 3. Mechanical arrangement for the slow rotation of component. 4. Component. 5. Outlet for solution.

Several methods of agitation similar to that used in the deposition of electrocomposites can be employed. Air agitation, bath circulation, ultrasonic agitation, and the two methods described by Kedward, Addison, and Tennet (1976)—the liquid-air system and the plate-pumper technique—have been used.

The type of movement to which the material is subjected and the suspension technique can both affect the result. Uneven flow conditions prevent a uniform dispersion of the second phase. This factor is more critical for electroless composites than for electro-deposited composites. Several research workers have found that air agitation with a slow rotation of the component yields the best result.

Contrary to the normal stipulation that electroless plating baths must be free from any debris, particles here are added, therefore the problem of bath stability is even higher than that of normal electroless plating baths. Stabilizers are invariably added to prevent spontaneous decomposition.

It is not uncommon to find that proprietary baths are used. Metzger (1972) used Kanigen, which is a proprietary electroless nickel bath employing hypophosphite, to develop a process for the deposition of Ni-P-SiC composites. This became the commercial process Kanisil. Dennis, Sheikh, and Silverstone (1981) and Hussain and Such (1981) used Ni Foss 80, a proprietary product of W. Canning Materials, Ltd.

All the normal precautions recommended for the maintenance of electroless plating baths must be followed in electroless composite plating baths. However, continuous filtration cannot be used. It is essential that periodic filtration, after settling the particulate matter, be used. Impurities might get into the bath from the particles. Therefore, the impurity level in the bath must be monitored.

A variety of materials are acceptable as particulate matter for electroless composites. These are given in Table 13. Normally particles less than 7 μm are used (Brown, 1985). The particulate matter should be insoluble or only slightly soluble in the bath. If it is slightly soluble, the dissolved ions should be very low in concentration and should not influence the deposition. Metal powders acting as nickel catalysts or that are more electropositive than nickel are unsuitable (Parker, 1972). Only purified particulate matter should be used. Otherwise, impurities in the particles may cause problems. For example, early attempts to prepare EN-diamond composites with artificial diamond were not successful. Homogeneous decomposition of the electroless plating bath occurred. Feldstein et al. (1983) attributed this decomposition to impurities in diamond, such as cobalt, nickel, copper, and iron, which are used as catalysts in the diamond synthesis.

TABLE 13   Materials Which Are Suitable for Incorporation into
Electroless Nickel

| | | |
|---|---|---|
| **A.** | Kaolin | glass | talcum |
| | graphite | plastics | diamond |

B.   Oxides, carbides, borides, nitrides, silicides, sulfides, silicates, sulfates, carbonates, phosphates, oxalates, and fluorides from

| | | |
|---|---|---|
| aluminum | tantalum | barium |
| boron | vanadium | strontium |
| chromium | tungsten | cerium |
| hafnium | zirconium | iron |
| molybdenum | manganese | nickel |
| silicon | magnesium | titanium |
| calcium | | |

C.   Metals and alloys from

| | | |
|---|---|---|
| boron | titanium | zirconium |
| chromium | tantalum | hafnium |
| vanadium | molybdenum | tungsten |
| stainless steel | | |

*Source*: From Metzger and Florian (1976).


## 3.2   Comparing Electroless and Electrocomposites

An important difference between the two techniques of preparing composites is that electroless deposition achieves, with a small concentration of particles in the bath, the same amount of incorporation obtained by using a large concentration of particles in the electroplating process. For example, an electroless nickel bath containing 10 g/L of SiC gives the same volume percentage of SiC in the deposit as is obtained by electroplating from a nickel bath containing 100 g/L of SiC. This behavior is not peculiar to incorporation of SiC. A similar trend has been noticed with $Al_2O_3$ incorporation. Figure 43 compares the extent of incorporation by the two methods from solutions containing the same concentration of particles. At a concentration of 10 g/L of alumina in the electrolyte, less than 1 wt.% of $Al_2O_3$ is incorporated into nickel, when the composite is prepared by electrodeposition from Watt's bath. On the other hand, electroless deposition from a bath containing 10 g/L alumina gives an incorporation of 8.8 wt.% in the deposit. To obtain this amount of incorporation by electrodeposition, we must use a concentration of 300 g/L of alumina in the bath (Gawrilov and Owtscharova, 1973a,b).

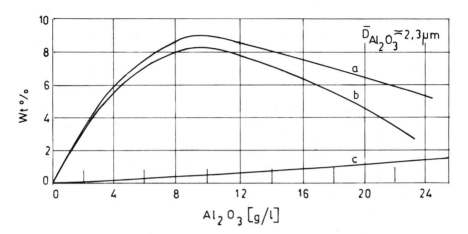

FIGURE 43 Dependence of amount incorporated on the concentration of $Al_2O_3$ in the solution. Electroless deposition of $Ni-P-Al_2O_3$ at (a) 83°C and (b) 93°C. (c) Results of Broszeit, Heinke, and Wiegand (1971) on electrodeposition of $Ni-Al_2O_3$. (Reproduced from *Chemical Nickel-Plating* by Gawrilov with the kind permission of Port Cullis Press Ltd., Surrey.)

Dennis, Sheikh, and Silverstone (1981) observed that the volume percent of chromium carbide in the electroless deposition of $Ni-P-Cr_3C_2$ is practically constant (about 25%) when the concentration of particles is varied from 10 to 200 g/L. This observation again points out that percentage incorporation is high, even at low concentration of particles in the bath, in electroless composites.

No explanation for the larger incorporation noticed in electroless plating has been given. Dennis, Sheikh, and Silverstone (1981) presented some calculations for the number of particles present in the solution and those that impinge on the substrate. If all the particles impinging on the surface from a 10-g/L solution get incorporated, the interparticle distance will be $4 \times 10^{-8}$ m. The experimentally observed value for interparticle distance is $6.5 \times 10^{-6}$ m. The ratio of the number of particles impinging to that incorporated is

$$\left( \frac{6.5 \times 10^{-6}}{4 \times 10^{-8}} \right)^3$$

which is nearly $10^6$. On the basis of this simple calculation, they conclude that particle concentration could be reduced by several

orders of magnitude before there is any reduction in particle concentration in the deposit. This conclusion is based on the assumption
that the number of particles incorporated is not a function of the
number of particles impacting. The basis for this assumption is
not clear. Besides, the calculations are not based on any specific
characteristic of electroless plating.

The average size of the particle, concentration of the particle
in the suspension, and the velocity of the suspension normal to
the surface of the substrate are used in calculating the above numbers. As such, the calculations should be valid for electrolytically
produced composites too. Since the experimental results show that
incorporation is more for electroless composites, the only inference
we can make is that a larger fraction of the impacting particles
gets incorporated into the electroless deposit.

The higher rate of incorporation obtained with electroless composite deposition has a practical significance. The cost of an electroless composite bath will be lower than that of an electrocomposite
bath. The lower cost is significant for expensive particulate matter,
such as diamond.

Another significant difference between the two techniques is
that in electrocomposites the amount of incorporation can be increased
by adding surfactants, whereas in electroless composites, surface
active agents do not affect the rate of particle incorporation.

On a component of complex shape, an electroless composite
gets coated uniformly. Thickness of the coating and the extent
of incorporation are uniform. In electrodeposited composites, both
thickness and rate of incorporation depend on current density,
which is bound to be nonuniform on an object of complex shape.
Consequently, the thickness and volume fraction of the particles
will be different at different points. It is difficult to "throw" the
deposit into deep recesses and holes with electrocomposites. Deep
recesses and holes do not pose serious problems as far as electroless
composites are concerned. On sharp corners, there is no excessive
buildup in the case of electroless composite which is not the case
with electrocomposites.

In nickel-based composites, one can expect differences in the
properties for electro- and electroless composites. The matrices
in the two composites are different; the electroless composite matrix
is Ni-P or Ni-B, whereas the electrocomposite matrix is Ni. Insofar
as the property contribution of the matrix is concerned, there will
be differences between the two.

## 3.3  Systems Studied

Composites of any metal that can be electroless plated can, in principle, be prepared. Electroless composites with Ni, Cu, Au, Co, and

Sn are therefore feasible. Of these, the most extensively investigated system is electroless nickel composites. Before we discuss this system, we shall briefly review the work on other systems.

Poniatowski and Clasing (1972) produced hardened gold, a composite containing Au and $TiO_2$. This was prepared by adding dropwise a strong acid, gold chloride solution mixed with $TiCl_4$ solution to ammoniacal hydrazine and heating in vacuum. The resulting material exhibited high hardness, tensile strength, and good creep resistance at high temperatures.

A couple of studies conducted in the USSR on $Cu-Al_2O_3$ from electroless plating baths are given in Metal Finishing Abstracts. Guseva, Mashchenko, and Borisenko (1977) studied $Cu-Al_2O_3$ deposition from electroless copper containing suspended $Al_2O_3$. They investigated the effect of $Al_2O_3$ concentration on deposition rate and coating composition. By comparing the properties of copper and $Cu-Al_2O_3$ composites, they found that the microhardness of the composite is more than that of copper by a factor greater than 2 and its electrical resistance is also higher than that of copper.

*Electroless Nickel Composites*

Particles of several materials can be used to make EN composites, and these are listed in Table 13. Basically all nonmetallic substances used for depositing electrocomposites are suitable for electroless composites. However, composites with SiC and diamond have been studied extensively. Recently EN-PTFE composites have attracted a growing interest.

A typical plating bath for electroless composite has the following composition (Brown, 1985):

| | |
|---|---|
| $NiSO_4 \cdot 6H_2O$ | 30 g/L |
| $Na\ H_2PO_2 \cdot H_2O$ | 20 g/L |
| Lactic acid | 25 g/L |
| Propionic acid | 5 g/L |
| Lead | 1–4 mg/L |
| pH | 4.4–4.8 |
| Particle concentration | 0.25–2% by weight |

In the above bath lead is present at a much higher concentration than in normal EN baths. The extra lead is needed to enhance the stability of the bath and has to be controlled by analytical estimation or stability tests. As a rule of thumb a particle concentration of 1% (by weight) or a little higher produces deposits containing 20–25% by volume of particles, whereas a concentration of 2% (by weight) leads to deposits with approximately 30% by volume of particles.

Gawrilov and Owtscharova (1973a,b) deposited $Ni-P-TiO_2$ using $TiO_2$ particles of diameter 0.2 μm and $Ni-P-Al_2O_3$ with 2.3-μm $Al_2O_3$

particles. They found that incorporation of particles depends on temperature and, to a certain extent, pH. In the Ni-P-Al$_2$O$_3$ system, the maximum extent of incorporation occurs at a concentration of 10 g/L Al$_2$O$_3$ in the suspension. They compared their results on the variation of incorporation rate with the concentration of Al$_2$O$_3$ with that of Broszeit et al. (1971) for electrodeposition of Ni-Al$_2$O$_3$ from Watt's bath at pH = 4.5 and 50°C. These results are shown in Fig. 43. The substantial increase in rate of incorporation in electroless plating is evident. Also notice that a temperature increase lowers the rate of deposition. The authors report a significant increase in hardness and wear resistance as measured on a Skoda-Sawin machine and a certain improvement in corrosion resistance for Ni-P-Al$_2$O$_3$ and Ni-P-TiO$_2$ deposits. Epifanova et al. (1975) and Hussain and Such (1981) have also investigated this system.

Gawrilov (1974) investigated the deposition of Ni-P-ZrO$_2$. The mean diameter of the particle was 0.7 µm. He obtained a maximum incorporation of 4.5 wt.% of ZrO$_2$. Stress, microhardness, and porosity were virtually unaffected by ZrO$_2$ content. Abrasion resistance of this composite was better under dry friction conditions. Coatings deposited from suspension containing 8 g of ZrO$_2$/L exhibited maximum corrosion resistance. He also showed that it is feasible to deposit Ni-B-ZrO$_2$. Guseva, Mashchenko, and Borisenko (1980) have also investigated EN-ZrO$_2$ composites.

The feasibility of incorporation of B and C up to 13-15% by volume in Ni-P deposits has been established. These coatings have better corrosion resistance and lower stress compared with Ni-P deposits (Kurtinaitene, Zhitkyavichyute, and Lunyatskas, 1983).

Some attempts have been made to incorporate fibers into EN. Two approaches have been made. Mitsubishi Heavy Industry (1976) proposes to realize this by depositing EN on nonmetallic fibers, chromium plating and, finally, bundling and heating them to form Ni-Cr alloy so as to get fiber-reinforced composites with Ni-Cr alloy matrix. We have not encountered practical applications of this rather unconventional approach. Guseva, Mashchenko, and Borisenko (1981) incorporated asbestos fibers into EN by suspending them in the EN bath. The resulting composite exhibited improved physico-mechanical properties.

Parker (1971,1972) has described a process for chemical deposition of alloy coatings by making use of electroless deposition of composites. In this process, he suspended metal powders in an electroless nickel bath and deposited an EN-metal composite. This composite was heated to form the alloy. Metallic particles could consist of Cr, Mo, W, Ti, V, Zr, Hf, Nb, Ta, or alloys of metals (stainless steel). The particle size could be from 0.1 to 50 µm. The patent specification (Parker, 1971) states that incorporation of metal particles can reach 65 vol.%.

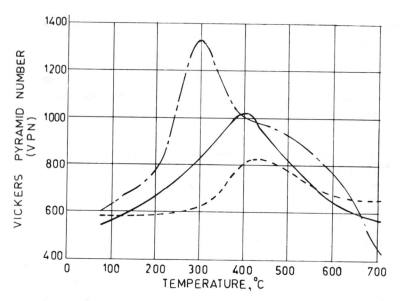

FIGURE 44 Effect of heat treatment on the Vickers hardness of electroless nickel containing chromium and tungsten. (From Parker, 1972; reproduced with the kind permission of Swiss Electroplating Society, Switzerland.)

Parker (1972) obtained, by codepositing Mo powder with EN, a deposit containing 70% Ni, 18% Mo, 4% Cr, and 8% P. He has studied the effect of heat treatment on the hardness of this composite and his results are shown in Fig. 44. Ni-P-Cr composite developed the highest hardness of 1330 kp/mm$^2$ (604.5 kg/mm$^2$) after heating for 1 hr at 300°C. Electroless nickel deposits exhibited under this condition a hardness of 820 kp/mm$^2$ (372.7 kg/mm$^2$). The maximum hardness of Ni-P-Cr composites was seen to be higher than the maximum hardness of EN, which is 1025 kp/mm$^2$ (466 kg/mm$^2$). Ni-P-Cr composites showed better corrosion resistance than conventional EN; a coupon with a 25-μm Ni-P-Cr coating had a corrosion rate of 22 mdd (heat treated) as compared with 91 mdd for a Ni-P coating. Wear resistance of Ni-P-Cr was also found to be better; the Taber wear index was 4.6 as plated and 2.1 after heat treatment at 600°C, compared with 18 for as-plated Ni-P and 8 for Ni-P heat-treated at 290°C for 10-16 hr.

Particles of the following materials have been incorporated into EN, BN (Prusov and Egorenkova, 1977), MoS$_2$ (Gawrilov and Butschkow, 1980); quartz, talc, CrB$_2$, WC, and graphite (Safina et al., 1978); carbides of silicon, titanium, and tungsten, oxides of aluminium

and titanium, $MoS_2$, graphite, and Teflon (Grund, Hofmann, and
Sobek, 1983); boron and silicon carbide (Parker, 1972); chromium
carbide (Dennis, Sheikh, and Silverstone, 1981).

The materials codeposited form quite an impressive list. Of
these, only a few have found commercial application. We will discuss
the systems which have found commercial applications or which
have been evaluated for industrial use.

*Electroless Nickel—SiC Composites.* Electroless composite with
SiC has been one of the most popular composites. This popularity
is due to SiC's inert nature, availability in a state of high purity,
and high hardness, which can impart good wear resistance to the
composite.

Metzger and Florian (1976) have investigated Ni-P-SiC electroless
composites. They found that a suspension of 10 g/L of SiC in an
EN bath gave the same percentage of particle incorporation as that
produced by electroplating, from a suspension of 100 g/L. Metzger's
work led to the commercial utilization of Ni-P-SiC composites under
the name Kanisil (Metzger, 1972). This system has also been investigated by Parker (1972), Gawrilov and Ernin (1975), Hubbell (1978a,b),
Ernin, Hussain, and Such (1981), and Broszeit (1982).

A 20-25% by volume of SiC (particle size 1-3 μm) in EN leads
to significant improvement in mechanical and physical properties.
Wear resistance of heat-treated Ni-P-SiC is better than hard chromium
or anodized aluminium. Corrosion resistance, as measured by salt
spray tests, of equal thickness of EN and electroless nickel composite
are nearly the same.

*Electroless Nickel—Diamond Composite.* This system has found
a great deal of practical application. Consequently it has also received
considerable attention from R & D groups. The pioneering work of
scientists at the Du Pont Company in the 1960s paved the way for
the evolution of a number of composite coatings for commercial application (Sharp, 1974; Graham and Gibbs, 1975; Editor, 1977). Work
on this composite has been reviewed by Feldstein et al. (1980, 1983)
and Barras et al. (1979). Diamond is available in several forms.
Broadly there are two types: monocrystalline and polycrystalline
shown in Fig. 45. Single crystals of diamond fracture along cleavage
planes, resulting in sharply defined edges and points. Natural
diamond and certain types of synthetic diamond belong to this category.

Shock-synthesized diamond, on the other hand, is polycrystalline.
It is an aggregate of diamond crystallites. It has an irregular surface
and is normally available in narrow size distribution.

The choice of the type of diamond that can be used for making
composites depends upon the end use of the composite. Monocrystalline

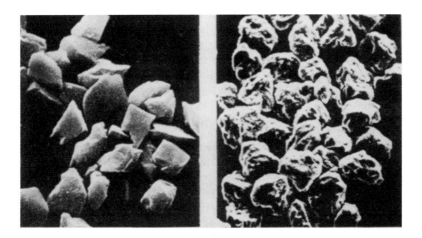

FIGURE 45  Diamond particles: left—monocrystalline; right—
polycrystalline. Reproduced from *Trans. IMF.*, 1978, *56*:118,
with the kind permission of Institute of Metal Finishing, London,
U.K.

diamond is suitable for grinding or abrasive application because
of their characteristic surface with sharp edges and points. Grinding
wheels, diamond files, etc., employing a composite layer as the
functional surface uses naturally occurring monocrystalline diamond.

There is a problem in using synthetic single-crystalline diamond
for making EN-diamond composites. They appear to catalyze the
homogeneous decomposition of the bath. Feldstein et al. (1983)
have suggested that this difficulty might be due to certain impurities
in the synthetic diamond, such as iron, nickel, cobalt, or copper,
which are used as catalysts in its manufacture. Even though this
has not been verified, it appears quite plausible. Interest in the
use of synthetic monocrystalline diamond has practically vanished
after the availability of shock-synthesized diamond.

For wear resistance, composites containing single crystals of
diamond will not be useful because their rough surface would abrade
the mating face. Shock-synthesized polycrystalline diamond is ideally
suited for this application. Under conditions of severe wear, it
tends to flatten, thereby bearing the load and reducing wear. For
less severe conditions, a composite of polycrystalline particles of
diamond may act as a surface of nonrolling "ball bearings."

Electroless nickel-polycrystalline diamond has good surface protection and antigalling properties. Though the earliest use of this
composite was for improving corrosion resistance (Odekerken, 1966),
its major application today is combating wear.

A few interesting practical details are discussed by Luckschandel
(1978). The plating baths used for coating this composite have to
be changed more often. Standard EN baths can be operated up
to 10 "turnovers," but the EN-diamond composite plating bath has
to be discarded (of course, after recovering the diamond) after
three turnovers. Even though the suspension may be quite uniform,
the inclusion rate is strongly dependent on orientation. If the component is not rotated, the surfaces facing upward would contain
twice as many particles as the vertically oriented surface and three
to four times more than surfaces facing down. Rotation or tumbling
in such a way that all surfaces of the component face upward in
a regular sequence is therefore essential. With intricate shapes
this could pose problems; the deposit thickness might be uniform
on a complex shape but distribution of the second-phase particles
may not be uniform.

Normally particles of diameter 6 μm are used for wear resistance.
Other standard diameters are 1, 3, and 4 μm. A standard coating
should contain 20-30% diamond by volume, and the thickness should
be around 20 μm.

*Electroless Nickel—PTFE Composite.* For dry lubrication PTFE
is very useful. It has a relatively high softening point (325°C)
and a low coefficient of friction. In fact, it has a lower coefficient
of friction than any other polymer. (The coefficient of friction of
steel on PTFE for low sliding speeds is 0.05, irrespective of whether
the sliding is under dry or wet conditions.) It is nonflammable
and does not support combustion. These properties make PTFE
an attractive material for composite.

The major hurdle in the preparation of EN-PTFE composites
is that PTFE is not wetted by water. Hydrophobicity has been the
main reason for the long delay in the realization of EN-PTFE composites (Feldstein et al., 1983). The problem of dispersion of this
polymer in an EN bath has been solved by using a concentrated
dispersion of the polymer in water. This concentrated dispersion
is prepared by dispersing the polymer in water, containing a suitable
blend of cationic and nonionic fluorocarbon surfactants. These surfactants make the dispersion stable. They are strongly adsorbed
on the surface of the polymer. Consequently, the concentration
of the free surfactant is so small that the addition of this suspension
to the bath does not introduce any significant amount of free sur-
factant (Tulsi, 1983).

The bath for composite deposition is prepared by adding the
requisite amount of PTFE dispersion to the EN bath. The following
composition and operating conditions are given by Tulsi (1983).

200 mL of EN concentrate diluted to 1 L

| | |
|---|---|
| PTFE dispersion (60% active) | 6.5-11 g/L |
| Particle size of PTFE | < 1 µm |
| Operating conditions | |
| Temperature | 88-92°C |
| pH | 4.7-5.2 |
| bath loading | 0.4-3.5 dm$^2$/L |
| agitation | Gentle |

Operating parameters and the EN solution composition are very similar to a normal EN solution.

The factors which influence the incorporation of PTFE are the concentration, size distribution, ratio of cationic to nonionic sur-factant, pH, and flow velocity of the solution. The amount of PTFE in the composite is normally 25-30% by volume (about 8% by weight). It can, however, be varied by altering the bath composition and the operating conditions. The phosphorus content of the deposit is between 6-12%.

The hardness of the deposit is 250 VPN in the as-plated condition and becomes 400 VPN after heat treatment at 400°C for 4 hr. The hardness is less than that of EN because the property of the composite is a linear combination of the matrix and dispersed phase.

*Electroless Nickel—Chromium Carbide Composite.* This composite has been investigated in detail by Dennis, Sheikh, and Silverstone (1981). These authors produced this composite by using Ni-FOSS 80 (proprietary product of W. Canning, Ltd.) and a concentration of 10 g/L of $Cr_3C_2$ particles of size 2-9 µm. They found that the proprietary solution could be employed without modifying it. The bath was operated at pH 4.5 and 92°C. Air agitation was used to keep the particles in suspension. The rate of deposition was about 18 µm/hr. The composite was found to contain 64.8% Ni, 7.2% P, and 27% $Cr_3C_2$. The EN deposit from this bath was analyzed to be 90.8% Ni and 9.1% P. The deposit contained 25% by volume of particles.

In the as-deposited state, the deposit has a hardness of 645 HV. After optimum heat treatment (500°C for 5 hr in vacuum) the composite attains a hardness of 1225 VHN. The values for EN are 515 VHN in the as-deposited state and 1155 VHN after optimum heat treatment (200°C for 5 hr in vacuum). The authors have tried to explain the improvement in hardness as due to the additional flow stress arising from the carbide particles. Therefore, they conclude

TABLE 14   Wear of Coated or Uncoated Dies After Forging 1000
EN3B Billets

| Deposit | Wear volume $(10^{-3}$ cm) | Comments |
|---|---|---|
| No deposit | 4.37 | Heavy, adherent scale |
| Brush plated Co-Mo | 1.01 | Very slight to zero scale |
| As-deposited electroless nickel | 2.39 | Local heavy wear very adherent scale |
| Fully hardened electroless nickel (5 hr at 200°C in vacuum) | 1.86 | Heavy scaling |
| As-deposited electroless nickel-chromium carbide | 2.03 | Heavy scaling |
| Fully hardened electroless nickel-chromium carbide (5 hr at 500°C in vacuum) composite | 1.35 | Slight scaling |

that the higher hardness of $EN-Cr_3C_2$ is caused by the additive
effects of the normal precipitation hardening mechanism (Silcock,
Heal, and Hardy, 1953-1954).

These authors have evaluated the suitability of this composite
coating for minimizing wear and erosion of die steels by a simulated
hot forging test. Their results are given in Table 14. It is seen
that the wear volume of the heat-treated composite is one-third
that of the die material. This compares favorably with the wear
volume of the Co-Mo layer, which is one-fourth of the die material.
The composite resists scaling.

## 3.4   Structure of Electroless Composites

The cross sections of several electroless nickel composites have
been studied by optical microscopy. These show that the particles
are uniformly dispersed in the deposit. Figure 46 shows a cross
section of a typical $EN-Al_2O_3$ composite. Microstructures of EN-SiC,
EN-diamond, and EN-PTFE are shown, respectively, in Figs. 47,
48, and 49. In all cases, the uniform distribution is clearly seen.

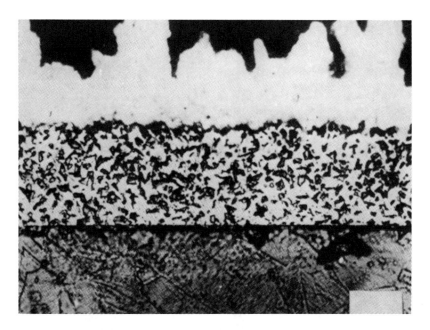

FIGURE 46  Optical micrograph of cross section of an EN-Al$_2$O$_3$ composite coating, 400 times.  Reproduced from *Trans. IMF.*, 1976, *56*:174, with the kind permission of the Institute of Metal Finishing, London, U.K.

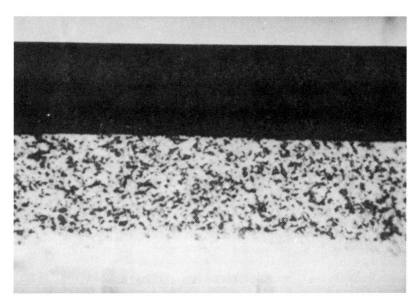

FIGURE 47  Optical micrograph of cross section of an EN-SiC composite coating, 600 times.  Reproduced from *Plating & Surface Finish.*, 1978, *65* (12):58 with the kind permission of Am. Electroplaters' Soc. Inc., U.S.A.

FIGURE 48   Optical micrograph of an EN-diamond composite containing
1.5-μm-diameter diamond particles.   Reproduced from *Metal Finish*,
1983, Aug. 35, with the kind permission of Metals and Plastics
Publications Inc., U.S.A.


In an electroless composite bath, where care has been taken
to ensure homogeneity in suspension, the rate of particle incorporation is the same at any height of the solution. However, it depends,
as pointed out earlier, on orientation. Hussain and Such (1981)
showed for a sample plated in an EN-SiC plating bath without rotation
that the coating on the surface facing upward had 11-13% by volume
of particles, whereas the layer on the downward-facing surface
contained only 5-6% by volume of particles. Figure 50 shows electron
micrographs of cross sections of EN-SiC coating on upward- and
downward-facing surfaces.
     It is well known that an electroless Ni-P deposit has a banded
structure. Composites do not appear to have a banded structure.
In the literature no one seems to have investigated this point,
except Dennis, Sheikh, and Silverstone (1981), who have shown
that the EN-$Cr_3C_2$ composite does not possess a lamellar structure.

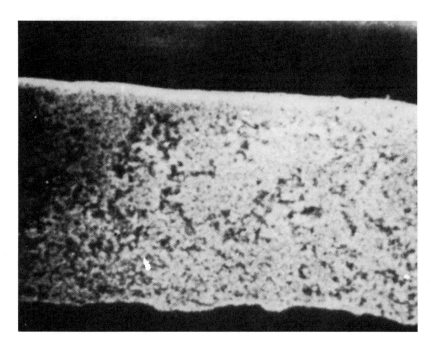

FIGURE 49   Optical micrograph of cross section of an EN-PTFE
composite coating.   Reproduced from *Metal Finish.*, 1983, Aug. 35,
with the kind permission of Metals and Plastics Publications, Inc.,
U.S.A.

## 3.5 Surface Finish

It is inevitable that surface roughness increases by incorporation
of particles into the electroless deposit. This is a consequence of
particles projecting out of the surface and results in a characteristic
appearance. A typical surface topography of an EN-SiC composite
is shown in Fig. 51. The actual roughness of an electroless composite depends on various factors, such as surface finish of the
substrate, particle size and distribution, extent of particle inclusion,
and coating thickness.

The effect of substrate roughness on the final finish of the
EN-SiC composite is shown in Table 15 (Hubbell, 1978a). Deposition
of the composite is seen to deteriorate the surface finish, when
the substrate roughness is poorer than 15 RMS. For larger substrate
roughness, plating improves the surface finish, which is perhaps
due to the leveling action of the deposit.

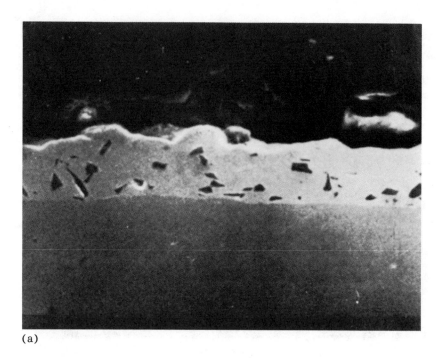

(a)

FIGURE 50   Electron micrographs of cross sections of (a) an upward-facing surface and (b) a downward-facing surface of an EN-SiC composite coating.   Reproduced from *Surf. Technol.*, 1981, *13*:119, with the kind permission of Elsevier Sequoia S.A., Switzerland.

Table 16 gives data on the surface finish of EN-diamond composites. Roughness of deposit is seen to increase with thickness, particle size, and substrate finish. Table 17 shows that the deposit becomes rougher when coarser particles are incorporated.

Improving the surface finish of composite by a postplating finishing operation is not easy. With EN-SiC, it can be accomplished by using diamond paste and sapphire. The mirrorlike surface resulting from such polishing is shown in Fig. 52.

Postplating finishing is extremely difficult in EN-diamond composites because of the high hardness of diamond. Wherever the surface finish is important, one should start with a substrate of low roughness and employ particles as small as permissible for the particular end use.

Recently, a novel smoothing technique has been developed (Feldstein, 1983) which appears promising. It consists of deposition of

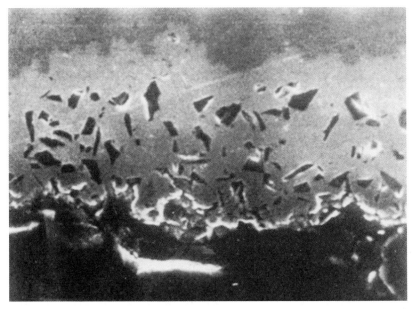

(b)

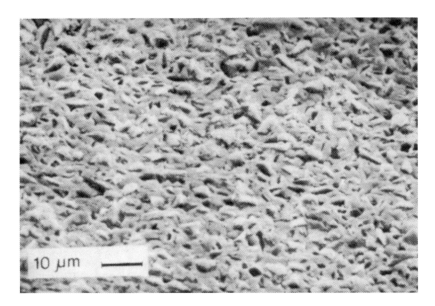

FIGURE 51   Electron micrograph of the surface of EN–SiC composite
with 1-3 μm SiC particles.   Reproduced from *Plating & Surf. Finish.*,
1978, *65* (12):58 with the kind permission of Am. Electroplaters
Soc. Inc., U.S.A.

TABLE 15  Effect of Substrate Roughness on the Surface Finish of EN-SiC Composites

| Sample[a] no. | Surface finish[b] before plating (RMS) | Surface finish[b] after plating (RMS) |
|---|---|---|
| 1.  8000 grit polish | 0.5-1.0 | 4.0-7.0 |
| 2.  1200 grit polish | 1.0-2.0 | 6.0-13.0 |
| 3.  320 grit emery cloth | 7.0-7.5 | 7.0-12.0 |
| 4.  280 stone | 12.0-15.0 | 14.0-16.0 |
| 5.  240 grit dry blast | 26.0-32.0 | 15.0 |
| 6.  24 grit dry | 160.0-190.0 | 58.0-68.0 |

[a]Coating thickness: 37.5-50 $\mu$m; size of SiC particles: 1 + 3 $\mu$m.
[b]Units are not given in the paper. Judging from the numerical values, we feel the surface finish is in microinches.
*Source*: From Hubbell (1978a).

TABLE 16  Effect of Substrate Roughness and Particle Size on Surface Finish of EN-Diamond Composite

| Particle size ($\mu$m) | Amount of diamond in the deposit (vol.%) | Coating thick-ness ($\mu$m) | Surface finish of composite AA ($\mu$m)[a] | |
|---|---|---|---|---|
| | | | Substrate roughness (AA 0.15 $\mu$m[a]) | Substrate roughness (AA 0.30 $\mu$m[a]) |
| 1.5 | — | 20 | 0.35 | 0.50 |
| 3.0 | — | 12.5 | 0.47 | 0.60 |
| 3.0 | — | 20 | 0.53 | 0.68 |
| 6 | 25 | 12.5 | 0.70 | 1.00 |
| 6 | 25 | 20 | 0.13 | 1.33 |
| 6 | Max. | 20 | 0.65 | 1.43 |

[a]Accuracy of surface roughness ± 10%.
Data of Barras et al. given in microinches is converted to micro-meters.
*Source*: From Barras et al. (1979).

TABLE 17  Effect of Particle Size and Postdeposition Smoothing on the Surface Finish of EN-Diamond Composite

| Diamond size (μm) | Surface finish of composite[a,b] | |
|---|---|---|
| | As plated (μm) | After smoothing by the new method (μm) |
| 6 | 1.00-1.13 | 0.38-0.46 |
| 4 | 0.66 | 0.34 |
| 3 | 0.48 | 0.30 |
| 1.5 | 0.18 | — |

[a]Surface roughness of substrates: 0.025-0.050 μm.
[b]Surface roughness characterizing parameter is not mentioned in the paper.
*Source*: From Feldstein et al. (1983).

FIGURE 52  EN-SiC-coated cast steel mold finished to mirrorlike surface. Surface roughness before deposition: 0.025-0.05 μm RMS; surface roughness after plating: 37.5 μm thick EN-SiC to 0.125-0.150 μm RMS; surface roughness after finishing: 0.025 μm RMS. Reproduced from *Plating Surf. Finish.*, 1978, *65* (12):58 with the kind permission of American Electroplaters Society Inc., U.S.A.

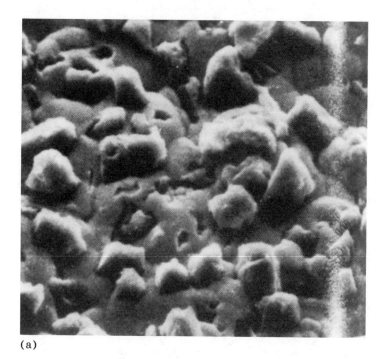

(a)

FIGURE 53   Scanning electron micrograph of EN-diamond composite
2000 times. (a) As-plated; surface roughness 87.5 μm AA. (b) After
smoothing by the new technique; surface roughness 40 μm AA.
(Reproduced from *Metal Finish.*, 1983, Aug., 35, with the kind
permission of Metals and Plastics Publications Inc., U.S.A.)

a second metallic layer which covers all exposed particles and subsequent partial removal of the secondary layer, taking care to leave
the particles still covered. Improvement arising out of this method
of smoothing is demonstrated by the scanning electron microscope
(SEM) pictures of surfaces shown in Fig. 53 and the data presented
in Table 17.

### 3.6   Hardness and Wear Resistance
of Electroless Composites

Data presented in a report published by the U.S. Department of
Energy in September 1985 has been quoted by Roos and Celis (1987),
which brings out the importance of wear-resistant coatings. Based
on actual experimental or production data, losses due to friction

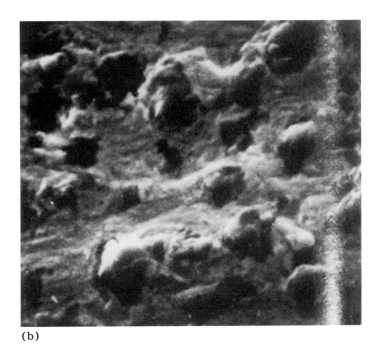

(b)

are estimated to be $6 \times 10^9$ kWh/yr. Through advanced surface
modification technologies to reduce friction, 9% of the energy loss
could be saved. The technologies considered are ion implantation,
laser surface hardening, electron beam surface hardening, and
deposition of wear-resistant coating.

The loss due to wear is estimated to be $2.3 \times 10^9$ kWh/yr.
This figure takes into account indirect losses related to replacement
cost of tools, downtime, and repair. About 71% of the wear loss
could be saved through surface modification technology. It is therefore clear that wear-resistant coatings do help to save energy.

The major application of electroless composites is combating
wear. In general, hardness and wear resistance are related: the
higher the hardness, the greater the wear resistance. Consequently,
hardness and wear resistance of these composites have received
maximum attention. Most of the studies pertain to composites based
on EN (Ni-P).

The hardness of a composite depends on the matrix and the
additional flow stress contributed by the particles. EN deposits
exhibit an increase in hardness after heat treatment. Similar behavior is to be expected for EN composites based on the arguments
given above. Such behavior has been found in all EN composites.

TABLE 18   Effect of Heat Treatment on the Microhardness
(measured with Tukon tester) of EN and EN Composite

|  | Vickers hardness (kg/mm$^2$) |
| --- | --- |
| Electroless nickel matrix as plated | 500 |
| Electroless nickel matrix heat treated | 870[a] |
| Silicon carbide composite as plated | 700 |
| Silicon carbide composite heat treated | 1300[a] |

[a]Baked 5 hr at 290°C.
*Source*: From Hubbell (1978b).

Table 18 and Fig. 44 show the improvement in hardness arising
from heat treatment of EN composites.

By and large, wear resistance of composite coatings has been
measured by three methods: (1) Taber abrasive wear test, (2)
Dow-Corning alpha wear test, and (3) DuPont's accelerated yarnline
wear test.

In the Taber abraser test two rotating abrasive wheels abrade
a rotating disk coated with the deposit, whose wear is to be measured. A sketch of the Taber abraser is shown in Fig. 54. The
contact pressure between the wheel and the sample is normally
100 psi (0.6895 MPa). In this method, wear rate is measured by
finding the weight loss. The wear rate is expressed by the Taber
wear index (TWI) which is equal to the loss of weight in milligrams
per 1000 cycles.

Dow-Corning's alpha wear tester, LFW-1, measures both friction
and wear. Figure 55 shows a sketch of LFW-1. A test block is pressed
against a rotating hardened steel ring, and the resulting weight
loss or the dimensions of the wear scar are measured. The coefficient
of friction is measured by a friction load pickup.

DuPont's accelerated yarnline wear tester is schematically shown
in Fig. 56. A pin coated with a deposit to be tested is rotated at
1 rpm, and a nylon monofilament smeared with 1 μm $Al_2O_3$ powder
passes round the coated surface at a speed of 300 ft/min and at
a tension of 10 g. The wear rate is measured by measuring the
depth or volume of the wear scar.

The Taber abraser measured predominantly the abrasive wear,
while LFW-1 essentially measures adhesive wear. The accelerated
yarnline wear test more closely approximates the entire spectrum
of wear mechanisms (abrasive, adhesive, and erosive wear).

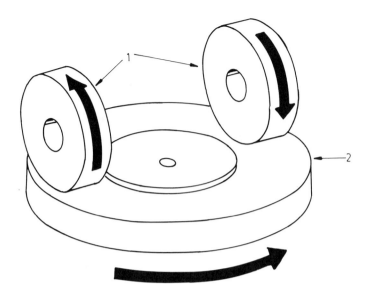

FIGURE 54  Schematic of Taber abraser. 1. Abrasive wheels weighted on test specimen. 2. Test specimen.

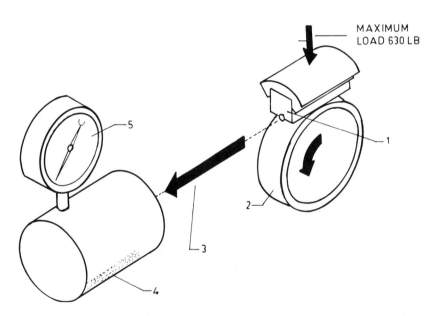

MAXIMUM
LOAD 630 LB

FIGURE 55  Schematic of alpha friction wear tester LFW-1. 1. Test block cylindrically seated for uniform load distribution over entire line of contact area. 2. Test ring. 3. Friction force at line of contact directly transmitted to load pickup. 4. Friction load pickup. 5. Friction load indicator.

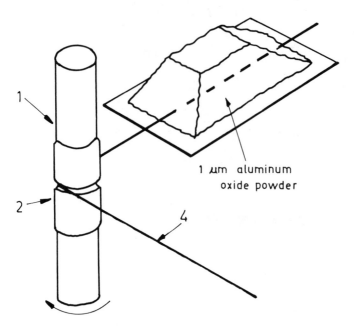

1 μm aluminum
oxide powder

Pin rotates at 1 rpm

FIGURE 56 Schematic of accelerated yarnline wear tester. 1. 25.4-mm-diameter pin. 2. Coating. 3. Wear scar. 4. Nylon monofilament (0.38 mm dia.) at 10 g tension moving at 1.52 m/s (3000 ft/min).

    Though all three methods have been used to measure the wear resistance of composites, the Taber wear test has been used far more often. Hard chrome and anodized coatings have been evaluated by it. When the TWI values of different composite coatings are compared, caution is needed.
    Martin, Leech, and Lowery (1979) have pointed out that while testing composite coatings, the standard procedure for determining TWI values should be modified to achieve a high degree of reproducibility. They found that an abrasive wheel with a harder matrix must be used to minimize tear-out of included particles. Further, they noticed that the standard 1000-2000 cycles test results suffer from lack of consistency and reproducibility. Such factors as heat generated during the test, atmospheric moisture, and test panel cleanliness have a much greater effect on TWI values of composites than on noncomposite materials. They suggest the following test cycle.

1. A clean-up run of 1000 cycles
2. Dehumidified cooling to ambient temperature with precise weighing
3. A sequence of three runs, each of 500 cycles duration, with dehumidified cooling and weighing between each run
4. Average the results of the three runs to establish the TWI value

This modified procedure gives a TWI value of 0.18-0.22 for EN-SiC composite, whereas the previous value for this composite with 2-μm size SiC particle was 1.7. Thus it is clear that comparing results of different authors may not be meaningful. We shall therefore discuss the conclusions on wear characteristics that could be drawn from the results of each group separately.

Wear resistance of composites depends on several factors, particularly particle size and volume fraction of the dispersed phase, hardness of the particle and matrix, and compatibility between the dispersed phase and the matrix. In addition, coefficient of friction, lubrication, surface finish, etc., also play a role.

The mechanism of wear of composites is rather complex and not well understood. Abrasive, adhesive, and erosive wear mechanisms appear to be operative. There is a mutual interaction between these which complicates the picture. Though a good deal of work has been done on the effect of various parameters that govern wear, the wear mechanism has not been investigated. We shall discuss the results of investigations on the wear of electroless composites.

Particle size has a significant influence on wear; larger particles produce a smaller wear rate. Table 19 shows the results obtained in the yarnline test, which illustrates that composites containing larger particles resist wear better. It is also seen that the wear rate is less for composites containing a larger amount of particles.

Taber wear indices of several composites are given in Hubbell's paper (1978b), and are shown in Table 20. Wear resistance is seen to be better for larger particles. Composites containing harder particles exhibit a lower wear rate. Likewise, a harder matrix, obtained by heat treatment of the composite, also contributes to lower wear rate. It is also seen that incorporation of any of the carbide particles reduces the wear rate of EN by a factor of 3 or more. Improvement in wear resistance arising out of heat treatment is much larger in EN than in electroless composites. Wear rate of most composites in the as-plated condition is slightly lower than that of heat-treated EN deposit. Parker's results on Taber wear tests on composites (Table 21) and yarnline wear test results (Table 22) support our general conclusions.

LFW-1 test results reveal interesting features. Tables 23 and 24 show Parker's (1974) data. The composite coating is more wear

TABLE 19   Effect of Particle Size and Percentage Loading of
Particles on Wear Resistance Measured by Yarnline Wear Test

| Test no. | Dispersed phase data Average particle size ($\mu$m) | Vol. % | Time (min) | Wear rate ($\mu$m/hr) |
|---|---|---|---|---|
| 1 | 12-22[a] | 20 | 85 | 3.4 |
| 2 | 9 | 20 | 85 | 5.1 |
| 3 | 5 | 20 | 85 | 6.2 |
| 4 | 3 | 29 | 30 | 11.6 |
| 5 | 3 | 5 | 10 | 65.0 |
| 6 | 1 | 20 | 2 | 216.0 |

[a]Particles were selected and tested in the range of 12-22 $\mu$m.
*Note*: Ni-B matrix was used with the codeposited polycrystalline
diamond.
*Source*: Data of Christini et al. quoted by Feldstein (1983).

TABLE 20   Wear Resistance of Electroless Composite as Measured
by Taber Abraser

| Type of particle | Particle size ($\mu$m) | Particle hardness (Knoop) | Hardness of matrix material (Vickers kg/mm$^2$) | Taber wear index |
|---|---|---|---|---|
| None | — | — | 500 | 10.75 |
| None | — | — | 870 | 3.75 |
| Silicon carbide | 10-15 | 2480 | 500 | 2.6-3.6 |
| Silicon carbide | 10-15 | 2480 | 870 | 1.5-1.9 |
| Silicon carbide | 1-3 | 2480 | 500 | 2.9-3.3 |
| Silicon carbide | 1-3 | 2480 | 870 | 1.7-2.1 |
| Boron carbide | 7-10 | 2600 | 500 | 2.1-2.3 |
| Boron carbide | 7-10 | 2600 | 870 | 1.0 |
| Tungsten carbide | 5 | 1850 | 500 | 3.0 |
| Tungsten carbide | 5 | 1850 | 870 | 2.0 |

Taber wear index = weight loss in mg/1000 cycles using CS-10
wheels - 1000-g load - average of 4000 cycles. (The lower Taber
wear index numbers indicate the best abrasion resistance.)
*Source*: From Hubbell (1978b).

TABLE 21   Wear Resistance of EN Composites and Other Coatings Measured by Taber Abraser

| | | Deposit Taber wear index | |
| Particle | Hardness (Knoop) | As plated | Heat treated[a] |
|---|---|---|---|
| None | — | 18 | 8 |
| Graphite | Soft | 15 | 8 |
| Chromium carbide | 1,735 | 8 | 2 |
| Tungsten carbide | 2,080 | 3 | 2 |
| Aluminum oxide | 2,100 | 10 | 5 |
| Titanium carbide | 2,470 | 3 | 2 |
| Silicon carbide | 2,500 | 3 | 2 |
| Boron carbide | 2,800 | 2 | 1 |
| Diamond (Du Pont) | 7,000 EST | 2 | 2 |
| Aluminum hardcoat | — | 2 | — |
| Hard chromium | 1,000 | 3 | — |

[a]10-16 hr at 290°C.
Taber wear index = weight loss mg/1000 cycles (average of 5000 cycles, CS-10 wheels 1,000-g load). Volume fraction of particles = 15-20%.
*Source*: From Parker (1972).

TABLE 22   Effect of Particle Size and Percentage Loading of Particles the Wear Rate of Electroless Ni-B-Diamond Composite as Measured by Accelerated Yarnline Wear Test

| Average particle size (μm) | Vol.% incorporated | Wear rate (μm/hr) |
|---|---|---|
| 17 | 20 | 3.4 |
| 5 | 20 | 6.2 |
| 3 | 30 | 12 |
| 3 | 5 | 65 |
| 1 | 20 | 216 |

*Note*: Dispersed phase—explosively formed diamond.
*Source*: From Graham and Gibbs (1975).

TABLE 23   LFW-1 Friction and Wear Test, Blocks Plated

| Electroless nickel coating | Particle diameter (μm) | Block surface hardness (HV$_{100}$) | Friction conflictions | | | | Block loss (mg) | Ring loss (mg) | Wear scar (mm) |
|---|---|---|---|---|---|---|---|---|---|
| | | | Static | | Kinetic | | | | |
| | | | Initial | Final | Initial | Final | | | |
| As-plated | — | 585 | 0.193 | 0.133 | 0.165 | 0.128 | 9.0 | 0.6 | 3.5 |
| 260°C/1 hr | — | 724 | 0.247 | 0.140 | 0.180 | 0.132 | 8.8 | 0.8 | 3.5 |
| 260°C/10 hr | — | 988 | 0.267 | 0.160 | 0.133 | 0.133 | 2.8 | 0.6 | 2.9 |
| 400°C/1 hr | — | 1064 | 0.180 | 0.117 | 0.178 | 0.120 | 2.3 | 0.5 | 1.5 |
| 540°C/1 hr | — | 892 | 0.160 | 0.100 | 0.187 | 0.103 | 1.7 | 0.7 | 1.5 |
| Composites | | | | | | | | | |
| + Teflon[a] | — | 900 | 0.213 | 0.140 | 0.160 | 0.113 | 2.8 | 0.9 | 2.4 |
| + SiC[b] | 2 | 724 | 0.180 | 0.137 | 0.133 | 0.124 | 1.0 | 1.0 | 1.0 |
| + B$_4$C[b] | 5 | 724 | 0.177 | 0.112 | 0.113 | 0.107 | 2.2 | 0.8 | 0.7 |
| + B$_4$C[c] | 5 | 724 | 0.120 | 0.113 | 0.113 | 0.103 | 2.0 | 0.8 | 0.7 |
| + WC | 1 | 707 | 0.167 | 0.160 | 0.133 | 0.133 | 3.2 | 3.1 | 0.7 |
| Diamond[a] | 3 | 1100 | — | 0.120 | 0.126 | 0.120 | 1.2 | 4.5 | 0.6 |
| No coating | — | 400 | 0.233 | 0.133 | 0.140 | 0.123 | 3.8 | +0.3 | 2.6 |

Test conditions: speed = 72 rpm, max. load = 68 kg, ring = Rc 65, lubricant = white oil, duration = 5000 cycles, block = 25 μm plated.
[a]Heat treated.
[b]As-plated.
[c]Lapped surface, as-plated.
Source: From Parker (1974).

TABLE 24 LFW-1 Friction and Wear Test, Rings Plated

| Electroless nickel coating | Particle diameter (μm) | Ring surface hardness ($HV_{100}$) | Kinetic friction coefficient | | Ring wt loss (mg) | Block | |
|---|---|---|---|---|---|---|---|
| | | | Initial | Final | | wt loss (mg) | Wear scar (mm) |
| As-plated | — | 623 | 0.108 | 0.086 | 102 | 0.2 | 1.80 |
| 260°C/1 hr | — | 743 | 0.117 | 0.091 | 50 | 0.2 | 1.87 |
| 260°C/10 hr | — | 1010 | 0.133 | 0.113 | 13 | 0.6 | 0.86 |
| 400°C/1 hr | — | 1060 | 0.108 | 0.106 | 1 | 0.0 | 1.43 |
| 440°C/1 hr | — | 950 | 0.167 | 0.126 | 9 | 0.1 | 1.80 |
| Composites | | | | | | | |
| + SiC[a] | 2 | 1090 | 0.134 | 0.127 | 0.1 | 5.1 | 3.20[b] |
| + B$_4$C | 5 | 670 | 0.133 | 0.094 | 3 | 6.1 | 3.13 |
| + WC | 1 | 700 | 0.167 | 0.121 | 96 | 2.9 | 2.30 |
| No coating | — | 900 | 0.183 | 0.143 | — | 3.4 | 2.40 |

Test conditions: speed = 72 rpm, max. load = 284 kg, ring = 25 μm plated, block = Rc 60, lubricant = white oil, duration = 25,000 cycles.
[a]Baked at 260°C/10 hr.
[b]Test continued to 101,408 cycles, wear scar 4.5 mm.
Source: From Parker (1974).

TABLE 25   Friction Coefficient and Wear of EN Composites as Measured by LFW-1

| Test ring | | Test block | | Friction coefficient | | | | Weight loss (mg) | |
| | | | | Static | | Kinetic | | | |
| Coating | Surface hardness (Vickers, kg/mm$^2$) | Coating | Surface hardness (Vickers, kg/mm$^2$) | Initial | Final | Initial | Final | Ring | Block |
|---|---|---|---|---|---|---|---|---|---|
| Silicon-carbide composite | 1300 | Electroless nickel | 870 | 0.16 | 0.13 | 0.1 | 0.1 | 0.0 | 6.4 |
| Silicon-carbide composite | 1300 | Hard chrome | 1000 | 0.15 | 0.10 | 0.07 | 0.07 | 0.5 | 1.1 |
| Silicon-carbide composite | 1300 | Silicon-carbide composite | 1300 | 0.1 | 0.07 | 0.09 | 0.06 | 0.2 | 0.2 |

Test conditions: load = 68 kg, speed = 72 rpm, duration = 5000 cycles, lubricant = mineral oil, ring material = SAE 4620 (ASTM STD D2614-68), block material = SAE 01 (ASTM STD D2614-68).
Source: From Hubbell (1978b).

resistant than EN, as evidenced by lower weight loss of a composite-coated block or ring. However, the weight loss of the mating surface is higher for composites than for EN, indicating that composites abrade the surface more than EN does. Composites containing WC exhibit more wear than other composites. Though the weight loss of an EN-WC-coated block is about the same as for other composites, the weight loss of an EN-WC-coated ring is abnormally high, nearly the same as for an EN-coated ring. This result is difficult to understand. The diamond containing composite coating is seen to produce even greater weight loss of the mating surface than other composites do.

Hubbell's (1978b) data given in Table 25 point out the importance of the proper choice of the mating surface for an EN-composite deposit. It is surprising that the coefficient of friction is lowest for the silicon carbide composite running against itself! Consequently, wear is low for this combination. This result is particularly interesting because abrasion of a material against itself generally exhibits large friction and high wear. Both EN and hard chrome appear to be compatible with silicon carbide composite, from the point of view of friction.

The importance of surface finish is shown by mercury Marine's LFW-1 test results, quoted by Hubbell (1978b). Steel rings coated with EN-SiC composite are run against pearlitic cast iron blocks (Knoop hardness approximately 250), and the volume loss on the block and ring is measured.

The results (Table 26) show that wear of the composite layer is negligible and not significantly affected by surface roughness. However, wear on the mating surface is enhanced by a poorer surface finish. The unpolished composite coating with 10-15-$\mu$m particles aggressively grinds off the mating surface. After polishing, it

TABLE 26   LFW-1 Wear Test. EN-SiC Composite-Coated Ring Versus Cast Iron Block

| Size of particles in the composite ($\mu$m) | Roughness of composite coated ring AA ($\mu$m) | | Ring wear | Block wear volume loss ($cm^3 \times 10^{-5}$) |
|---|---|---|---|---|
| 10-15 | As-plated | 0.41-0.50 | None | 346 |
| 10-15 | Polished | 0.15-0.20 | None | 100 |
| 1-3 | Polished | 0.25-0.28 | None | 24 |

*Source:* From Hubbell (1978b).

TABLE 27  Effect of Surface Finish and Particle Size on the Coefficient of Friction of EN-Diamond Composite as Measured by Yarnline Wear Test

| No. | Coating | Postplating finishing | Surface roughness after finishing AA ($\mu$m) | Coefficient of friction |
|---|---|---|---|---|
| 1 | EN | None | 0.15-0.13 | 0.46 |
| 2 | +1.5-$\mu$m diamond | GBB | 0.36-0.27 | 0.40 |
| 3 | +1.5-$\mu$m diamond | BF | 0.27-0.19 | 0.35 |
| 4 | +3-$\mu$m diamond | GBB | 0.65-0.53 | 0.41 |
| 5 | +3-$\mu$m diamond | BF | 0.44 | 0.36 |
| 6 | +6-$\mu$m diamond | GBB | 0.79-0.70 | 0.42 |
| 7 | +6-$\mu$m diamond | BF | 0.53-0.56 | 0.38 |
| EN composite-coated matte chrome | | Over | | |
| 8 | matte chrome | — | 1.91-2.29 | 0.21 |
| 9 | +1.5-$\mu$m diamond | BF | 2.11 | 0.30 |
| 10 | +3-$\mu$m diamond | BF | 1.98 | 0.32 |
| 11 | +6-$\mu$m diamond | BF | 1.80 | 0.33 |

GBB = glass bead blasting; BF = brush finishing.
*Source*: From Barras et al. (1979).

becomes less abrasive. The wear of the block is dramatically reduced by decreasing the particle size from 10-15 $\mu$m to 1-3 $\mu$m, even though the surface finish is not as good as the polished 10-15-$\mu$m particle composite. The wear is about one-fourth that of the polished 10-15-$\mu$m particle composite.

Accelerated yarnline wear test results of Barras et al. (1979) bring out the importance of particle size and surface finish in EN-diamond composites through the coefficient of friction (Table 27). For composites containing particles from 1.5 to 6 $\mu$m, the friction coefficient decreases with decreasing roughness. Friction coefficient of the composite is less than that of EN. Another interesting result is that a diamond composite coated on matte chromium shows a lower friction coefficient than when coated on steel substrates. This is all the more surprising because the composite coating on matte chromium is more than four times rougher than the coating on steel.

The above results demonstrate the ability of composite coatings to control the frictional property of a surface. This characteristic of composite coatings is exploited in the production of friction texturing disks to impart controlled amounts of twist into yarnlines.

Diamond-bearing composites exhibit interesting wear characteristics that are peculiar to themselves. It has been pointed out that natural diamond would not be suitable for wear-resistance applications since its surface exhibits sharp edges and points due to cleavage. Shock-synthesized polycrystalline diamond is best suited for wear-resistant coatings. The results (Table 28) of Graham and Gibbs (1975) provide an experimental proof of the poorer wear resistance of composites containing natural diamond.

The yarnline wear groove formed on composites with different diamonds was examined by SEM to get an idea of the number of diamond particles pulled out during wear. The results of this study have been quoted by Feldstein et al. (1983) and are given in Table 29.

The above data reinforces the conclusion that in composites containing diamond, the type of diamond plays an important role, since it determines the compatibility of the interaction of particles and matrix.

Graham and Gibbs (1975) investigated the difference in behavior of composites with natural and shock-synthesized diamond

TABLE 28  Wear Resistance of Electroless Ni-B Diamond Composite as Measured by Accelerated Yarnline Wear Test

| Coating[a] | Test time (min) | Wear rate ($\mu$m/hr) | Normalized wear resistance[b] |
|---|---|---|---|
| Ni-B (as-plated) | 0.03 | 23,000 | 1.0 |
| Ni-B 10 vol.% 9 $\mu$m natural diamond | 85 | 10.2 | 2,250 |
| Ni-B 10 vol.% 9 $\mu$m explosively formed diamond | 85 | 5.1 | 4,500 |

[a]Boron content of Ni-B matrix: 0.5 wt.%.
[b]Reciprocal of wear rate normalized with respect to as-plated electroless Ni-B.

TABLE 29   Data on Diamond Particles Pulled Out from Accelerated Yarnline Wear Test Grooves on NI-B Diamond Composites

| Test no. | | Area $I^a$ | Area $II^b$ | Percent difference | Comment |
|---|---|---|---|---|---|
| 1 | Diamond count for diamond $B^c$ | 50 | 34 | -32 | Difference primarily caused by diamond removal from the bottom of the wear grooves |
| 2 | Diamond count for natural diamond | 60 | 45 | -25 | |
| 3 | Diamond count for diamond $A^d$ | 42 | 46 | +9.5 | Difference because of the randomness and exposure of diamonds on the bottom of the wear groove, just underneath the surface of Ni-B matrix |

[a]Side of wear groove and adjacent as-plated composite surface.
[b]Bottom of the wear groove, where most of the wear occurs.
[c]A is polycrystalline diamond prepared as per Cowan (U.S. patent 3,401,019).
[d]B is diamond synthesized as per U.S. patents 2,947,608 through U.S. 2,947,611.
*Source:* From Feldstein et al. (1983).

by examining the yarnline wear groove with the help of SEM. Their
observations are briefly summarized below. Explosively formed diamond
is irregular and rough. The alloy Ni-B encapsulates the entire area
by growing into and around the irregularities. Natural diamond
being smooth, the mechanical anchorage arising from such encapsulation is rather small. Thus mechanical binding of explosively formed
diamond to the matrix is much stronger. Electron microscopic examination also reveals nucleation of electroless nickel at several places
on the surface of explosively formed diamond. This would further
enhance the binding between particle and matrix. Such nucleation
centers are rarely seen in natural diamond. Thus, bonding due
to chemical factors is higher in explosively formed diamond. As
a result of the enhanced mechanical and chemical binding of explosively formed diamond, particles do not get pulled out during
abrasion, and hence wear resistance is better.

The matrix also appears to play an important role. Table 30
gives the wear rate data for natural and shock-synthesized diamond
composites using a Ni-P alloy as the matrix. In this matrix, natural
diamond is inferior to polycrystalline. The wear rate of a shock-
synthesized diamond composite is practically half that of a Ni-P
composite with natural diamond. However, comparing the results
of Table 30 and Table 28, we see that the wear rate of polycrystalline
diamond in a Ni-P matrix is 3.78 μm/hr as compared with 5.1 μm/hr
in a Ni-B matrix. No attempt has been made to explain this large
difference in the wear rate arising from matrix influence. The reduction is too large to be accounted for on the basis of increased
hardness of the Ni-B matrix. The hardness of heat-treated Ni-P
is about the same as that of as-plated Ni-B. If the improvement
is due to hardness, then the wear rate of heat-treated Ni-P composite
should exhibit substantial reduction in wear rate. Parker's data

TABLE 30   Wear Rate of Ni-P Diamond Composite

| Test no. | Composite coating | Test time (min) | Wear rate[a] (μm/hr) |
|---|---|---|---|
| 1 | Electroless Ni-P 1-μm diamond A[b] | 2 | 378 |
| 2 | Electroless Ni-P 1-μm natural diamond | 2 | 732 |

[a]Results of accelerated yarnline wear test.
[b]Explosively formed polycrystalline diamond.
*Source:* Data from U.S. patent 3,936,577 and reissue 29,285 quoted
by Feldstein (1983).

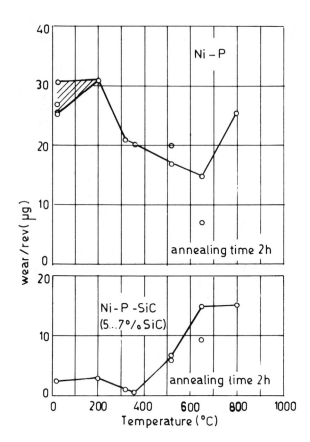

FIGURE 57  Effect of heat treatment on wear of Ni-P and Ni-P-SiC
(Broszeit, 1982) under mild abrasive conditions. (Reproduced with
the kind permission of Elsevier Sequoia S.A., Switzerland.)


(Table 21) shows that as-plated and heat-treated Ni-P-diamond
composites have similar wear rates. Broszeit's results (Fig. 57)
on wear of Ni-P-SiC show a similar trend. Therefore one might
infer that increasing the hardness of the matrix from 500 to 800
VPN cannot really produce a dramatic reduction, as is seen in
changing the matrix of a diamond composite from Ni-P to Ni-B.
Whether a similar effect would be observed with other particulate
matter is yet an unanswered question.

  Wear measurements on EN-PTFE composites have been carried
out by Tulsi (1983). His results are summarized in Table 31. In
the as-plated state, the wear rate is marginally higher for EN-PTFE

TABLE 31   Coefficient of Friction and Wear Rate of EN-PTFE
Composite as Measured by Pin and Ring Machine

| Coating | Pin Hardness (Vickers, kg/mm$^2$) | Coating on chrome steel ring | Friction coefficient | Specific wear rate ($10^{-6}$ mm$^3$/NM) |
|---|---|---|---|---|
| Ni-P (as-plated) | 600 | None | 0.6-0.7 | 35 |
| Ni-P-PTFE[a] (as-plated) | 250 | None | 0.2-0.3 | 40 |
| Ni-P-PTFE[a] (as-plated) | 250 | Ni-P-PTFE composite | 0.1-0.2 | 1 |
| Ni-P-PTFE[a] (heat treated)[b] | 400 | None | 0.2-0.5 | 20 |
| Ni-P-PTFE (heat treated) | 400 | Ni-P-PTFE composite | 0.1-0.7 | 2 |

[a]Volume percent of PTFE-25-30.
[b]400°C for 4 hr.
Source: From Tulsi (1983).


than for EN. However, heat treatment is reduced by half. The most
dramatic reduction in wear rate is found when EN-PTFE runs against
itself. Even if the mating surface is different, after "running-in,"
it will get coated with PTFE by transfer of particles from the composite so that the mating surface becomes virtually PTFE, thereby
making the wear rate quite low.

A comparison of wear resistance of different composites with
themselves and with other coatings will be useful. Broszeit (1982),
employing the Taber abraser, compared the effect of heat treatment
on the wear of Ni-P and Ni-P-SiC (Fig. 57). The former suffers
greater wear at lower annealing temperatures (200°C) and progressively decreases with temperature, exhibiting a minimum at 650°C,
beyond which it again increases. Ni-P-SiC shows little reduction
of wear up to 350°C, beyond which it increases, and deterioration
of wear properties is noticed at higher temperatures. The Ni-SiC
reaction (580°C) and partial melting of Ni-P eutectic (880°C) are
responsible for poorer wear characteristic at higher temperatures.
Throughout the temperature range, the composite resists wear better.

Figure 58, taken from Metzger and Florian (1976), shows wear
of Ni-P-SiC and Ni-SiC. Unfortunately, there is no discussion of

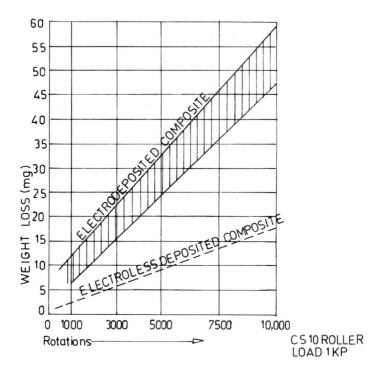

FIGURE 58  Comparison of wear of silicon carbide composite deposited by electroless and electroplating methods.


this important comparison. The authors have not even indicated the source of the data. The wear of electroless composite is seen to be smaller than the electrolytically deposited one. What is even more important is that the rate of increase of wear with the number of revolutions is greater for electrocomposites than for electroless composites. The present authors feel that the reduced wear of electroless composites is probably due to increased hardness of the matrix and better bonding between particles and matrix. Such a stronger interaction may be due to nucleation of electroless nickel at several points on the particle, in addition to encapsulation of the particle by electroless nickel. For electrocomposites, the particles are held only by the growth which penetrates into the irregular surface topography. Nucleation on the particle is unlikely since it is not a conductor. Hence the particle-matrix interaction can be expected to be stronger in electroless composites than in electro-composites.

Tables 32 and 33 compare the wear resistance of several coatings and materials. Wear resistance of EN shock-synthesized diamond is seen to be much better than other well-known wear-resistant materials like cemented WC, sintered $Al_2O_3$, flame-sprayed $Cr_2O_3$ electrodeposited chromium, and heat-treated tool steel.

From an applications point of view certain salient features of wear characteristics of electroless composites should be borne in mind. To keep the wear of the composite and mating surface low, 1-3-$\mu$m particles are used at a loading of about 20-25% by volume.

In service, the matrix surrounding each particle wears off to a greater extent, making the particles partly project from the mean plane of the surface and thereby making them take the load. This demands that the particles should be hard, tough, nonfriable, and well bonded to the matrix, if they are to enhance wear resistance. Otherwise, loose particles may agglomerate and aggressively abrade the mating surface.

No technology is free from limitations, and electroless composites are no exception. High impact, bending, and shear stress may damage the coating. In such situations, electroless composites should not be used. Due to embedding of particles, the composite coating is likely to have a lower ductility. Hence, they are to be avoided under loads which might deform the substrate and damage the coating.

Diamond-containing composites pose a problem of compatibility with the mating surface when they are used for wear resistance. Diamond, being an allotropic modification of carbon, can react or

TABLE 32 Wear Rate of Coatings and Materials as Measured by Taber Abraser

| | Wear rate | |
| --- | --- | --- |
| Coating or material | Per 1000 cycles ($10^4$ mil$^3$) | Relative to EN-diamond composite |
| 1. EN-polycrystalline diamond (20-30% by volume of 3-$\mu$m diamond particles) | 1.159 | 1.00 |
| 2. Cemented WC-grade C-9 (88% WC, 12% Co) | 2.746 | 2.37 |
| 3. Electroplated hard chromium | 4.699 | 4.05 |
| 4. Tool steel hardened to 62 $R_C$ | 12.815 | 13.25 |

*Source*: From Feldstein et al. (1983).

TABLE 33  Comparison of Wear Rates of Different Coatings or
Materials by Accelerated Yarnline Wear Test

| Wear-resistant coating | Wear rate | |
| of material | $mil^3/hr$ | Relative rate |
|---|---|---|
| Synthesized polycrystalline diamond coating[a] | 1.26 | 1.0 |
| Natural diamond composite coating[a] | 3.24 | 2.5 |
| Single-crystal diamond composite coating[a] | 4.20 | 3.3 |
| Cemented tungsten carbide grade C-9 (88% WC, 12% Co) | 43.74 | 35 |
| Sintered 99.5% aluminum oxide | 59.40 | 47 |
| Flame-sprayed chromium oxide | 106.86 | 85 |
| Flame-sprayed aluminum oxide | 173.56 | 138 |
| Silicon-carbide composite coating[a] | 349.80 | 278 |
| Electroplated hard chromium | 966.60 | 767 |
| Tool steel hardened, $R_C$ 62 | 3,478.80 | 2,761 |

[a]All composite coatings contained 20-30% (volume) of a 3-μm grade
particle in an electroless nickel matrix.
*Source*: From Feldstein et al. (1980).

dissolve in the mating surface if it is made out of carbide-forming
elements such as iron, chromium, and titanium, especially under
conditions of high loads and/or dry friction. Consequently, the
diamond particles may disintegrate and cause nickel to contact the
mating surface, which has a tendency to get cold-welded to a number
of metals. Hence attention must be given to the compatibility of
the mating surface when diamond-bearing composites are used.
Nonmetals and nonferrous materials are suitable for use with diamond
composites even under dry friction conditions. Ferrous materials
should be used with this composite only with a lubricant.

A composite containing particles produces relatively greater
wear on the mating surface than do hard coatings of metals and
alloys. This need not necessarily be viewed as a limitation, because
not all wear applications are concerned with wear of the mating
surface. Use of an EN-particle composite for molds and metal-forming
tools is an example of coating for wear resistance, where the wear
of the mating surface does not come into the picture.

The real limitation of a composite-based Ni-P matrix is its poor performance at higher temperatures (500°C). Kenton et al. (1983) have reported that codeposition of 20-30% by volume of calcium fluoride particles will give significant improvements in high-temperature wear performance. This suggestion has the potential to broaden the application of Ni-P composites.

## 3.7  Applications of Electroless Composites

Electroless composite coating on a production scale and its consequent use for industrial applications are relatively recent. Of the several particles embedded into electroless nickel, SiC, diamond, and recently PTFE have found commercial application. Though several matrices can be used, most applications use only Ni-P. The applications can be broadly classified into three categories, providing (i) wear resistance, (ii) a surface with a desired friction coefficient, and (iii) a hard surface for machining and finishing tools.

*Electroless Composite Coatings for Improved*
*Wear Resistance*

The useful life of molds for plastics, rubber, etc., has been improved by coating them with NI-P-SiC.

Die cast molds employed for the manufacture of components for irons from glass-fiber-reinforced thermosetting plastics normally lasted for about 10,000 moldings. Electroless Ni-P coating of this mold improved its life by a factor of 3. A 50-μm coating of Ni-P-SiC, after heat treatment, increased the life by 15 times (Metzger and Florian, 1976). Molds for switchboard parts containing 102 holes of 1.778-mm diameter were effectively protected by a 45-μm Ni-P-SiC layer. As a result, it could be made out of unhardened steel. In this application, this coating outperformed any other coating or alloy tried (Hubbell, 1978b).

In the plastics industry, increasing use of flame retardants and reinforcing glass fibers causes accelerated corrosion and abrasion of molds. A Ni-P-SiC coating has been found to be extremely useful for combating these problems. A 50-μm coating of an extruder screw, used for extrusion of materials like GE's Noryl (30% glass-filled polymer), has proved to be better than chrome plating. The use of this coating (37 μm thick) on a large compression mold, weighing 8000 lb (3636.4 kg), used for producing the reinforced plastic front-end piece for an automobile, is an interesting application. It demonstrates that the process of coating an EN composite is practical even for large parts. Figure 59 shows this mold (Hubbell, 1978b).

Forming and drawing of sheet metal poses the problem of pickup and galling. Ni-P-SiC coating has been used for getting over these problems. To mention an example: an automotive door hinge die

FIGURE 59   A large compression mold used for producing reinforced
plastic front-end piece of an automobile, weighing 8000 g (3636.4 kg),
coated with 37 μm of Ni-P-SiC for protection against wear. (Reproduced from *Trans. IMF.*, 1978, *56*:65 with the kind permission of
Inst. of Metal Finishing, U.K.)

required, on an average, 1 hr for repair, after producing 3000
parts. When the die was coated with 37 μm of EN-SiC, it needed
attention only after producing 400,000 parts (Hubbell, 1978b).
    A Ni-P-SiC coating has been used in foundries for resisting
wear and helping to release sand cores, without breakage, from
core boxes. Because of the intricate shape of many foundry patterns,
the Ni-P-SiC composite coating turns out to be an ideal coating
for improving the surface property needed by the foundry. For
instance, an epoxy pattern got eroded, especially at the angular
edges and rounded portions (due to blowing of sand before every
compaction) after 2000 molds were made. The use of this coating
permitted more than 25,000 impressions to be made before the pattern
wore out (Hubbell, 1978b).
    The special advantages of EN-SiC coatings are particularly
well demonstrated with a high-pressure valve.
    Figure 60 shows an inner thread of the working part. The
×50 magnification [Fig. 60 (b), left] shows the uniformity of coating
on the inner thread. The ×250 magnification [Fig. 60 (b), right]
demonstrates the good particle dispersion and the high embedding

rate in the electroless nickel coating even on such exposed surface parts of the high-pressure valve (Metzger and Florian, 1976).

Ni-P-diamond composite coatings are applied to improve the wear resistance of tools. Some examples are reamers used with highly abrasive aluminum alloys, contact heads of honing heads, broaching tools for graphite, valves for viscous rubber masses, and thread guides for use in textile machines and friction texturizing disks (Luckschandel, 1978).

Even though PTFE-containing composites are of very recent origin, commercial applications have already started. These applications have properties such as lack of adhesion, lack of galling, high dry lubricity, and good wear and corrosion resistance.

Molds for certain rubber compositions, especially those containing epichlorohydrin, suffer from severe attacks of hydrochloric acid, due to in situ release. Conventional release agents do not protect these molds. Besides, it is difficult to apply release agents uniformly on a complex contour, especially in recesses which are more vulnerable to attack. Ni-P-PTFE composite coatings of these molds protects them and helps to release the cured product (Tulsi, 1983).

Carburetor parts are coated with this composite to provide a nonstick, dry lubrication property, which helps minimize the buildup of gummy deposits on the choke shaft. This coating is superior to simple PTFE coatings (Tulsi, 1983).

The internal surface of a long aluminum air cylinder, being worn out by the wearing action of rubber pistons, could be protected by a NI-P-PTFE coating. A normal PTFE or hard anodized coating has a life-span of 10,000 to 30,000 cycles, whereas a 5-$\mu$m layer of this composite has a life-span in excess of 3,000,000; i.e., the improvement is at least 100 times (Tulsi, 1983).

Tulsi (1983) mentions the use of this coating for stainless steel butterfly valves working against a Nimonic ring, standard bolts and nuts requiring a precise amount of torque to exert the correct tension.

*Electroless Composite Coating to Provide a Surface with a Desired Friction Coefficient*

Slipless transmission of very high rotational speed (1 million rpm) is needed for textile fibers. Such a transmission requires a good-grip moderately rough surface. Ni-P-diamond can be tailored for this purpose. Other applications in this category are yarn brakes, infinitely variable gears, and friction clutches (Luckschandel, 1983).

*Electroless Composites for Coating Machining and Finishing Tools*

Diamond-bearing composites with particles with diameters of 6-12 $\mu$m diameters (see Section 3.6) caused greater wear to the mating

(a)

FIGURE 60   (a) Inner thread of a high-pressure valve. The whole
piece is coated with a 50-μm Ni-P-SiC layer and heat treated. (b)
Cross section of inner thread of the high-pressure valve. The equal
thickness of the composite coating should be mentioned (×50 magnification, left). At ×250 magnification (right) the uniform dispersion
of SiC can be noticed.   Reproduced from *Trans. IMF.*, 1976, 54:174,
with the kind permission of Inst. Metal Finishing, U.K.


surface. This property is exploited in making profiled diamond
tools used for microfinishing screw threads, ball-guiding grooves,
and other profile sections. For this application, the accuracy demanded
is so exacting that conventional electroplating methods cannot be
used, since the surface profile cannot be corrected after coating.
For this application, it is almost mandatory that electroless composite
plating should be used (Luckschandel, 1983).


3.8   Conclusions

Though suspending micron-size solid particles in an electroless
bath posed a threat to the stability of the bath, this has been
overcome and reliable coatings for wear resistance applications,
such as Kanisil and Nye-Carb, Nidiament, are available. Several
industrial applications of this coating, especially Ni-P-SiC, Ni-P-

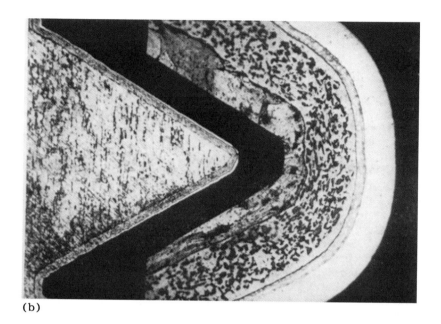

(b)

diamond, and NI-P-PTFE, have been established and more applications are in the offing. Since these applications are mainly for combating wear, a great deal of data on wear of these composites have been published. However, little work appears to have been done on the mechanism of wear.

In these composites the rate of embedding, for a given concentration of particles in suspension, is much higher than that observed in electrolytic deposition of composites. This phenomenon and the mechanism of codeposition of particles in electroless composite deposition have received no significant attention.

Most of the studies on properties of composites are with the Ni-P matrix. Very little work has been done on Ni-B matrix composites.

### References

Barras, R., Spencer, R., Bailey, N., Feldstein, N., and Lancsek, T. (1979). Proc. Electroless Nickel Conference, Gardner Publications, Ohio, p. 255.

Broszeit, E. (1982). Thin Solid Films, 95:133.

Broszeit, E., Heinke, G., and Wiegand, H. (1971). Metall. (Berlin), 25, 470:1110.

Brown, L. (1985). Trans. IMF, 62:139.

Dennis, J. K., Sheikh, S. T., and Silverstone, E. C. (1981).
     Trans. IMF, 59:118.
Editor (1977). Machine Design, Nov.:24.
Epifanova, V. S., Golovushkina, L. V., Prusov, Yu, V., and
     Flerov, V. N. (1975). Zashchita Metallov, 11:634.
Feldstein, N., Lancsek, T., Barras, R., Spencer, R., and Bailey
     (1980). Prod. Finish, July:65.
Feldstein, N., Lancsek, T., Lindsay, D., and Salerno, L. (1983).
     Metal Finish, Aug.:35.
Gawrilov, G. (1974). Galvanotechnik, 65:858.
Gawrilov, G. and Butschkow, D. (1980). Galvanotechnik (10):1088.
Gawrilov, G. and Ernin, E. (1975). Galvanotechnik, 66:397.
Gawrilov, G. and Owtscharova, E. (1973a). Galvanotechnik, 64
     (1):23 idem, (1973b). Metalloberfl, Angew. Electrochem, 27
     (2):41.
Graham, A. H. and Gibbs, T. W. (1975). Chapter 16 in Properties
     of Electrodeposits. Their Measurements and Significance (Sard,
     R., Leidheiser, Jr., H., and Ogburn, F., eds.). The Electro-
     chemical Society.
Grund, H., Hofmann, K., and Sobek, E. (1983). Schmierungstechnik,
     14 (3):75, idem (1983). R. Zh. Mash., 713:246.
Guseva, I. V., Mashchenko, T. S., and Borisenko, A. I. (1977).
     R. Zh. Korr. i, Zashch, ot. Korr., 12K:502 (from Met. Fin.
     Abst.), Idem (1980). ibid., 2K:269 (from Met. Fin. Abst.).
Guseva, I. V., Mashchenko, T. S., and Borisenko, A. I. (1981).
     R. Zh, Mash., 11B:710.
Hubbell, F. N. (1978a). Plating Surf. Finish, 65 (12):58.
Hubbell, F. N. (1978b). Trans. I.M.F., 56:65.
Hussain, M. S. and Such, T. E. (1981). Surface Technology, 13:119.
Kedward, E. C., Addison, C. A., and Tennet, A. A. B. (1976).
     Trans. IMF, 54:8.
Kenton, D. J. (1983). Paper presented at the ASM Surtech and
     Surface Coating Exposition, Dearborn, MI, May.
Kurtinaitene, V., Zhitkyavichyute, I., and Lunyatskas, M. (1983).
     R. Zh. Korr. I, Zashch. ot. Korr., 6K:404.
Lukschandel, J. (1978). Trans. IMF, 56:118.
Martin, W. B., Leech, G. R., and Lowery, C. H. (1979). Proc.
     Electroless nickel Conference, Gardner Publications, Ohio, p. 99.
Metzger, W. (1972). Galvanotechnik, 63:722.
Metzger, W. and Florian, Th. (1976). Trans. IMF, 54:174.
Mitsubishi Heavy Industry KK (1976). Jap. Patent 52/104405, March:1.
Odekerken (1966). Brit. Pat., 1,041,753, US Pat., 3,644,183 and
     DDR Pat., 41, 406.
Parker, K. (1971). US Patent, 3,562,000.
Parker, K. (1972). Prof. "Interfinish-72", SGT, Basel, Switzerland,
     p. 202.

Poniatowski, M. and Clasing, M. (1972). *Gold Bull.*, *5* (2):34.

Prusova, Yu, V. and Egorenkova, S. I. (1977). *R. Zh. Korr,
   i, Zashch, ot. Korr.*, *11K*:204.

Roos, J. R. and Celis, J. P. (1987). Proc. INCEF-86, The Electrochemical Soc. India, Bangalore (in Press).

Safina, F. K., Saifullin, R. S., Tremasov, N. V., and Sairanova,
   A. A. (1978). *Zashchita Metallov*, *14*:504.

Sharp, W. F. (1974). Proc. Diamond - Partner in Productivity,
   p. 121.

Silcock, J. M., Heal, T. J., and Hardy, H. K. (1953-4), *J. Inst.
   Metals*, *82*:239.

Tulsi, S. S. (1983). *Trans. I.M.F.*, *61*:147.

## 4   LAYERED OR LAMINATED COMPOSITES

Fine Damascus and Samurai swords are classical examples of laminated
composites. These swords are made from different steels and irons
by folding and forging them. The result is an extremely hard,
keen-edged sword with a relatively soft and tough body (Fox, 1974;
Wadsworth, Kum, and Sherby, 1986). These composites have their
modern technological counterparts in plywoods, bimetallic elements
for thermal actuation, clad metals for corrosion protection, metal
honeycomb sandwich panels, and a host of others. They are all
laminated structures in which the constituent materials integrate
their properties from layer to layer.

Several examples of deposits exhibiting a laminated or banded
structure are known. Copper deposited in the presence of benzotriazole, nickel deposited from bright nickel baths, alloy deposits
such as Fe-Ni, electroless Ni-P, and electroless Ni-B are some of
the examples of such deposits. Studies on deposits with banded
structures have been referred to by Brenner (1963), Raub and
Muller (1967), and Cohen, Koch, and Sard (1983). Even though
these investigators discussed in detail the cause of formation of
laminated structures, they did not examine them from the point
of view of composites.

### 4.1   Koehler's Strong Solid—An Example
####      of Layered Composite

Koehler (1970), from theoretical considerations, proposed a new
type of composite. It consisted of alternate layers of crystals of
A and B. The materials of A and B were chosen to satisfy the
following conditions.

The lattice parameters and thermal expansions of A and B are
nearly the same at the operating temperature, so there are no strains

at the interface of A and B. The line energy of dislocations should differ by a large value. The thickness of low-line-energy material, say A, should be so thin that dislocation generation cannot occur inside the A layer. This condition will be fulfilled if the thickness of A is about 100 atomic layers. The thickness of B layer should also be about the same. Adhesion between lamellae should be excellent.

In a layered composite of this type, Frank Reed sources cannot operate inside one layer, and a very large external stress will be needed to drive the dislocations from low-line-energy layer A to high-line-energy layer B. The result will be a strong solid.

It is readily seen that Koehler's strong solid is only a special type of laminated composite. Broadly speaking, a composite made from sufficiently thin alternate layers of a low- and high-elastic-constant material must behave like Koehler's strong solid. To realize maximum strength, we should choose materials A and B very carefully. According to Koehler, some of the useful combinations are Cu-Ni, Pt-Ir, Rh-Pd, and MgO-LiF.

Koehler suggested epitaxial growth as a method of preparing this type of composite. Perhaps this suggestion was made because epitaxial growth results in minimum strain at the substrate-deposit interface. Electrodeposition is probably a better technique for making this composite. It is well established that electrodeposits exhibit epitaxial growth in the initial layers, adhesion is very good, and thickness is easily controllable. These characteristics make electrodeposition a versatile technique for forming this type of layered composite.

## 4.2 Some Examples of Electrodeposited Laminated Composites

George U. Rose, Jr., was probably the first to produce copper-nickel laminate by electroplating alternately copper and nickel from two different baths. This material was developed to provide a stronger printing plate than an electrodeposited copper 6 mm thick, which got bent in service due to lack of adequate strength. His patented process of plating this laminate was used by the U.S. Bureau of Engraving and Printing (Blum and Slattery, 1921).

Blum (1921) made a scientific study of Rose's method of improving the strength of copper by interposing thin layers of nickel during the deposition of copper. His results on the mechanical properties are summarized in Table 34. It is seen that for a given ratio of thickness of copper to nickel, the strength increases as the number of layers of copper in 1 mm of the composite increase. A laminate with 63 $\mu$m of copper alternating with 6.3 $\mu$m of nickel exhibits 71% higher strength than does pure copper, without loss of ductility.

TABLE 34 Mechanical Properties of Electrolytic Copper and Copper–Nickel Plates

| Plate no. | Current density (A/cm²) | Copper, Thickness of each layer (μm) | Nickel, Thickness of each layer (μm) | Thickness ratio | Ultimate tensile strength (MPa) | Percentage improvement in UTS over copper | Elongation (%) |
|---|---|---|---|---|---|---|---|
| 2 | 0.06 | — | — | — | 165.7 | — | 10 |
| 4 | 0.06 | 250 | 25 | 10:1 | 219.7 | 33 | 8 |
| 5 | 0.06 | 125 | 12.5 | 10:1 | 243.2 | 47 | 9 |
| 6 | 0.06 | 63 | 6.3 | 10:1 | 283.4 | 71 | 13 |
| 7 | 0.06 | 125 | 25 | 5:1 | 295.2 | 79 | 7 |

Source: From Blum (1921).

TABLE 35

| Metal | Thickness ($\mu$m) | UTS (MPa) | Elongation (%) |
|---|---|---|---|
| Copper | 500 | 196.12 | 8 |
| Nickel | 500 | 392.24 | 4 |
| Copper-nickel 1:1 | 500 | 490.3 | 6.5 |
| Layer thickness 500 Å | | | |

Blum found that the grain size of copper was smaller when the nickel layer was interposed. He attributed the property improvement to the finer grain size. He has reported similar effect for layers of copper and silver as well as layers of copper deposited from sulfate and cyanide solutions.

Indira Rajagopal, Rajam, and Rajagopalan (1980) prepared a laminated copper-nickel composite by alternately plating 500-Å thick layers of the two metals. Standard acid copper and nickel sulfamate baths were used. To ensure a coherent deposit in each layer at such a small thickness, a pulse current with a duty ratio of 1:1 was used. Their results are given in Table 35. The tensile strength data was obtained by applying a load parallel to the laminae. It is seen that the UTS of the laminated composite is 490.3 MPa, which is greater than the 294.18-MPa value calculated by the mixture rule. Figure 61 shows the cross section of the deposit.

Recently Cameron Ogden (1986) has investigated in greater detail laminated composites prepared by plating alternate layers of copper and nickel from two different electrolytes, proprietary copper sulfate and nickel sulfamate baths. Deposition was carried out at 20 mA/cm$^2$ on to stainless steel by air agitation. Composites with layer thickness varying from 0.4 to 8.2 $\mu$m (150 to 3000 atomic layers) were investigated. Both tensile strength in the direction parallel to the layers and resistance to penetration perpendicular to the layers were measured.

Ogden's results on tensile strength are given in Fig. 62. The main conclusions are as follows:

1.  Tensile strength (TS) obeys the empirical relationship

$$TS = \frac{600}{t^{0.18}}$$

FIGURE 61 Cross section of laminated copper-nickel composite deposited from two different electrolytes using pulsed current. Carbon replica mag ×5000.

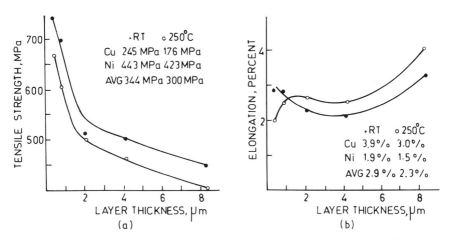

FIGURE 62 (a) Tensile strength and (b) elongation of Cu-Ni laminated composite as functions of individual layer thickness. Data for the as-plated deposit and heat treated in air at 250°C are shown. Tensile strength values for unlayered copper and nickel standards and average values expected on the basis of mixture rule. (From Cameron Ogden, 1986; reproduced with the permission of P & SF.)

where TS in megapascals and t is the individual layer thickness in micrometers.

2.  The strength of the composite is greater than what is predicted by the mixture rule at all layer thickness both at room temperature and at 250°C.
3.  For composites made from thinnest layers (0.4 $\mu$m), the strength is 75% greater than that of nickel (the stronger metal) and twice the average strength of Cu and Ni.
4.  The enhancement in strength occurs without sacrifice in ductility.
5.  For layer thicknesses greater than 2 $\mu$m, the stress at rupture is nearly four times the average for copper and nickel and three times that of copper, which is the stronger of the two from the point of view of penetration resistance.
6.  The strain at rupture is nearly constant for layer thicknesses greater than 2 $\mu$m. It is 80% higher than the copper-nickel average and about one-third higher than that of copper alone.

It is seen from the above that laminated composites prepared by electroplating exhibit significant improvement in mechanical properties.

In the investigations discussed so far the laminated composite has been prepared by depositing thin layers of the two metals from two different electrolytes. Though one could think of automated cathode movement, use of flow cells, etc., for deposition of layered composites on industrially useful components, a more viable commercial system would involve deposition of such materials from a single electrolyte.

Cohen, Koch, and Sard (1983) deposited laminated Ag-Pd composite from a single bath by using a modulated current. The plating bath consisted of

| | |
|---|---|
| LiCl | 500 g |
| AgCl | 12 g |
| $PdCl_2$ | 3 g |
| HCl | 20 mL |
| $H_2O$ | 1000 mL |

It was operated at 85° ± 0.5°C with moderate stirring (about 10 cm/s flow rate). Deposition was carried out with square-wave- and triangular-wave-modulated current. Due to the variation of current, the cathode potential varied from a lower value to a higher value periodically and caused the deposition of alternate layers of Ag-rich and Pd-rich phases. The authors called this laminated composite a *cyclic multilayered alloy* (CMA) coating.

SEM and AES studies of the deposit established that it is a laminate of Ag-rich and Pd-rich layers. Since Ag-Pd alloy is a

potential substitute for hard gold plating on electrical contacts,
the authors compared the contact resistance of the composite and
dc plated Ag-Pd alloy and found that there is no significant difference between the two but both are superior to the wrought alloy
R-156. However, the CMA coating exhibited an increasing brightness
with frequency of the square or triangular waveform.

Many alloy deposits exhibit a laminated structure. Of these,
electrodeposited Ni-Fe alloy foils have been studied. Levy (1969)
found Ni-Fe deposits with 25-50% Fe to be the strongest, exhibiting
strengths over 1372.8 MPa. MacInnis and Gow (1971) investigated
the tensile strength and hardness of 25-$\mu$m Ni-Fe foil, electrodeposited from a sulfate electrolyte at 0.1 A/cm$^2$, 60°C, and pH
of 2.5. The deposits containing 20-25% Fe exhibited a maximum
tensile strength of 1716 MPa and a maximum hardness of 650 KHN
(Knoop hardness number). The tensile strength of pure nickel
electrodeposit was found to be 1304.2 MPa, and that of commercially
available Ni-Fe (25%) alloy prepared by conventional techniques
was only 617.8 MPa. The electroplated alloy showed a laminar structure. Each layer was 0.5 $\mu$m thick with an average grain size of
0.5 $\mu$m. The higher strength of the electrodeposited NI-Fe alloy
as compared to the commercially available alloy was ascribed to
the fine laminar structure of the former.

Vedanayaki and Rajagopalan (1986) made a detailed study of
the cause of higher hardness for copper deposited from a copper
sulfate bath containing benzotriazole (BTA). Factors such as locked-
in stress, cold-working, small grain size, dispersion of particles,
and laminated structure were considered. These deposits exhibited
a laminar structure due perhaps to the incorporation of cuprous
benzotriazolate. They showed that grain size contributes substantially
to the hardness of copper deposited in the presence of BTA, but
it did not account for it fully. They attributed the higher hardness
to the small grain size and the laminated structure.

## 4.3  Conclusion

Electrodeposited laminated composites are just around the corner.
The feasibility of making such composites with improved mechanical
properties has been demonstrated by alternate deposition of metals
from two different electrolytes.

For the commercial exploitation of this type of composite, it
is essential that such a material should be deposited from a single
electrolyte. The feasibility of such a method has been demonstrated
by Cohen, Koch, and Sard (1983) and by Cameron Ogden (1986).

Enhancement of strength has been established for Fe-Ni laminated
films. Perhaps alloy deposits have high potential for making this
type of composite. Their composition varies significantly with current

density and mass transfer. These may be exploited to obtain a desired laminated composite.

By the vapor deposition technique, laminated composites with enhanced magnetic (Thaler, Ketterson, and Hillard, 1978) and electrical properties (Ruggiero, Barbu, and Beasly, 1980) have been produced. This broadens the scope of this class of composites, and probably such composites may be produced by electrochemical methods in the future.

References

Blum, W. (1921). *Trans. Am. Electrochem. Soc.*, *40*:307.
Blum, W. and Slattery, T. F. (1921). *Chem. and Met. Eng.*, *25*:24.
Brenner, A. (1963). *Electrodeposition of Alloys: Principles and Practice*, Academic Press, New York.
Cameron Ogden (1986). *Plating and Surf. Fin.*, *73*:1930.
Cohen, U., Koch, F. B., and Sard, R. (1983). *J. Electrochem. Soc.*, *130*:1987.
Fox, S. A. (1974). *Metals and Materials*, April:230.
Indira Rajagopal, Rajam, K. S., and Rajagopalan, S. R. (1980). Unpublished results.
Koehler, J. S. (1970). *Phys. Rev.*, *B2*:547.
Levy, E. M. (1969). *Plating*, *56*:903.
MacInnis, R. D. and Gow, K. V. (1971). *58*:135.
Raub, E. and Muller, K. (1967). *Fundamentals of Metal Deposition*, Elsevier, New York.
Ruggiero, S. T., Barbu, T. W., and Beasly, M. R. (1980). *Phys. Rev. Litt.*, *45*:1299.
Thaler, B., Ketterson, K., and Hillard, J. (1978). *Phys. Rev. Litt.*, *41*:336.
Vedanayaki, B. K. and Rajagopalan, S. R. (1986). *J. Electrochem. Soc. India*, *35*(4):293.
Wadsworth, J., Kum, D. W., and Sherby, O. D. (1986). *Metal Progress*, June:61.

## 5  OPTICAL COMPOSITES

Colored finishes are needed for one reason or another. For decorative applications, different colored coatings are desired. Black coatings are required on optical instruments, printing machinery, and military equipment to minimize reflection of light. A comparatively recent engineering application of composites for black coatings is the selective coating employed on solar collectors to improve their efficiency.

These composites will be referred to as optical composites since they have special optical properties, to which they owe their application. The composites described in other sections find application because of their mechanical properties. They can be referred to as mechanical composites. These names are based on the dominant property of the composite. The other nomenclature introduced earlier is based on the nature of the constituents of the composites.

Both fibers and particles improve the mechanical properties of composites. Optical properties of composites are enhanced only by particles.

## 5.1  Theory of the Optical Properties of Composites

By their very nature, composites are heterogeneous. In particle-reinforced composites, even though the matrix may be continuous (especially at low volume fraction of the particles), the particles represent an inhomogeneity. If they are larger than the wavelength of light, in which case the intensity of light reflected or transmitted by such a composite will vary from one region to another. Such composites exhibit no special optical property, and they are not considered to be optical composites.

On the other hand, if the particles are much smaller than the wavelength of light, the particles cannot directly participate in the propagation of electromagnetic waves, since the electric field does not vary appreciably over the sphere at any instant of time. However, the particle can alter the field in the medium surrounding it and thereby alter the polarizability, which in turn will influence the optical constant of the surrounding medium. Consequently, the optical property of the composite will be different from that of the matrix and particle. Such a composite is considered to be an optical composite.

An optical composite is thus an inhomogeneous medium from the point of view of propagation of light. The best way of describing such a medium is to characterize it by an effective optical constant, the value of which must be calculated from the optical constants of the particle and the matrix. For this purpose two theories have been proposed. One was developed by Maxwell Garnett (1904, 1906) and another by Bruggemann (1935).

Maxwell Garnett assumed the inhomogeneous medium to consist of spherical particles randomly distributed in a matrix which forms a continuous phase. He further assumed that the particles are much smaller than the wavelength of light. For this model, he derived the following expression for the effective refractive index ($\hat{n}_e$):

$$\frac{\hat{n}_e^2 - 1}{\hat{n}_e^2 + 2} = q\frac{\hat{n}_d^2 - 1}{\hat{n}_d^2 + 2} + (1 - q)\frac{n_m^2 - 1}{k_m^2 + 2} \qquad (5.1)$$

where $\hat{n}_x$ is the complex refractive index. ($\hat{n}_x = n_x - iX_x$; $n_x$ = refractive index and $X_x$ = extinction coefficient of the xth component.) X defines the absorption of light in the medium. Intensity of light falls to $\exp(-4\pi X)$ of its initial value by traveling a distance equal to the wavelength of the light in that medium. The extinction coefficient X is related to $\alpha$, the absorption coefficient of the exponential law of absorption ($I_x = I_0 \exp(-\alpha x)$ by $X = \alpha\lambda/4\pi$ where d = dispersed particle, m = matrix, q = volume fraction of the dispersed phase.

Bruggemann (1935) modeled the inhomogeneous medium as consisting of grains of the matrix phase and the particle phase. In a given element the probability of the dielectric constant (equal to the square of the refractive index) being $\varepsilon_m$ is I - f, while the probability of it being $\varepsilon_d$ is f.

On the basis of this model he showed that

$$f\left\{\frac{\hat{\varepsilon}_d - \hat{\varepsilon}_e}{\varepsilon_d + 2\hat{\varepsilon}_e}\right\} + 1 - f\left\{\frac{\hat{\varepsilon}_m - \hat{\varepsilon}_e}{\hat{\varepsilon}_m + 2\hat{\varepsilon}_e}\right\} = 0$$

These theories have been improved to take into account the nonspherical nature of particle, percolation, etc., and these have been reviewed by Berthier and Lafait (1981) and Granquist (1981). In this communication our interest is to broadly explain the theory of optical properties of composites. Hence we shall explain the dominant optical properties with the help of Maxwell Garnett's theory.

Indira, Rajagopalan, Siddiqui, and Doss (1964) were the first to realize that electrodeposited black nickel is an optical composite and to explain its optical properties by using the theory of Maxwell Garnett (Rajagopalan, Indira, and Doss, 1965). Reflectivities of electrodeposited black chromium (Berthier and Lafait, 1979) and black electrolytically colored aluminum (Granquist, 1981) have been explained in terms of the theories of Maxwell Garnett and Bruggemann.

We shall describe briefly the dominant properties of optical composites. A metal is characterized by a large reflectivity, which is a consequence of a high value of X. Consequently, it has an extremely low transparency. Even at very small thicknesses the transmitted light is absorbed. A dielectric material, on the other hand, has no absorption and exhibits low reflectivity and large transmission. In thin layers it will exhibit interference colors. An optical composite containing metal plus dielectric will act as a medium with a low $X_e$ value which will increase as the volume fraction q of metal increases. The variation of $X_e$ with q, calculated from Eq. 5.1, is shown in Fig. 63. It is seen that $X_e$ is low even at fairly large values of q.

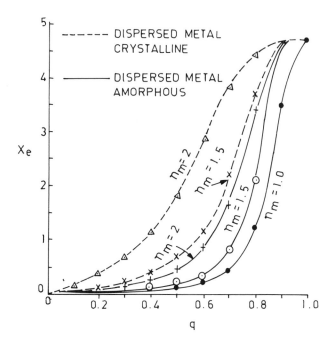

FIGURE 63   $X_e$ versus q curves for zinc dispersed in a dielectric matrix.

Since $X_e$ of the composite is small, the composite will have a low reflectivity and a significant amount of transmission. Therefore, in thin layers, it will exhibit, like a dielectric, interference colors, and in thick layers it will appear gray or black. Because $X_e \neq 0$, the transmitted beam will be partly absorbed. The interference colors will be more saturated (i.e., deeper) because partial absorption of the transmitted beam will tend to make the interference more complete by making the intensity of the beam that has traveled through the film nearly equal to that reflected at the air–film interface.

Thus, optical composites can provide "metallic" coatings which will exhibit saturated colors or have a gray to black appearance. Colored metallic finishes can therefore be tailored through optical composites.

## 5.2   Preparation of Optical Composites

From Section 5.1, it is clear that optical composites cannot be deposited in the same way as particle-dispersed composites. This is because of the requirement that the size of the dispersed phase should be smaller than the wavelength of light. It will be almost impossible to keep such particles in satisfactory suspension. Hence this class of composites is deposited by a different way. The particles are formed in situ at the electrode.

Black nickel, for example, is a composite containing a Ni-Zn alloy distributed in a matrix of $Ni_3S_2$ (Indira et al., 1964). This is deposited on the cathode, when the following solution is electrolyzed at a current density of 20 $mA/cm^2$.

| | |
|---|---|
| $NiSO_4$ $(NH_4)_2SO_4$ $6H_2O$ | 40 g/L |
| $NiSO_4$, $6H_2O$ | 80 g/L |
| $Zn SO_4$, $7H_2O$ | 40 g/L |
| KCNS | 25 g/L |
| pH | 5 g/L |

Black chromium is deposited on the cathode when an aqueous solution of pure chromic acid (free from sulfate) is electrolyzed in the presence of catalysts like acetate, nitrate, and borate. Normally a high current density and a lower temperature are used. The various bath compositions and conditions of formation can be found in the comprehensive review of Vasudevan, Grips, and Indira Rajagopal (1981).

Coating optical composites on aluminum is done by a different process, called electrolytic coloring of aluminum. The process consists of anodizing in sulfuric acid, followed by anodizing in phosphoric acid and then carrying out ac electrolysis in nickel sulfate or stannous sulfate or any other suitable metal salt solution with the anodized aluminum as one of the electrodes. Depending on the voltage and duration of phosphoric acid anodization as well as the time of ac electrolysis, aluminum acquires colors. All the processes described in the literature utilize basically the steps mentioned above. They differ with respect to the concentration of electrolytes employed and the conditions employed. Details with respect to these conditions have been given by Sheasby et al. (1980), Toshihiko Sato (1980), and Koichi Kuroda (1980).

## 5.3  Structure and Properties of Optical Composites

Black nickel was studied in detail by Indira et al. (1964), and they showed that it consists of zinc-rich nickel-zinc alloy and fine crystallites of $Ni_3S_2$. The deposit in the as-deposited condition did not give any x-ray diffraction pattern, and its electron diffraction revealed only the presence of $Ni_3S_2$. It was therefore concluded that black nickel films are composites containing a probably amorphous nickel-zinc alloy and fine crystallites of $Ni_3S_2$.

In thin layers black nickel exhibited vivid colors, and in thick layers it was black. Indira et al. (1964) and Rajagopalan et al. (1965) proposed for the first time that black nickel was an optical composite. Using Maxwell Garnett's theory, they explained the optical property of black nickel. They showed that the colors of black nickel were due to an interference phenomenon in absorbing material. They demonstrated (Rajagopalan and Indira, 1964) that black nickel was transparent in thin layers and became opaque in thick layers due to absorption of light. The black color of black nickel was due to the low reflectivity of the composite and the absorption of the light transmitted.

Black chromium is an optical composite of Cr and $Cr_2O_3$. These have been extensively studied from a theoretical point of view by Granquist et al., the Cornell group, and Ignatiev (Herzenberg and Silberglitt, 1982). Brethier and Lafait (1981) have also investigated this composite. In this composite, $Cr_2O_3$ appears to be amorphous, and Cr is a fine crystallite (Indira et al., 1982).

Unlike black nickel, black chromium appears to have a composition that varies with thickness (Brethier and Lafait, 1981). The metallic content decreases with thickness. As a result of the variation in composition, the optical constant also varies. Consequently, the surface layers are highly absorbing in the visible region, and the inner layers provide high infrared reflectivity. Thus the coating exhibits high absorption and low emissivity, thereby making it an ideal solar selective coating.

Electrolytically colored aluminum is an interesting example of an optical composite. It consists of aluminum oxide with cylindrical pores having a slightly wider bottom. Finely divided metal particles are present at the bottom (Fig. 64. Sheasby et al., 1980).

Thick layers of electrolytically colored aluminum are black. They are used as solar selective coatings. The reflectivity of such thick layers has been explained by Granqvist (1981), using Bruggemann's theory. Though Sheasby et al. (1980) proposed hypothetically that the colors of electrolytically colored aluminum are due to interference, they did not provide any quantitative explanation. Recently, Rajagopalan et al. (1987) have investigated this composite from a theoretical point of view by using Bruggemann's theory. They find that the colors of this composite could be understood in terms of a composite of three layers: a dielectric film of $Al_2O_3$ in contact with Al, a layer of composite containing metal + $Al_2O_3$ + air, and a layer of composite containing $Al_2O_3$ + air.

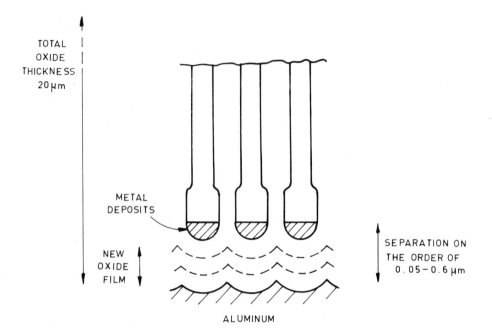

FIGURE 64   Interference effects produced by anodizing beneath the metal deposits.


## 5.4   Mechanism of Formation of Optical Composites

The mechanism of deposition of optical composites has not received much attention. Only some preliminary work on this topic has been done. This work is briefly reviewed here.

It has been pointed out that the property of optical composites is sensitive to the volume fraction of metal (q). For a black nickel optical composite (i.e., the one that exhibits vivid colors in thin layers), it is deposited at potentials more negative than -0.5 V with respect to SHE. In this region $CNS^-$ also gets reduced, probably to sulfide, which may react with nickel and form sulfide. If this hypothesis is correct, then q should decrease with increasing cathodic polarization. This conclusion is borne out by the results of an electrochemical study carried out by Indira et al. (1986). Typical results of this study are shown in Fig. 65.


## 5.5   Applications

By far the most important engineering application of optical composites is their use as selective coatings on collectors for utilization

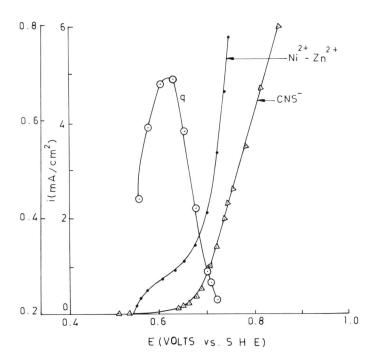

FIGURE 65 q (calculated from i-E curve) versus E and i-E curve for $Ni^{2+}$-$Zn^{2+}$ and $CNS^-$ systems.

of solar energy. Most of the energy from the sun is in the spectral region 0.1-2 μm. When a collector is used, the radiation normally is in the range of 5-10 μm. A selective coating should be an ideal absorber for the region 0.2-2 μm and an ideal reflector for wavelengths > 3 μm. Black chromium and black nickel nearly fulfill this requirement. However, corrosion resistance of black nickel is poor, while that of black chromium is good. Hence, black chromium has been extensively used as a selective coating.

Though electrolytically black colored aluminum has been investigated for use as a solar selective coating, it has not found practical application because of its high emissivity. Electrolytically colored coatings have found extensive applications as colored decorative finishes.

Black nickel and black chromium have been used as antireflection coatings on components of optical instruments as well as on certain parts of printing machinery.

References

Berthier, S. and Lafait, J. (1979). *J. Physique, 40*:1093.
Berthier, S. and Lafait, J. (1981). *J. De Phys., Coll. C1, 42*:285.
Bruggemann, D. A. G. (1935). *Ann. Phys., Leipzig, 24*:636.
Granquist, C. G. (1981). *J. De. Phys., Coll. C1, 42*:247.
Herzenberg, S. A. and Silberglitt, R. (1982). SPIE *324*:92.
Indira, K. S., Rajagopalan, S. R., Siddiqui, M. I. A., and Doss,
    K. S. G. (1964). *Electro Chim. Acta, 9*:1301.
Indira Rajagopal, Lakshmeswar, N. B., and Rajagopalan, S. R.
    (1982). (Unpublished).
Koichi Kuroda (1980). *Advanced Metal Finishing Technology in Japan*
    (Nobugoshi Baba et al., ed.) Technocrat, p. 120.
Maxwell Garnett, J. W. C. (1904). *Phil. Trans., 203*:385 (1906).
    Ibid 205:237.
Rajagopalan, S. R. and Indira, K. S. (1964). *J. Inst. Metals, 93*:129.
Rajagopalan, S. R., Indira, K. S., and Doss, K. S. G. (1965).
    *J. Electroanal. Chem., 10*:465.
Rajagopalan, S. R., Monica, V., and Indira Rajagopal (1986). (Un-
    published).
Sheasby, P. G., Patrie, J., Badia, M., and Cheetam, G. (1980).
    *Trans. Inst. Met. Fin., 58*:41.
Toshihiko Sato (1980). *Advanced Metal Finishing Technology in*
    *Japan* (Nobuyoshi Baba et al., ed.) Technocrat, p. 79, 102,
    107.
Vasudevan, N., Grips, V. K. W., and Indira Rajagopal (1981).
    *Surface Technology, 14*:119.

# 2
# Chemical Vapor Deposition

DEEPAK G. BHAT   *GTE Valenite Corporation, Troy, Michigan*

## 1   INTRODUCTION

The application of protective coatings on various types of substrates is an activity whose origin can probably be traced back to the beginning of civilization. The reasons for changing the characteristics of a given surface vary, depending upon the function and the desired durability of the surface. Thus, we can find examples of techniques of surface modification which cover an enormously wide range, from the application of a protective skin lotion, to applying a coat of paint on the wall of a living room, to spraying a variety of layers of various kinds of protective chemicals and pigments on an automobile body. In all these and other cases, the purpose is to change certain characteristics of the surface to enhance its appearance or useful life.

From an engineering point of view we are concerned, in this chapter, with the techniques of applying protective coatings in various industrial applications. These applications require various surface characteristics, such as corrosion resistance, wear resistance, and oxidation resistance. Numerous techniques are available to achieve these goals. We will examine those techniques which involve the application of coatings by deposition from the vapor phase.

## 2   CLASSIFICATION OF COATINGS

A coating may be defined as a layer of any substance used as a cover, protection, decoration, or finish. Therefore, by definition,

TABLE 1   Examples of Overlay Coatings

---

Deposition on the atomic scale
  Electrolytic techniques
    Electroplating
    Electroless plating
    Fused salt electrolysis
    Chemical displacement
  Vacuum techniques
    Vacuum evaporation
    Ion beam deposition
    Molecular beam epitaxy
  Plasma techniques
    Sputter deposition
    Activated reactive evaporation
    Plasma polymerization
    Ion plating
  Chemical vapor deposition techniques
    Deposition
    Reduction
    Infiltration
    Decomposition
    Spray pyrolysis
    Plasma-assisted
    Laser-assisted
    Metal organic

Particulate deposition
  Thermal techniques
    Plasma spraying
    Detonation-gun spraying
    Flame spraying
  Fusion techniques
    Thick-film ink coating
    Enameling
    Electrophoretic coating
  Impact plating technique

Bulk techniques
  Painting
  Dip coating
  Electrostatic printing
  Electrostatic spin coating
  Explosive cladding
  Roll bonding
  Weld overlay coating
  Laser overlay coating

---

TABLE 2   Examples of Conversion Coatings

| Surface modification | Technique |
|---|---|
| Chemical conversion | Electrolytic conversion<br>Anodization<br>Fused salt techniques<br>Pack cementation<br>Leaching techniques |
| Chemical vapor deposition | Thermal techniques<br>Plasma techniques<br>Laser techniques |
| Mechanical methods | Shot peening<br>Ion implantation |
| Thermal methods | Heat tinting |
| Surface enrichment | Diffusion from bulk |
| Physical vapor deposition | Sputtering<br>Arc deposition |

a coating has some properties that are *different* from the surface on which it is applied. There are three basic types of coatings:

1. *Overlay coatings* are deposited by the application of a new material onto the surface of a component.
2. *Conversion coatings* are applied by a modification of the composition of a surface by an in situ process.
3. *Combination coatings*, in which an *interface* is formed, that has properties which are different from those of the substrate and the coating.

The various methods of applying overlay coatings are given in Table 1. Table 2 gives examples of conversion coating techniques.

## 3   TECHNIQUES OF VAPOR DEPOSITION

The techniques of vapor deposition are broadly classified into two categories: chemical vapor deposition (CVD) and physical vapor deposition (PVD). Within these two general categories are many coating techniques, based on the principles of mass transfer from one source to another. Some techniques combine certain characteristics of two or more basic techniques, thus allowing greater versatility in the process or in the product. We will discuss the basic principles of CVD, the types of coatings deposited by CVD, and industrial applications of CVD. In addition to the conventional CVD, the newly emerging technologies of plasma CVD and laser CVD will also be briefly discussed.

## 4  CHEMICAL VAPOR DEPOSITION: DISCUSSION

Chemical vapor deposition may be defined as a technique in which a mixture of gases interacts with the surface of a substrate at a relatively high temperature, resulting in the decomposition of some of the constituents of the gas mixture and the formation of a solid film or coating of a metal or a compound on the substrate. The modern history of CVD probably dates back to the caveman, who used the soot formed by the incomplete oxidation of firewood to paint the walls of his cave and to draw figures on the wall. This was perhaps the first application of pyrolytic carbon (Blocher, 1966).

In its most fundamental form, CVD requires a source of precursor gases, a heated reaction chamber, and a system for the treatment and disposal of exhaust gases. The precursor gases include inert gases such as nitrogen and argon, reducing gases such as hydrogen, and a variety of reactive gases such as methane, carbon dioxide, water vapor, ammonia, chlorine, and others. Some of the precursors are in the form of a high-vapor-pressure liquid at room temperature, e.g., titanium tetrachloride ($TiCl_4$), silicon tetrachloride ($SiCl_4$), and methyl trichlorosilane ($CH_3SiCl_3$). These are heated to a relatively moderate temperature (typically below about 60°C), and the vapor is carried into the reaction chamber by bubbling a carrier gas (hydrogen or argon) through the liquid. Some of the precursors are formed by converting a solid metal or compound into a vapor, as in the case of aluminum chloride formed by a reaction of aluminum metal with chlorine or hydrochloric acid gas.

The gas mixture is carried into a reaction chamber heated to the desired temperature. The various techniques of heating include resistance heating using Kanthal, silicon carbide (Globar) or graphite heating elements, or induction. In some cases, the substrate may be directly heated by passing an electric current through it. In induction heating the part may be directly heated when it acts as a susceptor. In these two cases, the walls of the reactor are at a relatively low temperature, and therefore these techniques are described as *cold-wall* CVD. When the external heat source is used to heat the walls of the reaction chamber from which heat is radiated to the substrate, the technique is called *hot-wall* CVD. The various heating techniques are schematically illustrated in Fig. 1.

In conventional CVD, the reaction temperature ranges from about 900° to about 2000°C, depending on the coating to be deposited. In a variation of conventional CVD, called moderate temperature CVD (MTCVD), the reaction temperature is typically in the range of about 500° to about 850°C. This is usually achieved by using metal organic precursors which decompose at relatively lower temperatures. Therefore, it is also referred to as MOCVD (metal organic CVD). This technique is discussed further later. The reaction temperature can be further lowered by facilitating more energetic activation of the vapor phase reactions. The techniques in this category include plasma-assisted (or plasma-enhanced) CVD (PACVD or PECVD) and laser CVD (LCVD). As the names imply, the chemical reactions in the vapor phase are activated by the creation of a plasma in the gas phase or by shining a laser beam into the gas mixture. These techniques are described later.

The reaction gases exit the furnace chamber into a system of gas treatment devices. The purpose of these devices is to neutralize the obnoxious constituents of the exhaust gases, to remove solid particulates, and to cool the gases before they are exhausted to the atmosphere. These systems can take the form of a simple water scrubber tank, or they may involve elaborate neutralization and cooling towers, depending on the toxicity of the gas mixture and safety codes. Most CVD processes are carried out at subatmospheric pressures. Since the gases contain highly reactive species at high temperatures, the pumping equipment used to create the low pressure in the reaction chamber must be protected from these gases and from the particulates. This is usually done by using cold traps and nonreactive materials in the pump.

## 4.1 Chemical Reactions in CVD

The deposition of a coating by CVD occurs by one or more of a variety of chemical reactions, as given below.

1. Thermal decomposition or pyrolysis

$$CH_3SiCl_3 \rightarrow SiC + 3\ HCl$$

2. Reduction

$$WF_6 + 3\ H_2 \rightarrow W + 6\ HF$$

3. Oxidation

$$SiH_4 + O_2 \rightarrow SiO_2 + 2\ H_2$$

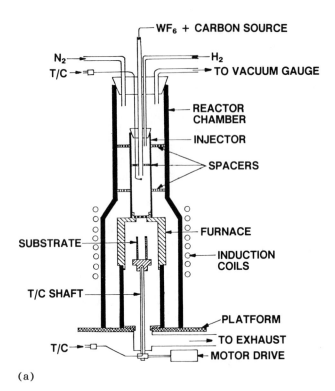

(a)

FIGURE 1   Heating techniques used in CVD. (a) Induction-heated hot-wall reactor. (From Bhat and DeKay, 1987. Copyright ASTM. Reprinted with permission). (b) Resistance-heated wire coating apparatus, and (c) radiation-heated fluidized-bed reactor. (Parts (b) and (c) from Powell, 1966; reprinted with permission of The Electrochemical Society.)

4.   Hydrolysis

$$2 \ AlCl_3 + 3 \ H_2O \rightarrow Al_2O_3 + 6 \ HCl$$

5.   Coreduction

$$TiCl_4 + 2 \ BCl_3 + 5 \ H_2 \rightarrow TiB_2 + 10 \ HCl$$

In the deposition of many of the coatings, the reactions proceed by one or more of the above basic principles. For example, in the deposition of refractory carbide or nitride coatings, a combination of thermal decomposition and reduction occurs; for example,

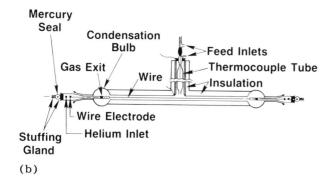

(b)

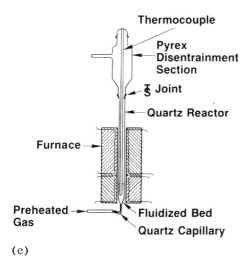

(c)

$$TiCl_4 + CH_4 \rightarrow TiC + 4\ HCl$$

$$AlCl_3 + NH_3 \rightarrow AlN + 3\ HCl$$

In these examples the deposits are formed by a reaction between the various precursor gases in the vapor phase. These are, therefore, examples of "overlay" coatings. In another situation, a coating may be formed by a reaction between a constituent of the vapor phase and the surface of the substrate. This is exemplified by the formation of a nickel aluminide coating on the surface of nickel by a reaction between $AlCl_3$ and hydrogen in the vapor phase and the nickel substrate (Sun et al., 1987). Another example is of a $Mo_2C$ coating on diamond particles by a reaction of molybdenum chlorides with diamond (Grishachev et al., 1984). Perhaps the most

TABLE 3   Characteristics of Chemical Vapor Deposition

---

1. Technique of deposition of solid via a vapor phase chemical reaction between precursor compounds in gaseous form at moderate to high temperatures.

2. The process can be carried out at atmospheric pressure as well as at low pressures.

3. Use of plasma and laser activation allows significant energization of chemical reactions, permitting deposition at very low temperatures.

4. Chemical composition of the coating can be varied to obtain graded deposits or mixtures of coatings.

5. Controlled variations in density and purity of the coating can be achieved.

6. Coatings on substrates of complex shapes and on particulate materials can be deposited in a fluidized-bed system.

7. Gas flow conditions are usually laminar, resulting in thick boundary layers at the substrate surface.

8. The deposits usually have a columnar grain structure, which is weak in flexure. Fine-grained, equiaxed deposits can be obtained by gas phase perturbation of chemical reactions by various techniques.

9. Control of vapor phase reactions is critical for achieving desirable properties in the deposit.

10. A wide variety of metals, alloys, ceramics, and compounds can be manufactured as coatings or as freestanding components.

---

widely known examples of this type of coating are gas carburizing, nitriding, and carbonitriding of steel. These coatings are then examples of *conversion coatings*. Thus, as one can see, CVD is a very versatile deposition technique for the formation of metals and compounds. Table 3 lists some of the characteristics of CVD.

## 4.2   Modeling of CVD Processes

The mechanisms by which deposits are formed in a CVD process involve various chemical processes which are influenced by several factors. Some of these factors relate to the CVD reactor system design, and others depend on the process parameters and properties of gases. Thus, any attempt to develop or describe a "model" for a CVD process requires some knowledge of these factors.

The CVD reactor system can be modeled by using the principles of heat transfer and fluid dynamics of gases. This is done by defining a series of "numbers" which depict the interrelationships between the various properties of gases and the geometry of the system. These numbers are listed in Table 4. They define the patterns of heat and mass flow of gases in a reactor of a given design and arrangement, allowing a reasonable prediction of deposition efficiencies. For example, Wahl and Hoffmann (1980) used these principles to calculate gas flow patterns and deposition rates of $Si_3N_4$ and tungsten in a CVD reaction chamber, and obtained a good correlation with experimental results.

In most CVD systems, the gas flow is laminar, although some local areas of turbulent flow may exist. Yet the gas flow pattern can be quite complicated, depending upon the reactor geometry, because it is determined by the relative dominance of forced convection (due to pressure differences in local regions) and free convection (due to gravity). This effect is described by the parameter $C = Gr/Re^2$. Forced convection dominates at small values of C. The exact value of C, or the transition between free and forced convection, depends upon the properties of the gas mixture and the size of the reactor.

In addition to the hydrodynamic parameters, the nature of the gas flow patterns in a CVD reactor is influenced by the kinetics of various chemical reactions which may occur in the bulk gas or on heated surfaces inside the reaction chamber. These factors are much more difficult to define precisely, since the exact concentrations of various species are difficult to predict. However, despite this interdependence between hydrodynamic characteristics of the gas mixture and the kinetics of the CVD process, it is possible to separate the two if the concentrations of reactant species or reaction rates are small. Recent papers by Jensen and Graves (1983), Wahl et al. (1987), Rosenberger (1987), Wang et al. (1987), and others have addressed these problems in considerable detail.

The approach for modeling a CVD process is based on equilibrium thermodynamics of reactions. Factors considered in defining the scope of a CVD reaction system include

1.  Process variables: temperature, pressure, and input gas composition
2.  System chemistry: compositions of possible chemical species formed during the reactions, their thermodynamic and kinetic properties, and reaction mechanisms
3.  Mass transport properties: diffusion, thermal and forced convection of gases

TABLE 4  Characteristic Numbers for Flow Conditions in a Chemical Reactor

| Number | Formula | Reference |
|---|---|---|
| 1. Reynolds (Re) | $\rho v l / \eta$ | |
| 2. Froude (Fr) | $v / \sqrt{gl}$ | |
| 3. Prandtl (Pr) | $\eta C_p / \lambda$ | Wahl and Hoffmann (1980) |
| 4. Schmidt (Sc) | $\eta / \rho D$ | |
| 5. Mach (M) | $v / v_c$ | |
| 6. Knudsen ($N_{Kn}$) | $l_m / d$ | Oxley (1966) |
| 7. Sherwood (Sh) | $\dot{m} l / D \, \Delta \rho$ | Wahl and Hoffmann (1980) |
| 8. Nusselt (Nu) | $\dot{Q} l / \lambda \, \Delta T$ | Wahl and Hoffmann (1980) |
| 9. Euler (Eu) | $2 \, \Delta P / \rho v^2$ | Fox and McDonald (1978) |
| 10. Grashof (Gr) | $(Re^2 / Fr) \cdot (\Delta \rho / \rho)$ | |
| 11. Raleigh (Ra) | $Pr \cdot Gr$ | |
| 12. Damkoehler for a surface reaction of nth order ($Da_S$) | $r_s l / DC_{As}^{\ n}$ | Wahl and Hoffmann (1980) |
| 13. Damkoehler for a homogeneous reaction in the bulk gas ($Da_b$) | $r_v l^2 / DC_{Ab}^{\ n}$ | |
| 14. Thiele (Th) | $\sqrt{Da}$ | |

$\rho$ = density of gas mixture
$\Delta \rho$ = density difference in the reactor
$v$ = velocity of gas mixture
$l$ = characteristic length of flow field
$\eta$ = dynamic viscosity of gas mixture
$g$ = gravitational constant
$C_p$ = specific heat of gas mixture at constant pressure
$\lambda$ = thermal conductivity of gas mixture
$D$ = coefficient of diffusion in gas
$v_c$ = velocity of sound
$l_m$ = mean free path of gas molecules
$d$ = characteristic diameter of flow field
$\dot{m}$ = mass deposition rate
$\dot{Q}$ = rate of heat transfer
$\Delta T$ = temperature difference
$\Delta P$ = pressure difference
$r_s = k_1 C_{As}^{\ n}$ = surface reaction rate
$r_v = k_2 C_{Ab}^{\ n}$ = bulk reaction rate
$C_{As}^{\ n}, C_{Ab}^{\ n}$ = concentrations of component A on the surface and in bulk gas, respectively

4.  Flow behavior: types of fluid flow, flow patterns, viscosities and velocities of gases.

The latter two depend, to a great extent, on the reactor arrangement and geometry and can be calculated through the use of the "numbers" defined in Table 4.

Although it is generally possible to identify the various process parameters in CVD with a remarkable degree of precision, it is still not feasible at this time to describe a complete "model" of a CVD process which takes into account all the factors. This is mainly due to the fact that there is very little or no information as yet available about many of the chemical species that can and do form during the high-temperature reactions. In addition, due to the paucity of kinetic data for many possible reactions, descriptions of processes based on thermodynamic considerations are often at variance with experimental results. Therefore, attempts to describe a model for a CVD process require the formulation of some simplifying assumptions. For example, Spear (1984) has described a simple and elegant model, as shown in Fig. 2.

The various steps are as follows:

1.  Forced flow of reactant gases into the system
2.  Diffusion and bulk (viscous) flow of reactant gases through the boundary layer
3.  Adsorption of gases onto the substrate

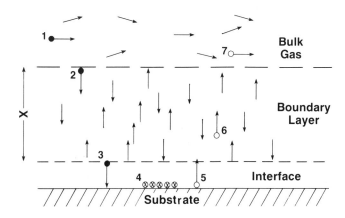

FIGURE 2  Schematic of the reaction sequence in a typical CVD process using the concentration boundary layer model (from Spear, 1984. This paper was originally presented at the 1984 Spring Meeting of The Electrochemical Society, Inc., held in Cincinnati, Ohio.)

4.  Chemical reactions of the adsorbed species or of adsorbed species
    and gaseous species
5.  Desorption of adsorbed species from the substrate
6.  Diffusion and bulk (viscous) flow of product gases through
    the boundary layer to the bulk gas
7.  Forced exit of gases from the system

   Another approach, due to van den Brekel (1977) and van den
Brekel and Bloem (1977), allows the formulation of a dimensionless
"CVD number" which defines the criteria for depositing a uniform
coating on a surface. The CVD number is the ratio of resistances
to diffusion flux and mass transfer flux in a system with a stagnant
boundary layer. When the CVD number is greater than 1, the process
is controlled by diffusion through the boundary layer, and when it
is less than 1 the process is controlled by the kinetics of surface
reactions. The growth rate is highly sensitive to the kinetics at low
temperatures, but is controlled by diffusion at high temperatures
when a stagnant boundary layer is present, as shown in Fig. 3.
Using such simplified models, we can predict the outcome of a CVD
reaction and design a process to obtain a tailored product.
   In recent years, computer calculations of multiple chemical reactions in a CVD process using equilibrium thermodynamic data have
been used extensively to develop CVD "phase diagrams." These

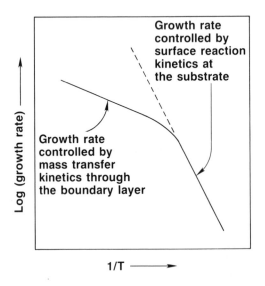

FIGURE 3  Schematic illustration of the effect of temperature and
kinetics on the growth rate of deposit in a CVD process.

**TABLE 5**  Thermochemical Data for Titanium Chloride

Titanium tetrachloride (TiCl₄)
(liquid)

$Cl_4Ti$
GFW = 189.712

$S°_{298.15}$ = 60.326 gibbs/mol

$\Delta Hf°_{298.15}$ = -192.2 ± 0.9 kcal/mol

$\Delta Hm°$ = 2.382 kcal/mol

$\Delta Hv°$ = 8.55 kcal/mol

$T_m$ = 249.05 K

$T_b$ = 409 K

| T, K | gibbs/mol | | | kcal/mol | | | Log $K_p$ |
|---|---|---|---|---|---|---|---|
| | $C_p°$ | $S°$ | $-(G°-H°_{298})/T$ | $H°-H°_{298}$ | $\Delta Hf°_{298}$ | $\Delta Gf°$ | |
| 0 | | | | | | | |
| 100 | | | | | | | |
| 200 | 34.584 | 46.498 | 63.496 | 3.400 | -193.496 | -181.719 | 198.573 |
| 298 | 37.704 | 60.326 | 60.326 | 0.000 | -192.200 | -176.226 | 129.177 |
| 300 | 34.709 | 60.541 | 60.327 | 0.068 | -192.127 | -176.127 | 128.308 |
| 400 | 34.936 | 70.557 | 61.691 | 3.547 | -190.972 | -170.961 | 93.408 |
| 500 | 35.153 | 70.376 | 64.274 | 7.051 | -189.820 | -166.091 | 72.598 |
| 600 | 35.370 | 84.805 | 67.176 | 10.577 | -188.698 | -161.452 | 58.809 |
| 700 | 35.586 | 90.273 | 70.095 | 14.125 | -187.592 | -156.999 | 49.017 |
| 800 | 35.803 | 95.093 | 72.921 | 17.694 | -186.505 | -152.704 | 41.717 |
| 900 | 36.020 | 99.269 | 75.618 | 21.285 | -185.430 | -148.542 | 36.071 |
| 1000 | 36.237 | 100.075 | 78.177 | 24.898 | -184.369 | -144.503 | 31.581 |

*Source:* From Stull and Prophet (1971).

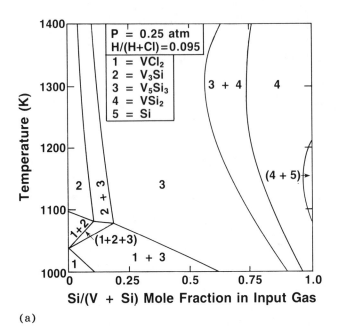

(a)

FIGURE 4   Typical CVD phase diagrams. (a) The V-Si-Cl-H system.
(From Spear, 1979. This paper was originally presented at the
Fall meeting of the Electrochemical Society, Inc., in Los Angeles,
CA.) (b) The Ti-B-Cl-H system. (From Randich and Gerlach, 1981.)

computer programs carry out several complex calculations in an
iterative manner such that the total free energy of the system is
minimized. The calculations take into account all possible chemical
species that can occur and for which appropriate thermodynamic
data are available. Several compilations of thermodynamic properties
of metals and compounds are available in the literature. See, for
example, Wicks and Block (1963), Schick (1966), Stull and Prophet
(1971), Barin and Knacke (1973), Chase et al. (1974, 1975, 1982),
and Barin, Knacke, and Kubaschewski (1977). A typical listing
is shown in Table 5.

    Typically, the various thermodynamic properties are determined
experimentally or, if experimental data are not available, by extrapo-
lation. Another acceptable method is to examine the data for similar
types of compounds of other species and to use these as guidelines
in predicting the properties of the species of interest. Errors can
occur in extrapolating from a given set of data if all experimental
parameters are not clearly or fully defined. It is also important

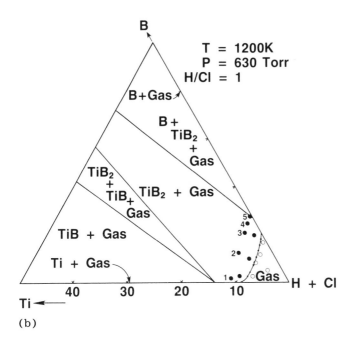

(b)

to examine the basis of various compilations, mainly with respect to the standard states used, if data for various species have to be obtained from different compilations. Uncertainties in thermochemical data can also produce significant differences in deposition rates calculated under equilibrium conditions (Rebenne and Pollard, 1985). Recently, Hunt (1987) showed that calculations of equilibrium states of a CVD system in which silicon is deposited showed significantly lower yields when more recent thermochemical data were used for silicon chlorides.

The results of computer calculations are expressed in terms of the concentrations of stable species at the given values of temperatures and pressures. Several such programs have been developed and used in recent years. See, for example, Eriksson (1971), Besmann (1977), Randich and Gerlach (1981), and Bernard (1981). Figure 4 gives some examples of the CVD "phase diagrams" obtained by these methods (Spear, 1979; Randich and Gerlach, 1981).

Another type of diagram that can be set up through a series of experiments is what may be called a "morphological diagram" (Bhat and Roman, 1987). An example of such a diagram is shown in Fig. 5 for CVD $Si_3N_4$. In such a diagram, the effects of key process parameters on the morphology of the deposit can be illustrated. These diagrams help in correlating the effects of process parameters

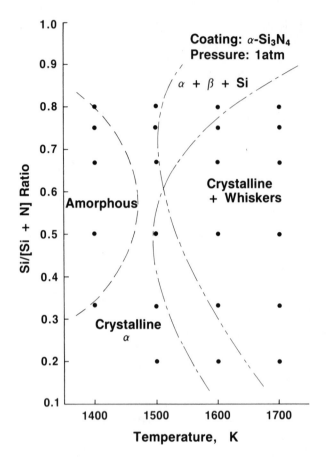

FIGURE 5   CVD morphological diagram for $Si_3N_4$ deposited at one atmosphere pressure. (From Bhat and Roman, 1987.)

on the nucleation and growth phenomena during deposition. Similar diagrams have also been given by Niihara and Hirai (1976) and Holzl (1978).

## 4.3   Materials Deposited by CVD

The various CVD techniques are used to deposit a wide range of materials, both in the form of freestanding bodies and as coatings for a variety of applications. These have been described in great detail in Powell et al. (1955, 1966), Holzl (1968), Yee (1978), and Pierson (1980).

The CVD technique is readily applicable for the deposition of refractory metals. The precursors for the deposition of metals, and most compounds, are the respective metal halides. For a successful application of CVD, it is necessary to be able to decompose these halides at relatively moderate temperatures (e.g., below about 1000°C). Thus, many metals whose halides are stable in this temperature range are difficult to use successfully. In many such cases, organometallic compounds (e.g., metal acetylacetonates, methyl or ethyl compounds, etc.) as well as metal carbonyls have been used successfully to deposit respective metals. Examples of metals which can be deposited by this approach include Cu, Pb, Fe, Co, Ni, Ru, Rh, Ir, Pt, and the refractory metals W and Mo (Powell et al., 1955). Further details about the use of organometallic compounds are given later. Most other metals of interest can be deposited by the decomposition or disproportionation of their halides. The most common metal halides are chlorides and, in some cases, fluorides or iodides. Table 6 shows typical deposition conditions for several metals (Holzl, 1968). Examples of chemical vapor deposited metal components are shown in Fig. 6.

Another interesting, and increasingly important, application of CVD is for making powders and whiskers of refractory materials. Whiskers, in particular, are becoming an important class of engineering materials because of their tremendous potential in the development of composites. The addition of whiskers of extremely fine dimensions, in the micron range, to ceramics has been shown to result in significant improvements in the toughness of such composites. A typical example of a whisker-reinforced ceramic is shown in Fig. 7(a). The material is a SiC whisker-reinforced alumina cutting tool which shows a much improved fracture toughness as compared with a conventional alumina cutting tool, as shown in Fig. 7(b).

Whiskers such as those used in the above example can be produced by CVD. Some examples of whiskers made by CVD, such as $Si_3N_4$ and TiC, are shown in Fig. 8. A number of compounds have been deposited in the whisker form by CVD. These include $Al_2O_3$ (Yamai and Saito, 1978; Hayashi et al., 1987), TiC (Takahashi et al., 1970; Hamamura et al., 1974; Kato et al., 1977; Wokulski, 1987), TiN (Kato and Tamari, 1975; Bojarski et al., 1981), $Cr_3C_2$ (Motojima and Kuzuya, 1985), SiC (Milewski et al., 1985; Cheng et al., 1987), $Si_3N_4$ (Shlichta and Holliday, 1985), ZrC and ZrN (Kato and Tamari, 1980), and $ZrO_2$ (Egashira et al., 1987). It is expected that many more materials will be studied in whisker form for their potential applications in composites.

Since the typical deposition temperatures in the conventional CVD are above about 800°C, it is necessary to select the substrate material judiciously. Most steels, for example, would be unsuitable

TABLE 6   Deposition Conditions for Several Metals

| Deposit | Metal reactant | Other reactants | Temperature (°C) | Pressure (torr) | Deposition rate (μm/min) |
|---------|----------------|-----------------|-------------------|-----------------|---------------------------|
| Tungsten | $WF_6$ | $H_2$ | 250–1,200 | 1–760 | 0.1–50 |
|  | $WCl_6$ | $H_2$ | 850–1,400 | 1–20 | 0.25–35 |
|  | $WCl_6$ | — | 1,400–2,000 | 1–20 | 2.5–50 |
|  | $W(CO)_6$ | — | 180–600 | 0.1–1 | 0.1–1.2 |
| Molybdenum | $MoF_6$ | $H_2$ | 700–1,200 | 20–350 | 1.2–30 |
|  | $MoCl_5$ | $H_2$ | 650–1,200 | 1–20 | 1.2–20 |
|  | $MoCl_5$ | — | 1,250–1,600 | 10–20 | 2.5–20 |
|  | $Mo(CO)_6$ | — | 150–600 | 0.1–1 | 0.1–1 |
| Rhenium | $ReF_6$ | $H_2$ | 400–1,400 | 1–100 | 1–15 |
|  | $ReCl_5$ | — | 800–1,200 | 1–200 | 1–15 |
| Niobium | $NbCl_5$ | $H_2$ | 800–1,200 | 1–760 | 0.08–25 |
|  | $NbCl_5$ | — | 1,880 | 1–20 | 2.5 |
|  | $NbBr_5$ | $H_2$ | 800–1,200 | 1–760 | 0.08–25 |
| Tantalum | $TaCl_5$ | $H_2$ | 800–1,200 | 1–760 | 0.08–25 |
|  | $TaCl_5$ | — | 2,000 | 1–20 | 2.5 |
| Zirconium | $ZrI_4$ | — | 1,200–1,600 | 1–20 | 1–2.5 |
| Hafnium | $HfI_4$ | — | 1,400–2,000 | 1–20 | 1–2.5 |
| Nickel | $Ni(CO)_4$ | — | 150–250 | 100–760 | 2.5–35 |
| Iron | $Fe(CO)_5$ | — | 150–450 | 100–760 | 2.5–50 |
| Vanadium | $VI_2$ | — | 1,000–1,200 | 1–20 | 1–2.5 |
| Chromium | $CrI_3$ | — | 1,000–1,200 | 1–20 | 1–2.5 |
| Titanium | $TiI_4$ | — | 1,000–1,400 | 1–20 | 1–2.5 |

*Source*: From Holzl (1968). Reprinted with permission of John Wiley & Sons, New York.

in this regard because of the solid-state phase transformation and the resultant dimensional changes which occur at the eutectoid transformation temperature of steel. Second, the difference in the coefficients of thermal expansion of steel and various coatings will set up considerable tangential compressive stresses at the coating-substrate interface at room temperature. This can result in a loss of coating adhesion either by buckling or microcracking, depending

FIGURE 6 Metallic components deposited by CVD. (a) Niobium nozzle, 200-mm length × 1.25-2.25-mm wall thickness. (b) Rhenium thrust chamber for liquid rockets, 75-mm major diameter × 175-mm length × 0.75-mm wall thickness. (c) Tungsten crucible, 325-mm diameter × 575-mm height × 1.5-mm wall thickness. (d) Tungsten manifold, about 175 mm long. (Photographs courtesy of Ultramet Corporation, Pacoima, CA.)

on the radius of curvature of the surface. In addition, the phase transformation also alters the microstructure and properties of steel which cannot be easily regained after coating. Another factor which should be considered is the reactivity of the steel surface with the furnace gases, which may cause the formation of undesirable phases at the interface or embrittlement of the steel. As shown earlier, the furnace gases typically contain hydrogen and halides, such as HCl, which are formed during the deposition reaction.

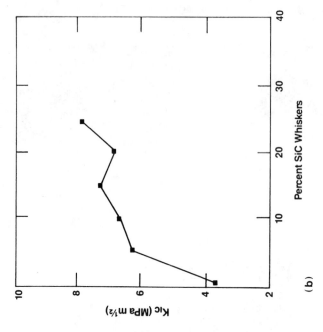

(b)

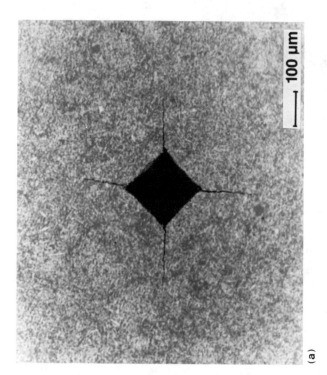

100 μm

(a)

FIGURE 7 Effect of whisker addition on the toughness of a ceramic material. (a) Photomicrograph of a SiC whisker–reinforced alumina composite cutting tool (WG–300), showing the ability of whiskers to deflect a propagating crack produced by microindentation. (b) Graphical representation of the effect of SiC whisker addition in improving fracture toughness of alumina ceramic. (Data courtesy of Dr. P. N. Vaidyanathan, Greenleaf Corporation, Saegertown, PA.)

FIGURE 8  Photomicrographs of whiskers grown by CVD: (a) $Si_3N_4$. (b) TiC.

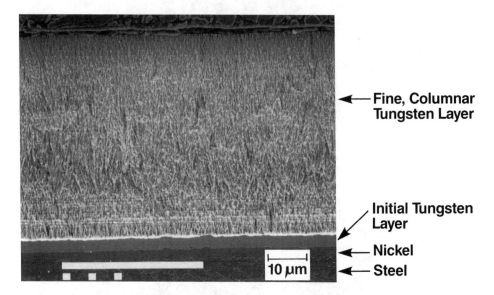

FIGURE 9    Polished cross section of a tungsten-coated 1018 steel.
The surface of steel is protected from the corrosive fluoride gases
during tungsten deposition by a layer of electroless nickel.

The commonly used substrates often include various refractory
metals (most notably Mo), graphite, mullite, and other ceramics
which are not readily attacked by the reactive gases at high temperatures. Steels have been used successfully as substrates when the
deposition temperature is below 700°C. In some cases, the surface
of steel must be protected from the attack of halide vapors by
depositing a film of nickel by electrolytic or electroless methods
(Archer, 1975; Archer and Yee, 1978; Bhat and DeKay, 1987).
Figure 9 shows the cross section of a fine-grained tungsten coating
on mild steel in which the substrate was first coated with a thin
film of nickel to protect the surface from the reactive fluoride gases
formed during deposition.

The use of the CVD technique is more widespread for depositing
compound coatings such as oxides, carbides, nitrides, borides,
and silicides, as summarized in Tables 7 through 11. Other compounds
such as phosphides, sulfides, aluminides, and other materials have
also been deposited by CVD.

4.4    Applications of CVD Coatings

The coatings deposited by CVD are used in various applications
requiring wear resistance, oxidation resistance, corrosion resistance,

TABLE 7  Deposition Conditions for Some Carbides

| Carbide | Precursors | Temperature (°C) | Pressure (torr) | Reference |
|---|---|---|---|---|
| $B_4C$ | $BCl_3-CH_4-H_2$ | 700–1,000 | 760 | Cochran and Stephenson (1970) |
| $B_xC_y$ | $BBr_3-CH_4-H_2$ $BI_3-CI_3$ | 900–1,800 | 0.001–760 | Ploog (1974) |
| $B_4C$ | $BCl_3-CH_4-H_2$ $BCl_3-CH_4-H_2$ | 1,025–1,225 850–1050 | 50 10–100 | Jansson and Carlsson (1985) Hannache et al. (1985) |
| $Cr_3C_2$ | $CrCl_3-C_4H_{10}-H_2$ | 900–1,200 | 760 | Motojima and Kuzuwa (1985) |
| $HfC$ | $HfCl_x-CH_4-H_2$ $HfCl_x-CH_4-H_2$ | 1,250–1,450 1,000–1,300 | 20 760 | Kaplan (1972) Hakim (1975) |
| $Mo_2C$ | $MoCl_x$–Diamond | 700–1,000 | 0.75–67.5 | Grishachev et al. (1984) |
| $NbC$ | $NbCl_5-CH_4$ $NbCl_5-CH_4-Ar$ | 600–900 1,000–1,200 | 760 760 | Funke et al. (1969) |
| $SiC$ | $SiCl_4-CH_4-H_2$ $Si_2Cl_6-C_3H_8-H_2$ | 950–1,150 450–1,175 | 220–360 760 | Bhat and Panos (1981) Motojima et al. (1986) |
| $TaC$ | $TaCl_5-CH_4$ $TaCl_5-CH_4-Ar$ | 850–1,000 1,200–1,300 | 760 760 | Funke et al. (1969) |
| $TiC$ | $TiCl_4-CH_4-H_2$ $TiCl_4-CH_4-H_2$ $TiCl_x-CH_4-H_2$ | 1,000–1,200 1,000 950–1,075 | 760 50 4–725 | Funke et al. (1969) Gass et al. (1977) Stjernberg et al. (1977) |

(continued)

Table 7 (continued)

| Carbide | Precursors | Temperature (°C) | Pressure (torr) | Reference |
|---|---|---|---|---|
| [TiC] | $TiCl_4$–$CH_4$–$H_2$ | 950–1,100 | 760 | Cho et al. (1982) |
| | $TiCl_4$–$C_3H_8$–$H_2$ | 850–1,200 | 760 | Baik et al. (1984) |
| | $TiCl_4$–$C_3H_8$–$H_2$ | 875–975 | 50–760 | Kim et al. (1986) |
| | $TiCl_x$–$C_4H_{10}$–$H_2$ | 750–850 | 760 | Piton et al. (1987) |
| WC | $WCl_6$–$CH_4$–$H_2$ | 900–1,150 | 5–100 | Mantle et al. (1975) |
| $W_2C$ | $WF_6$–$C_6H_6$–$H_2$ | 500–700 | 5–500 | Archer and Yee (1978) |
| $W_3C$ | $WF_6$–$C_6H_6$–$H_2$ | 300–500 | 5–500 | Archer and Yee (1978) |
| | $WF_6$–$CH_3OH$–$H_2$ | 300–600 | 200–250 | Holzl et al. (1984) |
| | $WF_6$–$C_2H_5OH$–$H_2$ | 300–600 | 200–250 | |
| $W_8C$ | $WF_6$–$C_3H_8$–$H_2$ | 400–900 | 760 | Demyashev et al. (1985) |
| ZrC | $ZrCl_4$–$CH_4$–$H_2$ | 1,550–2,100 | 5–200 | Lepie (1964) |
| | $ZrCl_4$–$CH_4$–$H_2$ | 1,300–1,700 | 760 | Funke et al. (1969) |
| | $Zr$–$CH_2Cl_2$–$H_2$–$CH_4$ $Zr$–$Cl_2$–$H_2$–$CH_4$ | 1,000–1,400 | 760 | Ikawa (1972) |
| | $ZrCl_4$–$CH_4$–$H_2$ | 1,540–1,840 | 760 | Driesner et al. (1973) |
| | $ZrCl_4$–$CH_4$–$H_2$ | 1,125–1,390 | 760 | Wallace (1973) |
| | $ZrCl_4$–$CH_4$–$H_2$ | 1,400–1,900 | 760 | Salles et al. (1987) |
| TiAlOC | $TiCl_4$–$AlCl_3$ –$H_2$–$CO$ | 1,000 | 760 | Gates (1986) |

TABLE 8  Deposition Conditions for Some Oxides

| Oxide | Precursors | Temperature (°C) | Pressure (torr) | Reference |
|---|---|---|---|---|
| $Al_2O_3$ | $AlCl_3-H_2O$ | 1,000–1,500 | 0.5–5 | Wong and Robinson (1970) |
|  | $AlCl_3-H_2-CO_2$ |  |  |  |
|  | $AlCl_3-O_2$ |  |  |  |
|  | $AlF_3-H_2O$ | 1,400 | 760 | Yamai and Saito (1978) |
|  | $AlCl_3-H_2-CO_2$ | 1,050 | 50 | Altena et al. (1983) |
|  | $AlCl_3-H_2-CO_2$ | 1,000–1,200 | 100 | Park et al. (1983) |
|  | $AlCl_3-H_2O$ | 250–1,000 | 2–7.5 | Mantyla et al. (1985) |
|  | $AlX_3-H_2-CO_2$ | 1,050–1,500 | 760 | Colombier and Lux (1986) |
|  | $AlX_3-CO_2$ |  |  | (X = Cl, Br, I) |
| $CeO_2$ | $CeCl_3-O_2-N_2$ | 1,200 | 40 | Taylor and Trotter (1972) |
| $SiO_2-P$ | $SiH_4-PH_3-O_2-N_2$ | 350–450 | 0.6 | Logar et al. (1977) |
| $SnO_2$ | $SnCl_4-O_2$ | 800–1,000 | 760 | Tabata (1975) |
| $TaO$ | $TaCl_5-H_2-O_2$ | 600 | 400 | Hieber (1974) |
| $Ta_2O_5$ | $TaCl_5-H_2-O_2$ | 700 | 400 | Hieber (1974) |
|  | $TaCl_5-H_2-O_2$ | 600–900 | 1–375 | Hieber and Stolz (1975) |
| $Ti_xO_y$ | $TiCl_4-CO_2-H_2$ | 1,015 | 50 | Fredriksson and Carlsson |
|  |  | 930–1,075 | 5–22.5 | (1985, 1986) |
| $UO_2$ | $UF_6-H_2-O_2$ | 950–1,500 | 2–20 | Heestand et al. (1967) |
| $ZrO_2$ | $ZrCl_4-CO-H_2$ | 725–1,125 | 38 | Minet et al. (1987) |
| $Al_2O_3/Ti_2O_3$ | $AlCl_3-TiCl_4-H_2-CO_2$ | 1,000–1,100 | 760 | Fonzi (1976) |
| $MgAlO_4$ | $AlCl_3-MgCl_2-CO_2-H_2$ | 950 | 760 | Kawahara et al. (1987) |

TABLE 9  Deposition Conditions for Some Nitrides

| Nitride | Precursors | Temperature (°C) | Pressure (torr) | Reference |
|---|---|---|---|---|
| AlN | $AlC_3 \cdot NH_3-H_2$ | 700–1,400 | 5–10 | Lewis (1970) |
| | $AlCl_3-NH_3-H_2$ | 700–1,200 | 760 | Arnold et al. (1977) |
| | $AlCl_3-NH_3-H_2$ | 500–1,325 | 1.7–320 | Bhat and Panos (1981) |
| | $AlCl_3-NH_3-N_2$ | 650–1,500 | 0.25–50 | Suzuki and Tanji (1987) |
| | $AlBr_3-NH_3-H_2$ | 400–900 | 760 | Pauleau et al. (1982) |
| BN | $BCl_3-NH_3-H_2$ | 1,550–1,850 | 1.6–2.2 | Clerc and Gerlach (1975) |
| | $BCl_3-NH_3-N_2$ | 900–1,900 | 2.0 | Tanji et al. (1987) |
| | $B_2H_6-NH_3-H_2$ | 700–1,250 | 760 | Hirayama and Shohno (1975) |
| | $BF_3-NH_3$ | 1,150 | 15 | Hannache et al. (1983) |
| | $B_{10}H_{14}-NH_3$ | 300–1,150 | $2 \times 10^{-7}$ | Nakamura (1985) |
| | $BCl_3-NH_3-H_2$ | 1,200–2,000 | 5–60 | Matsuda et al. (1986) |
| | $B_3N_3H_3Cl_3$ | 1,100–1,500 | 760 | Gebhardt (1973) |
| | | 1,050 | $1 \times 10^{-3}$ | Singh (1987) |
| HfN | $HfCl_4-N_2-H_2$ | 1,200 | 760 | Kieffer et al. (1973) |
| | $HfCl_x-N_2-H_2$ | 1,000–1,300 | 760 | Hakim (1975) |
| $Nb_xN_y$ | $NbCl_5-H_2-N_2$ | 700–1,100 | 760 | Kieda et al. (1987) |
| $Si_3N_4$ | $SiCl_4-NH_3-H_2$ | 1,100–1,500 | 5–300 | Niihara and Hirai (1976) |
| | $SiCl_4-NH_3-H_2$ | 850–1,600 | 25–80 | Bhat (1980) |

| | | | | |
|---|---|---|---|---|
| | $SiF_4-NH_3-H_2$ | 1,450–1,600 | 40–80 | Motojima et al. (1986) |
| | $Si_2Cl_6-NH_3-H_2$ | 800–1,300 | 25–100 | Bhat and Roman (1987) |
| | $SiCl_4-NH_3-H_2$ | 925–1,425 | 760 | |
| TaN | $TaCl_5-N_2-H_2$ | 1,000–1,500 | 400 | Hieber (1974) |
| $TaN_x$ | $TaCl_5-N_2-H_2-Ar$ | 700–1,100 | 760 | Kieda et al. (1987) |
| $Ta_2N$ | $TaCl_5-N_2-H_2$ | 800–1,000 | 400 | Hieber (1974) |
| $Ta_3N_5$ | $TaCl_5-NH_3$ | 900–1,300 | 400 | Hieber (1974) |
| TiN | $TiCl_4-N_2-H_2$ | 1,040 | 185–190 | Lindstrom and Stjernberg (1975) |
| | $TiCl_4-N_2-H_2$ | 1,000–1,050 | 760 | Sjostrand (1979) |
| | $TiCl_4-NH_3-H_2$ | 650–700 | 760 | |
| | $TiCl_4-H_2-N_2$ | 930–1,080 | 760 | Lee (1982) |
| | $TiCl_4-NH_3$ | 400–700 | 760 | Kurtz and Gordon (1986) |
| $VN_x$ | $VCl_4-N_2-H_2-Ar$ | 900–1,100 | 760 | Kieda et al. (1987) |
| $W_2N$ | $WCl_6-NH_3-H_2$ | 800–900 | 760 | Landingham and Austin (1969) |
| ZrN | $ZrCl_4-N_2-H_2$ | 1,150–1,200 | 760 | Kieffer et al. (1973) |
| $Si_3N_4-TiN$ | $SiCl_4-TiCl_4-NH_3-H_2$ | 1,050–1,450 | 1–8 | Hirai and Hayashi (1982) |
| $Si_3N_4-BN$ | $SiCl_4-BCl_3-NH_3-H_2$ | 1,400 | 10 | Nakae et al. (1985) |
| TiN-BN | $TiCl_4-BCl_3-NH_3-H_2$ | 1,400 | 10 | Nakae et al. (1985) |

TABLE 10  Deposition Conditions for Some Borides

| Boride | Precursors | Temperature (°C) | Pressure (torr) | Reference |
|---|---|---|---|---|
| $HfB_2$ | $HfCl_4-BCl_3-H_2$ | 1,400 | 3 | Gebhardt and Cree (1965) |
| $NbB_2$ | $NbBr_5-BBr_3$ | 850–1,750 | 0.025–0.2 | Armas et al. (1976) |
| Ni–B | $Ni(CO)_4-B_2H_6-CO$ | 150 | 650 | Mullendore and Pope (1987) |
| $SiB_4$ | $SiH_4-BCl_3-H_2$ | 800–1,400 | 50–600 | Dirkx and Spear (1984) |
| $SiB_x$ | $SiBr_4-BBr_3$ | 975–1,375 | 0.05 | Armas and Combescure (1977) |
| $TaB_2$ | $TaBr_5-BBr_3$ | 850–1,750 | 0.025–0.2 | Armas et al. (1976) |
| | $TaCl_5-B_2H_6$ | 500–1,025 | 760 | Randich (1980) |
| $TiB_2$ | $TiCl_4-BCl_3-H_2$ | 1,200–1,415 | 3–15 | Gebhardt and Cree (1965) |
| | $TiCl_4-B_2H_6$ | 600–900 | 760 | Pierson and Mullendore (1980) |
| | $TiCl_4-BCl_3-H_2$ | 750–1,050 | 760 | Caputo et al. (1985) |
| | $TiCl_4-BCl_3-H_2$ | 1,200 | 50 | Desmaison et al. (1987) |
| $ZrB_2$ | $ZrCl_4-BCl_3-H_2$ | 1,400 | 3–6 | Gebhardt and Cree (1965) |

TABLE 11 Deposition Conditions for Some Silicides

| Coating system | Substrate | Substrate Temp. (°C) | Pressure (torr) | Chemical mixture | Reference |
|---|---|---|---|---|---|
| $Mo_xSi_y$ | Graphite | 700–1,400 | 760 | $MoCl_5$–$SiCl_4$–$H_2$ | Kehr (1977) |
| $MoSi_2$ | Alumina | 150–550 | 1 | $MoF_6$–$SiH_4$–Ar | West and Beeson (1987) |
| $Ni_xSi_y$ | Nickel | 400–900 | 760 | $SiCl_4$–$H_2$ $Si_2Cl_6$–$H_2$ | Motojima et al. (1987) |
| | Graphite | 800–1,100 | 760 | $SiCl_4$–$NiCl_2$–$H_2$ $Si_2Cl_6$–$NiCl_2$–$H_2$ | Motojima et al. (1987) |
| $NbSi_2$ | Graphite | 900–1,200 | 760 | $NbCl_5$–$SiCl_4$–$H_2$ | Kehr (1977) |
| $Ta_xSi_y$ | Graphite | 1,000–1,200 | 760 | $TaCl_5$–$SiCl_4$–$H_2$ | Kehr (1977) |
| $TaSi_2$ | — | 630–750 | 45–300 | $TaCl_5$–$SiH_4$ | Wieczorek (1985) |
| $TaSi_2$ | Si | 650–950 | 0.3–2 | $TaCl_5$–$SiH_4$–$H_2$ | Bouteville et al. (1987) |
| $TiSi_2$ | Si | 700–1,000 | 0.75 | $TiCl_4$–Substrate | Bouteville et al. (1987) |
| $TiSi_2$ | Graphite | 600–900 | 760 | $TiCl_4$–$SiH_4$ | Million-Brodaz et al. (1987) |
| $Ti$–$Si$ | Ni-base superalloy | 1,000–1,100 | 30–120 | $TiCl_4$–$SiCl_4$–$H_2$ | Wahl et al. (1981) |
| $VSi_2$ | Graphite | 800–1,200 | 190 | $VCl_4$–$SiCl_4$–$H_2$ | Wang and Spear (1984) |
| $V_5Si_3$ | Graphite | 900–1,200 | 190 | $VCl_4$–$SiCl_4$–$H_2$ | Wang and Spear (1984) |
| $W$–$Si$ | Graphite | 600–800 | 100 | $WF_6$–$SiH_4$ | Lo et al. (1973) |
| $WSi_2$ | Si | 330–450 | 0.2 | $WF_6$–$SiH_4$ | Monnig et al. (1984) |
| | | 320–400 | 0.1–0.65 | $WF_6$–$SiH_4$ | Rode et al. (1987) |

and electrical, optical, or tribological properties. Depending upon
the applications, the characteristics of coatings may be varied by
controlling process parameters and equipment. The purity of the
coating may be a very critical factor in some applications because
the impurities may significantly affect certain properties, such as
the electrical and optical properties of the coatings. In such cases,
very high purity gases and high-vacuum ($10^{-5}$ to $10^{-8}$ torr) equipment are necessary to deposit useful coatings. On the other hand,
coatings used in tribological or wear applications may not be as
critically sensitive to small amounts of impurities. In such cases,
the gas purity and/or vacuum requirements may not be as stringent
as in the previous example. Typical impurities include oxygen,
chlorine or other halides, hydrogen from the furnace gases, and
metallic impurities from the furnace hardware (Fe, Cr, Ni, C) or
insulation (Al, Si, Ca, Mg, etc.). In any case, the selection of
coatings for a particular application requires a consideration of
several factors.

## Coatings for Wear Resistance

Holleck (1986) and Sundgren and Hentzell (1986) have reviewed
the criteria for the selection of hard coatings and coating-substrate
combinations for applications involving wear. Typical coatings in
this category include refractory borides, carbides, nitrides, and
oxides. These compounds can be classified according to the chemical
bonding in these materials—metallic (borides, carbides, and nitrides
of transition metals), covalent (borides, carbides, and nitrides
of Al, Si, B, as well as diamond), and ionic (oxides of Al, Be,
Ti, Zr). Along with the important characteristics of the coating
layers, we must consider properties of the substrate and interface.
The important properties include adhesion of coating, interdiffusion
and thermal mismatch at the interface, and hardness, strength,
and toughness of the substrate. The fracture toughness of the
coating, as well as the coating-substrate combination, is an important
property in applications involving wear and impact. The grain size,
stoichiometry, and uniformity of coating are also important, since
these properties greatly influence the wear behavior.

　　Prominent in the category of wear-resistant coatings are those
for metal-cutting tools. The important properties of coatings in
cutting applications include hardness, chemical stability, wear
resistance, low coefficient of friction, favorable thermal conductivity,
and thermal stability (Bhat and Woerner, 1986). Typical coatings
which meet these requirements include TiC, TiN, $Al_2O_3$, and their
combinations. Other coatings such as TaC, HfN, and $TiB_2$ have
also been used. The selection of a given coating or coating combination depends upon the performance criteria required in a given
metal-cutting operation, as shown in Table 12.

TABLE 12  Selection Criteria for a Coated Cutting Tool

| Criterion | Optimum coating |
|---|---|
| High temperature stability<br>Chemical stability<br>Crater wear resistance } | $Al_2O_3$<br>TiN<br>TiC |
| Hardness—edge retention | TiC<br>TiN<br>$Al_2O_3$ |
| Abrasion resistance—flank wear | $Al_2O_3$<br>TiC<br>TiN |
| Coefficient of friction<br>Grain size } | TiN<br>TiC<br>$Al_2O_3$ |

*Source*: From Bhat and Woerner (1986). Reprinted with permission from *Journal of Metals*, vol. 38 (2), 1986, a publication of The Metallurgical Society, Warrendale, Pennsylvania.

These coatings are deposited on cemented tungsten carbide substrates. As in any application of coating on a surface, compatibility between the coating and substrate is important. For cemented carbide substrate, this requirement consists of prevention of cobalt and carbon migration from the binder phase of the substrate to the surface at high temperatures and their interaction with the coating furnace atmosphere as well as the precursor gases. If such an interaction occurs, it leads to the formation of a brittle carbide $Co_6W_6C$, called eta phase, at the coating-substrate interface. This is shown in Fig. 10 for a cemented carbide tool coated with a multi-layered TiC-Ti(C,N)-TiN coating. The dark layer at the coating-substrate interface is the brittle eta phase. This phase reduces the adhesion of the coating and may also cause microchipping at the cutting edge during machining (Cho et al., 1986). Therefore, a proper control of furnace atmosphere is essential (Bhat et al., 1986). Figure 11 shows the microstructure of a coated carbide cutting tool in which a uniform multilayered coating consisting of TiC, $Al_2O_3$, and TiN is applied in order to enhance the service life and performance of the tool in the machining of steels and cast irons. An example of how a judicious choice of substrate and coating can improve the performance of a tool is shown in Fig. 12 (Bhat and Woerner, 1986). As the data show, the wear and deformation resistance of the

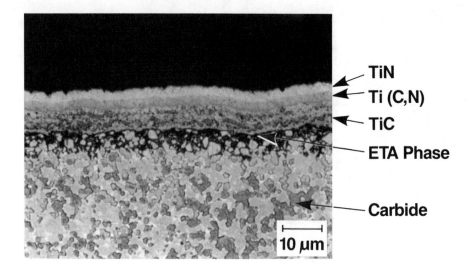

FIGURE 10    Photomicrograph of etched cross section of a coated cemented carbide cutting tool, showing the three coating layers (TiC, $TiC_xN_y$, and TiN) and the brittle eta phase. Eta phase is formed due to a loss of carbon from the substrate during TiC deposition.

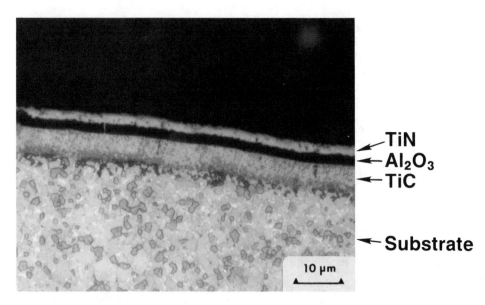

FIGURE 11    Microstructure of a multilayered CVD coating on a cemented carbide cutting tool insert. The first coating is TiC, followed by $Al_2O_3$ (dark layer), and TiN (outer layer). (Photograph courtesy of GTE Valenite Corporation, Troy, MI.)

**TEST CONDITIONS: 350 SFM (2 REV. CUT + 2 REV. DWELL PER CYCLE)**
**0.040 IPR FEED, 0.100 INCH DOC**

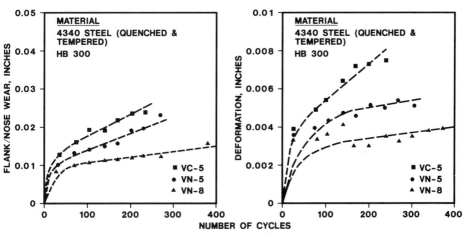

FIGURE 12   Effect of substrate composition and CVD TiN coating on the machining performance of a coated cemented carbide cutting tool. (From Bhat and Woerner, 1986, reprinted with permission from *Journal of Metals*, 32(2), 1986, a publication of The Metallurgical Society, Warrendale, Pennsylvania.)

uncoated carbide VC-5 is improved by a coating of TiN (VN-5). When the same TiN coating is applied to a different carbide substrate which has greater hardness and deformation resistance (VN-8), the overall tool life is further improved.

In recent years, ceramic cutting tools have become popular due to their greater deformation resistance at higher machining speeds required for improved productivity. One such class of ceramic tools includes silicon nitride and Si-Al-O-N materials. Coating applied to these tools have also shown improved performance in metal-cutting operations where chemical wear resistance is required, as for example in the machining of steel (Sarin and Buljan, 1983, 1984; Sarin et al., 1983; Bhat et al., 1987). An example of improvement in the tool life of a silicon nitride cutting tool after coating is shown in Fig. 13.

Another application of CVD coating is in areas involving erosion and abrasion, such as sandblast nozzles, slurry transport equipment, coal gasification equipment, mining equipment, etc. In these situations,

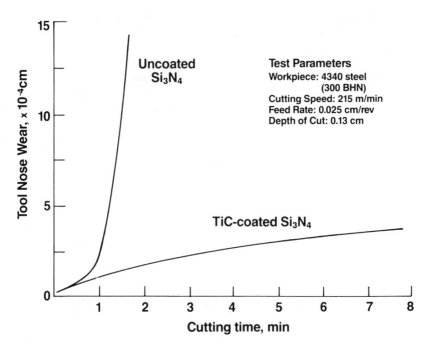

**FIGURE 13** Graph showing the effect of TiC coating on the wear resistance of a silicon nitride cutting tool in the machining of low-alloy steel. (From Sarin and Buljan, 1984.)

airborne or waterborne particles of sand, fly ash, or other particulate matter traveling at some velocity and under certain pressure cause abrasion and erosion of surfaces they come in contact with. In conventional practice, these surfaces are typically coated with abrasion-resistant materials by a variety of methods. Electrolytic chromium plating, thermal or plasma spraying, weld overlay coating or laser cladding are some of the commonly used techniques. Chemical vapor deposition has also been successfully used in many of these applications. As shown in Fig. 14, a CVD coating of a tungsten-carbon alloy CM 500L performed significantly better than the normally used Cr plating in a sand slurry abrasion test, and was used successfully to replace the latter (Bhat and DeKay, 1987). The performance of CM 500L was attributed in part to its unique, ultrafine-grained microstructure, shown in Fig. 15(a) (Bhat and Holzl, 1982). The development of this type of microstructure is a unique feature of the modified CVD technique called controlled nucleation thermochemical deposition (CNTD, Stiglich et al., 1980).

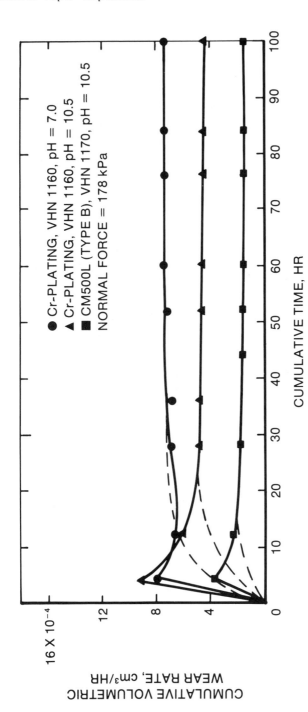

FIGURE 14 Comparison of slurry erosion behavior of electrolytic chromium plating and CVD CM 500L (tungsten–carbon alloy) coating in the modified Miller slurry abrasivity test. (From Bhat and DeKay, 1987. Copyright ASTM; reprinted with permission.)

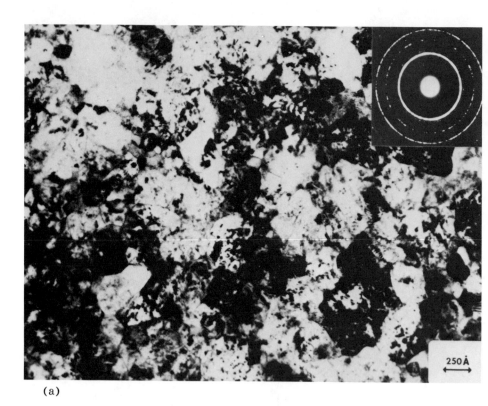

(a)

FIGURE 15   Microstructures of CVD coatings: (a) CM 500L. The
transmission electron micrograph shows the ultrafine-grained structure
of this alloy which is deposited by the CNTD process. (From Bhat
and Holzl, 1982, reprinted with permission of Elsevier Sequoia, S.A.,
Switzerland.) (b) Tungsten. The optical micrograph shows the
coarse, columnar structure which is typical of most of the CVD
coatings.


This microstructure may be compared to the typical, coarse, columnar
grain structure obtained in most CVD coatings, as shown in Fig.
15(b) for a tungsten deposit on a steel mandrel. The erosion resistance of CVD coatings was also compared with several other coating
techniques (Hickey et al., 1984; Qureshi et al., 1986) and was
found to be superior due to their fine-grained microstructures.
Figure 16 shows the steady-state erosion rate of CNTD W-C and
SiC coatings (CM500L and CM 4000, respectively) as compared with
other hard coatings against silicon carbide grit (Hickey et al., 1984),

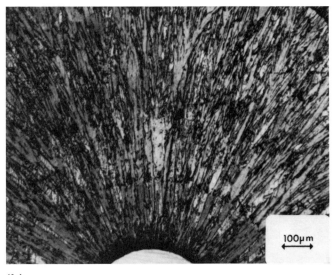

(b)

and Fig. 17 shows the relative slurry abrasion resistance of CM500L
in comparison with hard coatings applied by other techniques (Bhat
and Holzl, 1982). Hickey et al. (1984) concluded that the exceptional
wear resistance of CNTD coatings was related to their extremely
fine grain size and fracture toughness.

Chemical vapor deposition has also been used to apply wear-
resistant coatings for gun barrels. Haskell and Imam (1975) found
that the application of a coating of tungsten to the inner diameter
of a nickel-plated steel gun tube material resulted in a nearly 10-fold
increase in the erosive wear resistance in tests simulating the firing
of an explosive charge through the tube. A process feasibility study
was also made for depositing a Ta-W alloy coating (Glaski and Crowson, 1977; Bracuti et al., 1981) and a W-C alloy coating (Bracuti
et al., 1981) by CVD. Limited testing of the latter indicated that
the W-C coating showed a significantly higher erosion resistance
than the Ta-W coating.

## Coatings for Tribological Applications

The commonly used techniques for applying tribological coatings
include the various physical vapor deposition methods, such as
sputtering and evaporation, or methods such as flame or plasma
spray techniques. The purpose of coatings in the tribological applications is generally to improve the coefficient of friction between

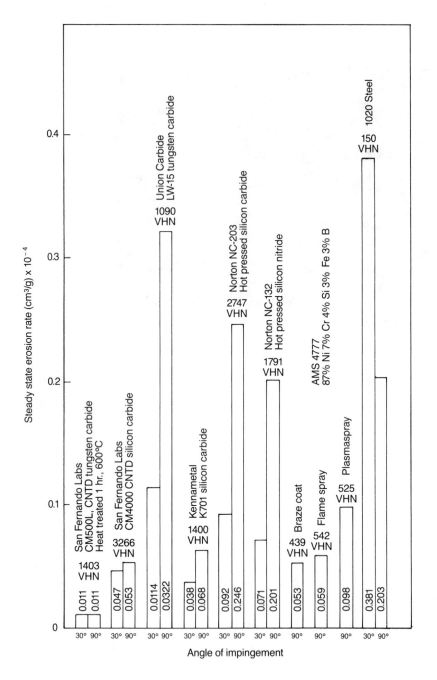

FIGURE 16 Steady-state erosive wear rates of ultrafine-grained CVD W-C alloy (CM500L) and SiC (CM4000) coatings and other hard facing materials, coatings and ceramics (Hickey et al., 1984; reprinted with permission of Elsevier Sequoia, S.A., Switzerland).

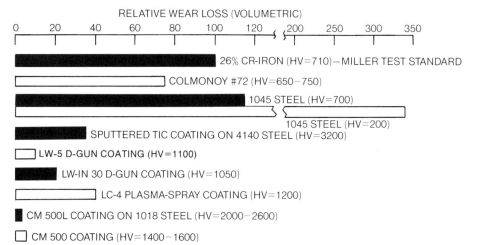

FIGURE 17  Slurry abrasion resistance of the ultrafine-grained
W-C alloy CM500L and various other hard facing materials and coatings using the Miller slurry abrasivity test (ASTM G 75-82). (Data
from Bhat and Holzl, 1982.)

sliding or rolling surfaces in contact, thereby reducing wear due
to adhesion, abrasion, or other causes. Typical coatings used in
these applications include refractory compounds such as carbides,
nitrides, and borides of transition metals. The frictional properties
of several CVD coatings are shown in Table 13. Other important
properties of coatings in these applications include hardness, elastic
modulus, fracture toughness, adhesion, grain size, and, to a certain
extent, chemical stability, depending upon the service environment.
Hintermann (1981) demonstrated the enhanced performance of steel
gyro bearings coated with CVD TiC, which was attributed to improved
resistance to adhesive and abrasive wear, and improved chemical
resistance against the byproducts of lubricant breakdown during
operation. Figure 18 shows comparative data for uncoated and TiC-
coated steel-bearing components.

An important factor in the application of tribological coatings
is the consideration of the nature of surfaces in contact. In some
instances, a friction "couple" may show a high friction coefficient
but exhibit a low wear rate. Other factors which can influence
the behavior include contact temperature, pressure, and the environment, i.e., the presence of moisture, surface oxidation, or lubrication. The details of these considerations in various applications
such as bearings, forming tools, abrasive and corrosive environments
at low and high temperatures have been discussed by Hintermann
(1981).

TABLE 13    Coefficient of Friction Data for Some CVD Coatings

| Coating system | Contact surface | Contact force | Coefficient of friction | Reference |
|---|---|---|---|---|
| $Cr_7C_3$ | Steel | — | 0.69 | Perry and Archer (1980) |
| TiC-coated steel race | Steel ball bearing | 2000 N | 0.14 | Boving et al. (1983) |
| TiC-coated carbide | TiC-coated carbide | 9 $kN/cm^2$ | 0.16 (RT) 0.34 (800C) | Hintermann and Boving (1981) |
| CM 500 (W-C alloy) | CM 500 | 2.8 MPa 15.0 MPa | 0.03 0.06 | Stiglich and Bhat (1980) |
| CM 500 | Cemented carbide | 2.8 MPa 11.7 MPa | 0.04 0.16 | |
| CM 4000 (CNTD SiC) | Cemented carbide | 2.8 MPa | 0.15-0.30 | Hintermann (1981) |
| $Cr_7C_3$ | Steel | 5 N | 0.79 | |
| $Fe_xB$ | Steel | 5 N | 0.76 | |
| SiC | Steel | 5 N | 0.23 | |
| TiC | Steel | 5 N | 0.25 | |
| TiN | Steel | 5 N | 0.49 | |
| SiC | TiC | 5 N | 0.2-0.26 | |
| TiC | TiC | 5 N | 0.22-0.32 | |
| TiN | TiC | 5 N | 0.16-0.18 | |
| SiC | SiC | 5 N | 0.27-0.47 | |
| TiC | SiC | 5 N | 0.25-0.35 | |

| Material | Counterface | Load | Friction coefficient | Reference |
|---|---|---|---|---|
| TiC | TiN | 5 N | 0.25–0.31 | Habig (1986) |
| TiN | TiN | 5 N | 0.19 | |
| TiC | Al$_2$O$_3$ | 20 N | 0.19–0.37 | |
| Cr$_7$C$_3$ | Cr$_7$C$_3$ | 5 N | 0.29 | |
| Fe$_x$B | Fe$_x$B | 5 N | 0.4 | |
| Tin-coated Pin | TiN-coated Disc | 2.5 N | 0.15–0.2 (Air) 0.35–0.6 (Vac) | Das and Kumar (1983) |
| Boron on Beryllium | Sapphire Diamond | 0.295 N 0.225 N | 0.23–1.05 0.06–0.32 | |
| W–C coated Steel | W–C coated Steel | — | 0.1–0.6 | Bergmann and Vogel (1986) (sputter-assisted CVD Coatings) |
| Cr–C coated Steel | Cr–C coated Steel | — | 0.15–0.45 | |
| TiC-coated Steel | TiC-coated Steel | — | 0.33–0.48 | |
| TiN-coated carbide | Diamond Cubic BN | ⩽ 10 N | 0.03 0.10 | Steinmann et al. (1987) |
| TiC-coated carbide | Diamond Cubic BN | ⩽ 10 N | 0.12 0.13 | |

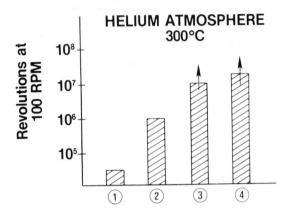

FIGURE 18   Effect of TiC coating on the frictional and wear properties of steel ball-bearing components. (1) Standard steel ball-bearing,
complete seizure. (2) Standard steel ball-bearing with $MoS_2$ spray,
complete seizure. (3) Ball-bearing with TiC-coated races, wear of
balls but no seizure. (4) Ball-bearing with TiC-coated balls, wear
of races but no seizure. (From Hintermann, 1981; reprinted with
permission of Elsevier Sequoia, S.A., Switzerland.)

## Coatings for High-Temperature Applications

The main requirement of coatings for high-temperature applications
is their thermal stability. Refractory metals and compounds having
low vapor pressures and high decomposition temperatures are generally suitable for high-temperature applications, depending upon
the service environment. Many refractory metals and ceramics can
be used in inert or vacuum atmospheres. For applications involving
ambient or reactive atmospheres, oxidation and/or chemical resistance
is also needed. Thus, many refractory oxides and oxide composites
are suitable candidates for these applications. In addition to the
resistance to the environment, these coatings are also required
to possess other properties, such as abrasion resistance, compatible
thermal expansion characteristics and strength to withstand thermal
shock often encountered in these applications. In this respect,
coatings of refractory metal silicides and transition metal aluminides
have been found to be useful.

Generally, these coatings are applied by a variety of techniques.
Thermal spray and plasma spray are most common. The use of CVD
techniques is fairly limited, although these have been used increasingly in recent years with the advent of carbon-carbon composites
in many high-temperature applications. Typical coatings used in

these applications include silicon carbide, silicon nitride, aluminum oxide, and various refractory metal silicides. Silicon carbide, for example, has a high hardness, strength, and chemical resistance, in addition to abrasion and chemical resistance at high temperature. The high-temperature properties of SiC depend on its purity and microstructure. As shown in Fig. 19, the room-temperature and high-temperature strength of CVD SiC is significantly higher than that of the conventional SiC made by bulk ceramic processing techniques. It was found that the SiC deposited by CVD had extremely fine grain size, which was believed to be the reason for its high strength. A comparison between the conventional CVD SiC and the fine-grained SiC is shown in Fig. 20. The fine-grained SiC was deposited by the CNTD technique mentioned earlier.

Typical high-temperature applications of these coatings include rocket nozzles, after-burner components, reentry cones, heat-exchanger components in high-temperature gas turbines and ceramic automotive engines, etc. One such example is shown in Fig. 21, in which a reaction-bonded silicon nitride (RBSN) turbine vane

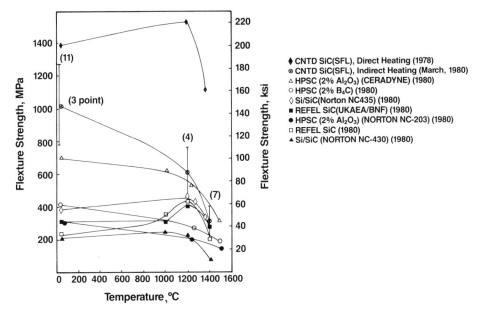

FIGURE 19 Temperature dependence of transverse rupture strength of CVD fine-grained SiC coating on graphite in comparison to bulk SiC made by conventional ceramic processing techniques. (From Panos and Bhat, 1980.)

FIGURE 20  Photomicrographs showing difference between (a) coarse-grained and (b) fine-grained CVD SiC. Coating is deposited on a resistively heated tungsten filament by the pyrolysis of methyl trichlorosilane.

FIGURE 21   Reaction-bonded silicon nitride (RBSN) turbine vane
coated with CVD alpha silicon nitride. The surface hardness, wear
resistance, and oxidation resistance of substrate are substantially
improved by the coating. (From Stiglich et al., 1980; reprinted
with permission of Ceramurgia International.)

was coated with CVD silicon nitride to improve its high-temperature
oxidation resistance, as well as wear resistance, since CVD $Si_3N_4$
is almost twice as hard as the bulk $Si_3N_4$ (Stiglich et al., 1980).
The ability of the CVD technique to achieve excellent microstructural
control can also be illustrated in the case of $Si_3N_4$, as shown in
Fig. 22 (Bhat, 1980). Figure 22(a) shows the fracture cross section
of a conventional $Si_3N_4$ coating in which the coarse columnar grain
structure can be seen. The refinement of this grain structure by
the incorporation of a small amount of Al in the deposit is shown
in Fig. 22(b). Thus, it is possible to refine the grain structure
of a CVD coating by setting up competitive reactions during deposition.

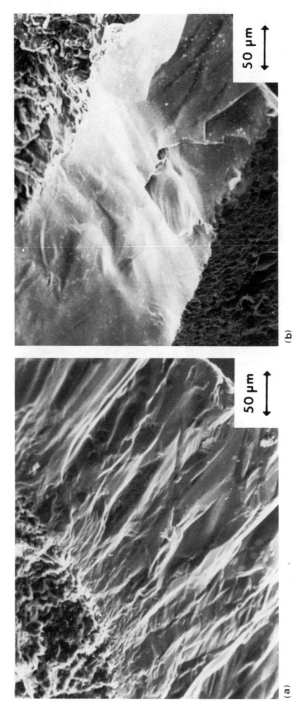

(b)

(a)

FIGURE 22  Photomicrograph showing the improvement in grain size of CVD silicon nitride deposit by codeposition of a second phase: (a) Conventional coarse-grained silicon nitride. (b) Fine-grained silicon nitride containing small amounts of codeposited AlN. (From Bhat, 1980.)

## 5  METAL ORGANIC CHEMICAL VAPOR DEPOSITION

A variation of the conventional CVD technique is metal organic
CVD (MOCVD). In this technique, organometallic compounds of
metals which decompose at relatively lower temperatures are used
as precursors. The advantages of MOCVD include the possibility
of depositing on thermally sensitive substrates and of depositing
multicomponent coatings. Some disadvantages of this technique are
decreased deposition rate, increased density of crystal defects and
impurities in the film, and greater care required in handling some
of the metal organic compounds because of their high reactivity
(Constant and Morancho, 1983). The main impetus for using organometallic precursors was provided by the need for epitaxial deposition
of semiconductors (Dapkus, 1982). In the microelectronics field,
this technique is also called organometallic vapor phase epitaxy
(OMVPE).

In this process, one or more constituents of the film are trans-
ported to the reaction zone in the form of metal alkyls while other
constituents may be in the form of hydrides. Other precursors
such as chlorine-substituted metal alkyls or coordination compounds
may also be used. Dapkus (1982) has summarized the types of semiconductors grown by MOCVD.

MOCVD technique has also been used in other fields of application. Coatings of metals have been deposited by MOCVD in cases
where metal halides are stable at high temperatures or where conventional CVD may be difficult, as mentioned earlier. In addition,
coatings of oxides, nitrides, carbides, and silicides have been
deposited using organometallic compounds. The purpose of these
applications is often the possibility of using lower deposition temperatures when the substrate cannot withstand high temperatures of
conventional CVD. Many organometallic compounds decompose at
moderate temperatures, allowing deposition on substrates such as
steels. Therefore, this technique has also been referred to as
moderate temperature CVD (MTCVD). Table 14 shows some of the
coatings deposited by MOCVD.

## 6  PLASMA-ASSISTED CHEMICAL VAPOR DEPOSITION

### 6.1  Introduction

This technique has recently come into prominence because it allows
deposition to take place at reduced substrate temperatures, typically
below about 600°C. Instead of requiring thermal energy to heat and
decompose gaseous precursors as in the case of the conventional
CVD processes, the kinetic energy of electrons in the plasma is
utilized to activate the chemical reactions in the vapor phase.

TABLE 14  Coatings Deposited by Metal Organic CVD

| Coating | Precursors | Temperature (°C) | Pressure (torr) | Reference |
|---------|-----------|------------------|-----------------|-----------|
| $Al_2O_3$ | $Al(OC_3H_7^i)_3$ | 700–800 | <10 | Hough (1972) |
| | Al-triisopropoxide | 270–420 | 100 | Saraie et al. (1985) |
| $B_7O$ | $B(C_2H_5O)_3-H_2$ | 800 | 0.76 | Michalski et al. (1987) |
| Co | $Co_2(CO)_8$ | 200–400 | — | Gross and Schnoes (1987) |
| Co, Fe, Ni | $M(C_2H_5)_2$ | 550 | $4\times10^{-5}$ | Stauf et al. (1987) |
| CoSi | $H_3SiCo(CO)_4$ | 670–700 | 0.4–2 | Aylett and Tannahill (1985) |
| $Cr_7C_3$ | $Cr[CH(CH_3)_2]_2]_2$ | 300–550 | 0.5–50 | Maury et al. (1987) |
| $\beta-FeSi_2$ | $(H_3Si)_2Fe(CO)_4$ | 670–700 | 0.4–2 | Aylett and Tannahill (1985) |
| $Mn_3Si$ | $H_3SiMn(CO)_4$ | 670–700 | 0.4–2 | Aylett and Tannahill (1985) |
| SiC | $CH_3SiCl_3-H_2$, $(CH_3)_2SiCl_2-H_2$ | 800–1,200 | 760 | Brutsch (1985) |
| | $CH_3SiCl_3-H_2$ | 900–1,200 | 760 | Motojima et al. (1986) |
| | Polycarbosilanes | 350–800 | 760 | Schilling et al. (1983) |
| | $CH_3SiCl_3-H_2$ | 1,150–1,450 | 70 | Tsui and Spear (1984) |
| | $CH_3SiCl_3-C_3H_8-H_2$ | 1,150–1,250 | 230 | Bhat and Panos (1981) |
| | $(CH_3)_4Si-H_2$ | 1,000 | 15 | Desmaison et al. (1987) |
| | $CH_3SiCl_3-H_2$ | 1,300–1,500 | 760 | Gyarmati et al. (1983) |

| | | | | |
|---|---|---|---|---|
| $Si_3N_4$ | $(CH_3)_4Si-NH_3$ | 525–1,500 | 1–760 | Lartigue et al. (1984) |
| $SnO_2$ | $(CH_3)_4Sn,$ $(C_2H_5)_4Sn,$ $(C_4H_9)_4Sn,$ $(C_4H_9)_2(CH_3COO)_2Sn$ | 400–500 | — | Kamimori and Mizuhashi (1981) |
| TiC | $(C_5H_5)_2TiCl_2-H_2$ | 825–1,050 | 1–7 | Stolyarov et al. (1983) |
| Ti(C,N) | $(CH_3)_3N-TiCl_4$ $CH_3CN-TiCl_4$ $CH_3(NH)_2CH_3-TiCl_4$ $HCN-TiCl_4$ | 560–950 | 15–720 | Bonetti-Lang et al. (1982) |
| $TiO_2$ | $Ti(C_3H_7O)_2$ | 190–550 | 760 | Komiyama et al. (1987) |
| $Y_2O_3$ | $Y_2(thd)_3$ | 430–490 | 7.5–22.5 | Brennfleck et al. (1985) |
| $ZrO_2$ | $Zr(OC_3H_7{}^i)_4$ | 700–800 | 2.7 | Hough (1972) |
| | $Zr(OC_5H_{11})_4$ | 750–950 | 760 | Hough (1972) |
| | $Zr(tfacac)_4-O_2$ | 450–750 | 760 | Balog et al. (1977) |
| | $Zr(thd)_4-O_2$ | 300–430 | 7.5–22.5 | Brennfleck et al. (1985) |
| | Zr 2,4 pentadionate | 450 | 760 | Dor et al. (1983) |
| | $Zr(tfacac)_4-O_2$ $Zr(C_3H_7O)_2$ | <425 | 760 | Komiyama et al. (1987) |

A plasma is an assembly of ions, electrons, neutral atoms, and molecules in which the motion of the particles is dominated by electromagnetic interaction. A relatively large amount of energy can be stored in the internal energy of the particles in the plasma. The plasma can be of two types: thermal plasma, as created in an arc discharge at atmospheric pressure, and cold plasma, as created in a low-pressure glow discharge. The electrons, ions, and neutral gas molecules in a thermal plasma are in local thermodynamic equilibrium. On the other hand, in the cold, nonequilibrium plasma the electrons, and, to a smaller extent, ions are considerably more energetic than the neutral gas molecules. It is this high energy of the plasma that causes activation of a chemical reaction at a relatively low temperature, typically less than 300°C. At the same time, because of its nonequilibrium nature, the plasma does not heat the bulk of the gas and the substrate. Therefore, "glow discharge" plasma-assisted CVD techniques have found major applications in a variety of fields, such as microelectronics, optical and solar coatings, and, increasingly, novel wear-resistant coatings.

A glow discharge plasma is defined as a region of relatively low-pressure and low-temperature gas which is ionized by applying a high-frequency field across the gas volume. The state of ionization of the plasma is sustained by the high-energy electrons in such a way that the plasma, as a whole, is quasineutral. When an electric field is applied to an ionized gas, energy is transferred more readily to electrons than to ions, because of the lighter mass of electrons. This mass difference between electrons and ions also limits the amount of energy transfer to the latter during subsequent elastic collisions. As a result of this, the kinetic energy of electrons is rapidly increased through successive collisions to a point where inelastic collisions can occur. At this point, the high-energy electrons begin to cause ionization and formation of free radicals by further interactions with neutral gas molecules (Thornton, 1982).

The electron temperature of the plasma is typically above 10,000 K while the gas temperature is less than 300°C. This ratio is maintained by keeping the pressure below about 10 torr (Bell, 1981). The primary function of the plasma is to generate chemically active ions and free radicals which react with other ions, atoms, and molecules in the gas phase or at the substrate surface to induce lattice damage and chemical reactions. The rate of production of active species is a function of electron density, reactant concentration, and rate coefficient. In other words, the rate of production of active species depends on the electric field strength, gas pressure, and the mean free path of particles between collisions. Thornton (1982) has described the features of glow discharge plasmas and the analytical relationships involved in the plasma-gas interactions.

In plasma-assisted CVD, the activation barrier to the chemical reaction is overcome by the dissociation of the reactive gases in the plasma due to the impact of the high-energy electrons. This allows the reaction gas temperature to be lowered considerably, which is a very important consideration in many applications such as microelectronics. A major difference between conventional CVD and plasma CVD is that the thermodynamic principles of chemical reactions which govern the former do not apply to plasma CVD. In other words, the dissociation of gaseous molecules by the plasma is nonselective. As a consequence, coatings formed in a plasma CVD reaction are quite different from those produced in conventional CVD. The compositions of phases produced in plasma CVD can be quite unique in that their formation is no longer dependent upon equilibrium thermodynamic constraints. These coatings are typically amorphous and may be considered as a generic class in themselves, having unique chemical bonding and properties (Thornton, 1980).

The apparatus used for plasma-assisted CVD is of a relatively simple design, as shown in Fig. 23 (Veprek, 1985). Another design of a plasma CVD chamber uses the substrate material itself to form the wall of the deposition chamber, as shown in Fig. 24 (Bicker, 1986). This design helps to reduce contamination and nonuniformity of the coating. Typical parameters for plasma generation include RF power in the range from low megahertz to microwave frequencies, and pressure in the range from 0.1 to 10 torr. Details of plasma deposition apparatus and techniques for CVD as well as PVD can be found in Bunshah (1982) and Bunshah and Deshpandey (1985).

## 6.2 Applications of Plasma CVD

An example of the effect of plasma in lowering the reaction temperature may be seen in Fig. 25 (Veprek, 1985). In the conventional CVD process, the deposition of TiC, Ti(C, N), and TiN typically occurs above 1200 K, 1000 K, and 900 K, respectively when the net free energy change for the reaction in question becomes negative. When a plasma is present, the reaction temperatures are considerably lowered, as shown in Fig. 25.

Several new studies have recently been published on the formation of ceramic coatings by plasma CVD. In one example, the researchers deposited a hybridized $Si_3N_4$-SiC film (Kamata et al., 1986). They found that the films were amorphous, with no agglomerated clusters of SiC or $Si_3N_4$, but a series of $SiN_xC_y$ compounds under various experimental conditions. This cannot be achieved by the conventional ceramic processing or CVD techniques. Plasma CVD techniques have also been used to prepare ultrafine powders of ceramic materials, such as SiC (Kijima and Konishi, 1985).

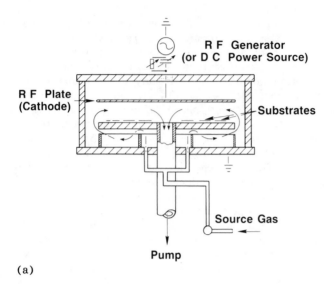

(a)

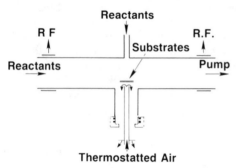

(b)

FIGURE 23   Schematic arrangement of apparatus used for plasma-assisted CVD. (a) Radial flow parallel plate reactor. (b) Discharge tube with the substrate located in the plasma column. (From Veprek, 1985; reprinted with permission of Elsevier Sequoia, S.A., Switzerland.)

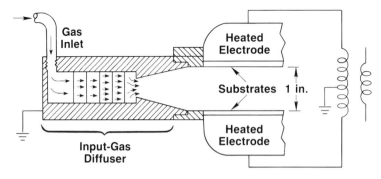

FIGURE 24 Schematic of a plasma CVD apparatus in which the substrate itself acts as the wall of the deposition chamber. (From Bickler, 1986.)

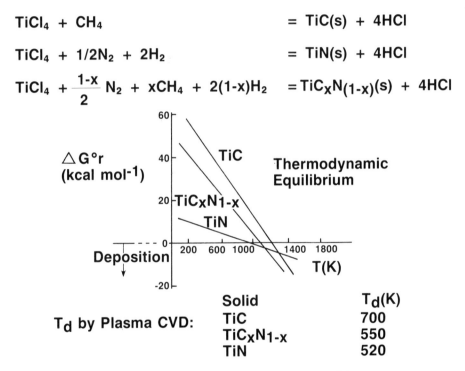

$$TiCl_4 + CH_4 = TiC(s) + 4HCl$$

$$TiCl_4 + 1/2N_2 + 2H_2 = TiN(s) + 4HCl$$

$$TiCl_4 + \frac{1-x}{2}N_2 + xCH_4 + 2(1-x)H_2 = TiC_xN_{(1-x)}(s) + 4HCl$$

$T_d$ by Plasma CVD:

| Solid | $T_d(K)$ |
|---|---|
| TiC | 700 |
| $TiC_xN_{1-x}$ | 550 |
| TiN | 520 |

FIGURE 25 Effect of plasma activation on the deposition temperature for TiC, TiN and $TiC_xN_y$ coatings deposited by CVD. The temperatures at which the standard free energy changes for the reactions become negative are significantly higher in conventional CVD than the temperatures at which the plasma assisted reactions occur. (From Veprek, 1985, reprinted with permission of Elsevier Sequoia, S.A., Switzerland.)

TABLE 15   Materials Deposited by Low Pressure Plasma CVD

| Material | Deposition temp. (K) | Deposition rate (cm/s) | Reactants |
|---|---|---|---|
| Amorphous Si | 523–573 | $10^{-8}$–$10^{-7}$ | $SiH_4$, $SiF_4$–$H_2$, $Si(s)$–$H_2$ |
| Microcrystalline Si | 523–673 | $10^{-8}$–$10^{-7}$ | $SiH_4$–$H_2$, $SiF_4$–$H_2$, $Si(s)$–$H_2$ |
| Amorphous Ge | 523–673 | $10^{-8}$–$10^{-7}$ | $GeH_4$ |
| Microcrystalline Ge | 523–673 | $10^{-8}$–$10^{-7}$ | $GeH_4$–$H_2$, $Ge(s)$–$H_2$ |
| Amorphous B | 673 | $10^{-8}$–$10^{-7}$ | $B_2H_6$, $BCl_3$–$H_2$, $BBr_3$ |
| Amorphous P Microcrystalline P | 293–473 | $\leq 10^{-5}$ | $P(s)$–$H_2$ |
| As | < 373 | $\leq 10^{-6}$ | $AsH_3$, $As(s)$–$H_2$ |
| Se, Te, Sb, Bi | $\leq$ 373 | $10^{-7}$–$10^{-6}$ | $Me$–$H_2$ |
| Mo, Ni | | | $Me(CO)_4$ |
| Diamond-like C | $\leq$ 523 | $10^{-8}$–$10^{-5}$ | $C_nH_m$ |
| Graphite | 1073–1273 | $\leq 10^{-5}$ | $C(s)$–$H_2$, $C(s)$–$N_2$ |
| CdS | 373–573 | $\leq 10^{-6}$ | $Cd$–$H_2S$ |
| GaP | 473–573 | $10^{-8}$ | $Ga(CH_3)_3$–$PH_3$ |
| $SiO_2$ | $\geq$ 523 | $10^{-8}$–$10^{-6}$ | $Si(OC_2H_5)_4$; $SiH_4$–$O_2$, $N_2O$ |
| $GeO_2$ | $\geq$ 523 | $10^{-8}$–$10^{-6}$ | $Ge(OC_2H_5)_4$; $GeH_4$–$O_2$, $N_2O$ |
| $SiO_2/GeO_2$ | 1273 | ~$3\times10^{-4}$ | $SiCl_4$–$GeCl_4$–$O_2$ |
| $Al_2O_3$ | 523–773 | $10^{-8}$–$10^{-7}$ | $AlCl_3$–$O_2$ |
| $TiO_2$ | 473–673 | $10^{-8}$ | $TiCl_4$–$O_2$; metal organics |
| $B_2O_3$ | | | $B(OC_2H_5)_3$–$O_2$ |
| $Si_3N_4$ | 573–773 | $10^{-8}$–$10^{-7}$ | $SiH_4$–$N_2$, $NH_3$ |
| AlN | $\leq$ 1273 | $\leq 10^{-6}$ | $AlCl_3$–$N_2$ |
| GaN | $\leq$ 873 | $10^{-8}$–$10^{-7}$ | $GaCl_4$–$N_2$ |
| TiN | 523–1273 | $10^{-8}$–$10^{-6}$ | $TiCl_4$–$H_2$+$N_2$ |
| BN | 673–973 | | $B_2H_6$–$NH_3$ |
| $P_3N_5$ | 633–673 | $\leq 5\times10^{-6}$ | $P(s)$–$N_2$; $PH_3$–$N_2$ |
| SiC | 473–773 | $10^{-8}$ | $SiH_4$–$C_nH_m$ |

Table 15 (continued)

| Material | Deposition temp. (K) | Deposition rate (cm/s) | Reactants |
|---|---|---|---|
| TiC | 673-873 | $10^{-8}-10^{-6}$ | $TiCl_4-CH_4(C_2H_2)+H_2$ |
| GeC | 473-573 | $10^{-8}$ | |
| $B_xC$ | 673 | $10^{-8}-10^{-7}$ | $B_2H_6-CH_4$ |

*Source:* From Veprek (1985). Reprinted with permission of Elsevier Sequoia S.A., Switzerland.

The most common applications of plasma-assisted CVD have been in the electronic industry. Table 15 lists various materials deposited in a low-pressure, plasma-assisted CVD process, along with reaction temperatures and deposition rates (Veprek, 1985). A wide variety of materials can be readily deposited by this technique.

An important new development in recent years in the application of plasma-assisted CVD is the deposition of diamond-like carbon films (Thompson, 1984; Bichler et al., 1987). These films have some unique application potential ranging from wear-resistant coatings for cutting tools, to coatings for laser mirrors, bearings, fiber-optic seals, dielectrics, p-n junctions, etc. These coatings are typically prepared by the decomposition of hydrocarbon gases by RF plasma and by ion beam deposition techniques. The films are amorphous, with a short-range tetrahedral bonding between the carbon atoms which gives rise to the diamond-like properties.

Moustakas et al. (1987) recently reviewed the various techniques for the growth of diamond from the vapor phase. These include various conventional and "assisted" thermal CVD techniques as well as the plasma techniques. Dischler and Brandt (1985) have reviewed the properties and applications of amorphous diamond-like carbon films. Table 16 compares the properties of amorphous carbon and diamond films. As described by these authors, as well as by Messier et al. (1987) and Setaka (1987), the mechanism of forming these films involves the formation, in a plasma, of atomic hydrogen and chemically active fragments of hydrocarbon ions and radicals. These species react on the surface of the substrate and form a surface film. Depending on the extent of this interaction, various carbon cross-linkages may form, giving rise to either an amorphous carbon film or a crystalline film showing the diamond structure. Haubner and Lux (1987) recently showed that the crystal structure

TABLE 16   Properties of Amorphous Carbon and Diamond Films

| Properties at 300 K | Amorphous carbon | Diamond |
|---|---|---|
| Density, $g/cm^3$, $\rho$ | 1.5-1.8 | 3.515 |
| Knoop hardness, $kg/mm^2$ | 1,250-1,650 | 10,300 {100} |
| | | 11,000 {111} |
| | | 11,500 {110} |
| Hydrogen content (H/C) | 0.15-0.60 | 0.001-0.010 |
| (a/o) | 13-38 | 0.1-1.0 |
| Electrical resistance, $\Omega$-cm | $10^{13}$ | $10^{16}$ |
| Optical energy gap, eV | 0.8-1.8 | 5.48 |
| Transmission bandwidth, $\mu$m | $0.5-\alpha$ | $0.225-\alpha$ |
| Infrared bands, $\mu$m | 3.4, 6-18 | 2.5-6.5 |
| Refractive index, n (to 1 $\mu$m) | 1.8-2.2 | 2.40 |
| Calibrated refractive index, $n/\rho$ | 1.2-1.22 | 0.68 |
| C-hybridization | 68% $sp^3$ | 100% $sp^3$ |
| | 30% $sp^2$ | |
| | 2% $sp^1$ | |

*Source*: Data from Dischler and Brandt, 1985. Reprinted from *Industrial Diamond Review*, vol. 45(508), 3/85.

of the diamond film depends on the concentration of hydrocarbon (methane), substrate temperature, and plasma intensity. Table 17 summarizes the properties of diamond films and their potential applications (Spear, 1987).

A variation of the plasma-assisted CVD has been recently reported in the literature (Meiners, 1982; Jackson et al., 1987). In this technique, called *plasma afterglow* CVD, the reactive constituent is introduced in the afterglow zone of the reactor, as shown in Fig. 26. The CVD reactor consists of three zones: the discharge zone, where the microwave plasma is generated in a glow discharge; the afterglow zone, in which a reactive constituent is introduced; and the processing zone where the deposition occurs. The distinction between the afterglow process and the conventional plasma processes is that in the afterglow technique only certain desirable "active" species are formed in the discharge. By contrast, in the conventional plasma processes all species are generated and react with the sub-

TABLE 17   Unique Properties of Diamond Films and Their Current or Potential Applications

| Properties | Applications |
|---|---|
| 1. High hardness | Abrasive coatings for cutting tools |
| 2. Low coefficient of friction | |
| 3. High thermal conductivity | Heat sinks for electronic devices |
| 4. Electric insulator | |
| 5. High thermal conductivity | Microwave power devices |
| 6. High heat resistivity | |
| 7. Large bandgap | |
| 8. Low dielectric constant | RF electronic devices |
| 9. High hole mobility | High-speed electronic devices |
| 10. High resistivity to acids | Electronic devices for severe |
| 11. High resistivity to radiation | environments such as in space |
| 12. High transparency | or in nuclear reactors |
| 13. Large refractive index | Electro-optical devices |

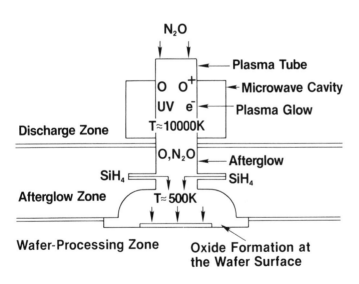

FIGURE 26   Schematic of the afterglow CVD reactor, showing the different zones. In the conventional plasma CVD, the reactions take place primarily in the discharge zone. (From Jackson et al., 1987; reprinted with permission of Solid State Technology.)

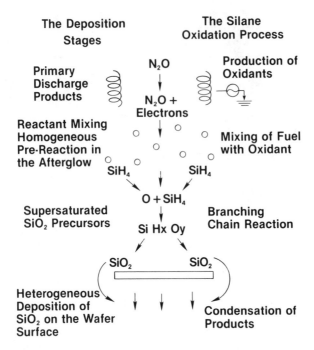

FIGURE 27   Schematic of the primary stages of the afterglow deposition process. (From Jackson et al., 1987; reprinted with permission of *Solid State Technology*.)

strate (Jackson et al., 1987). Thus, in the semiconductor wafer fabrication, the conventional process can cause undesirable plasma-substrate interactions, including incorporation of undesirable species in the film, or damage due to plasma bombardment. Many of these effects are avoided in the afterglow deposition technique. A typical sequence of reactions in an afterglow deposition process for $SiO_2$ using silane and nitrous oxide is shown in Fig. 27 (Jackson et al., 1987). It was shown by these authors that this technique resulted in better film properties, such as refractive index, higher deposition rates, and reduced hydrogen absorption.

## 7   LASER CHEMICAL VAPOR DEPOSITION

Laser chemical vapor deposition (LCVD) is a newly emerging technology in which the conventional CVD processes are enhanced by laser activation. In that sense, laser CVD is similar to plasma CVD

TABLE 18   Comparison between Laser CVD and Plasma-Assisted CVD

| Laser CVD | Plasma-assisted CVD |
|---|---|
| Narrow excitation energy distribution | Broad excitation energy distribution |
| Well-defined and controllable reaction volume | Large reaction volume |
| Highly directional light source allows precise localization of deposition | Can cause contamination from chamber walls |
| Gas phase reactions are reduced | Gas phase reactions are possible |
| Monochromatic light source allows selective excitation of desirable species | Excitation of gaseous species is nonselective in conventional plasma processes |
| Can be carried out at any pressure | Limited (low) pressure range |
| Irradiation damage is considerably reduced | Irradiation damage in insulating films is possible |
| Optical properties of gases and substrate are important in photolytic LCVD | Optical properties are not important |
| Laser sources include IR, visible, UV, and multiphoton wavelengths | Plasma sources include RF and microwave frequencies |

processes. However, there are some major differences between these two techniques. The energy distribution of electrons in a plasma discharge, for example, is much broader than that of the photons emitted from a laser (Solanki et al., 1985). This and other differences between the two techniques give rise to some unique advantages for laser CVD. Table 18 compares the characteristics of the PACVD and LCVD techniques.

Conventional CVD and the plasma CVD processes are thermally driven and therefore generally cause preheating of a large volume of reactants. This can lead to contamination of the deposit from heated surfaces. On the other hand, laser CVD processes are localized in the heated volume, thereby reducing contamination problems. There are two main categories of laser CVD: thermal and photochemical (Allen, 1984). These are also referred to as pyrolytic and

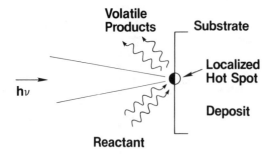

FIGURE 28  Schematic illustration of the pyrolytic (thermal) laser
CVD technique. (From Solanki et al., 1985; reprinted with permission
of Solid State Technology.)


photolytic laser CVD, respectively (Solanki et al., 1985). These
two methods are described briefly below.

Thermal or pyrolytic LCVD involves the selection of a laser
wavelength such that the reactant species transmit the light without
absorption, while the substrate is absorbent. This creates a localized
hot spot on the substrate where deposition can occur, as shown in
Fig. 28. Thus, the deposition reaction is driven locally by thermal
energy, similar to the conventional CVD, the only difference being
that the gas volume is not heated. This may be considered similar
to the cold-wall CVD where the substrate is heated by resistance
or direct induction. The difference between the cold-wall CVD and
pyrolytic laser CVD, in addition to the heat source, is that the
latter technique allows extreme versatility in localizing the heated
area by controlling the size of the laser beam. It also allows the
possibility of scanning the beam across the surface, as in the direct-
write laser CVD technique used in microelectronics applications
(Allen, 1986). The localization of heat allows much higher reactant
concentrations and pressures to be used with no significant gas
phase nucleation, since a large thermal gradient is present at the
deposition site. Continuous wave lasers, such as $CO_2$, Ar, and
Kr ion lasers, are preferred for pyrolytic laser CVD because pulsed
lasers have very short pulse durations (in the nanosecond range)
giving rise to short-duration thermal spikes instead of sustained
heating, which is important for a steady deposition rate.

Since the laser is used as a heat source, gaseous precursors
used in the conventional CVD can generally be used, as long as
they do not absorb laser radiation. In addition, the characteristics
of the substrate and deposited film, such as their optical and thermal
properties, become important (Allen, 1986). For example, the deposi-

tion of a highly reflective film on an absorbant surface reduces the absorbed laser energy, and consequently the substrate temperature and deposition rate are reduced. Similarly, a highly conductive film on a poorly conductive substrate causes a broadening of the temperature profile at the deposition site, again affecting the deposition rate.

Photochemical or photolytic laser CVD requires the gas phase to have a high-absorption cross section and the substrate to be transparent to the laser beam. Therefore, the wavelength of the laser becomes an important parameter. The laser may be continuous or pulsed. The gaseous species absorb the laser photons and reach unstable excited states. In the process of transition to the lower, more stable states, these molecules undergo fragmentation, thereby creating reactive species. Thus, chemical reactions take place in the excited gas volume near the substrate, which can be maintained at a much lower temperature than in conventional CVD. Another advantage of photochemical laser CVD is that, by appropriately selecting the wavelength of light, only selected gaseous species can be created. This permits a better control of the stoichiometry and purity of the deposited film, as compared, for example, to the plasma CVD technique. The principle of photochemical laser CVD is illustrated in Fig. 29.

Some of the versatile applications of laser CVD include laser photolithography, repair of VLSIC masks, laser evaporation-deposition, and metalization. Esrom and Wahl (1987) have recently proposed a model for laser CVD in which a series of dimensionless parameters are defined which characterize the heat transfer mechanisms at the surface. The purpose of the model is to evaluate techniques for maximizing the deposition rate by understanding the synergism between various processes occurring during laser CVD. The various materials deposited by laser CVD are summarized in Tables 19 and 20.

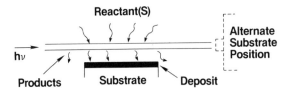

FIGURE 29 Schematic illustration of photolytic (photochemical) laser CVD technique. (From Solanki et al., 1985; reprinted with permission of Solid State Technology.)

TABLE 19  Materials Deposited by Photochemical Laser CVD

| Films | Substrates | Precursors | Lasers |
|---|---|---|---|
| Al | $SiO_2$, $Al_2O_3$, Si | $Al(isobutyl)_3$, $Al(CH_3)_3$ $Al_2(CH_3)_6$ | $2\times Ar+CO_2$, $H_2$ Lamp, ArF, KrF |
| Cd | $SiO_2$, p–GaAs, p–Si, InP | $Cd(CH_3)_2$ | $2\times Ar$ (257 nm) Ar, Kr |
| Cd–Zn | $SiO_2$, Si | $Cd(CH_3)_2$, $Zn(CH_3)_2$ Aqueous $CdSO_4$, $ZnSO_4$ | $2\times Ar$ He–Ne |
| Cr | $SiO_2$, Si | $Cr(CO)_6$ | Nd:YAG |
| Cu | Si | $Cu(F_6acac)_2$ | $2\times Ar$ |
| Fe–Cr–W | $SiO_2$ | $Fe(CO)_5$, $Cr(CO)_6$, $W(CO)_6$ | $2\times Ar$ (257 nm) |
| Ge | — | $GeH_4$ | KrF |
| Hg–Cd–Te | $SiO_2$ | $Hg(CH_3)_2$, $Cd(CH_3)_2$, $Te(C_2H_5)_2$ | Hg/Xe Lamp |

| | | | |
|---|---|---|---|
| InP | GaAs, $SiO_2$, InP | $(CH_3)_3In$, $P(CH_3)_3$ | ArF |
| Mn | ZnS | $Mn_2(CO)_{10}$ | Kr (UV) |
| Mo–Cr | $SiO_2$ | $Mo(C_6H_6)_2$, $Cr(C_6H_6)_2$ | Ar |
| Si | —, $SiO_2$ | $Si_2H_6 + Hg$, $SiH_4$ | Hg lamp, $CO_2$ |
| $SiO_2$ | $SiO_2$+Si | $SiH_4 + N_2O$ with Hg vapor | Hg lamp |
| $Si_3N_4$ | $SiO_2$, Si | $SiH_4 + NH_3$ | Hg lamp |
| Ti | $LiNbO_3$ | $TiCl_4$ | $2 \times Ar$ |
| W | $SiO_2$ | $WF_6 + H_2$, $W(CO)_6$ | $CO_2$, $2 \times Ar$ |
| ZnS | CdS | ZnS | Ar |

*Source:* Data courtesy of Dr. Susan Allen, Center for Laser Science and Engineering, University of Iowa, Iowa City, Iowa.

TABLE 20  Materials Deposited by Pyrolytic Laser CVD

| Films | Substrates | Precursors | Lasers |
|---|---|---|---|
| Al | $SiO_2$ | $Al(isobutyl)_3$ | $2 \times Ar$ (257 nm) + $CO_2$ |
| C | Graphite, $Al_2O_3$, Ceramics, W | $C_2H_2$, $C_2H_4$, $CH_4$ | Ar (488 nm) |
| Cr | $SiO_2$ | $Cr(C_6H_6)_2$ | Ar |
| Cu | Miscellaneous | $Cu(HCOO)_2$ | Nd:YAG |
| Fe | $SiO_2$, Si | $Fe(CO)_5$ | $CO_2$, Kr |
| p-GaAs | GaAs | $Ga(CH_3)_3 + AsH_3$ | Nd:YAG |
| Mo | $SiO_2$ | $Mo(C_6H_6)_2$ | Ar |
| Ni | $SiO_2$, Si | $Ni(CO)_4$ | $CO_2$, Kr |
| Si | Si | $SiH_4$ | Kr |
| Poly Si | CMOS gate array | $SiH_4 + B_2H_4$ | Ar |
| Poly Si (doped) | Si | $SiH_4-PH_3$, $B_2H_6$  $SiH_4+BCl_3$, $B(CH_3)_3$ or $Al(CH_3)_3$ | Ar  Ar |
| $SiO_2$ | — | $SiH_4 + N_2O$ | Kr |
| TiC | $SiO_2$, Steel | $TiCl_4 + CH_4$ | $CO_2$ |
| W | $SiO_2$, Si | $WF_6 + H_2$ | CO, Kr |

*Source:* Data courtesy of Dr. Susan Allen, Center for Laser Science and Engineering, University of Iowa, Iowa City, Iowa.

# 8 SUMMARY

A review of the chemical vapor deposition technique for the deposition of various types of coatings has been presented. An attempt was made to highlight the versatility of the conventional CVD technique in modifying the properties of surfaces to meet the demands of various service conditions. In addition to the classical CVD techniques, the general principles of plasma-assisted CVD and laser CVD were presented. These emerging technologies have considerably widened the range of applications of the CVD technique and allowed engineers to design new materials for the ever increasing demands of developing new applications. It is hoped that the interested reader will be well served by the extensive list of references at the end, which highlight the predominant publications and conferences which deal with this subject.

## ACKNOWLEDGMENTS

The author is indebted to the editor, Dr. T. S. Sudarshan, who asked me to write this chapter and then graciously allowed me to delay the deadline several times to accommodate my busy schedule. Further, the author wishes to acknowledge the many courtesies extended to him by numerous colleagues and peers in the preparation of this chapter. The author is also indebted to many researchers, whose works have been cited here, for providing original drawings and technical data and to the various publishing houses for permission to reprint copyrighted material. Finally, the author is grateful to GTE Valenite for permission to publish the work.

## REFERENCES

Allen, S. D. (1984). Laser Chemical Vapor Deposition (LCVD), *Emergent Process Methods for High Temperature Ceramics*, Materials Science Research, vol. 17 (R. F. Davis, H. Palmour III, and R. L. Porter, eds.). Plenum Press, New York, pp. 397-413.
Allen, S. D. (1986). *IEEE Circuits and Devices Magazine*, pp. 32-36.
Altena, H., Colombier, C., and Lux, B. (1983). "Growth of $\alpha$-Al$_2$O$_3$ on Single and Polycrystalline Substrates by CVD," Proceedings of 4th European Conference on Chemical Vapour Deposition, Eindhoven, The Netherlands, pp. 435-443.
Archer, N. J. (1975). "Tungsten Carbide Coatings on Steel," Proceedings of 5th International Conference on Chemical Vapor

Deposition, Fulmer Grange, England, The Electrochemical Society,
Inc., Princeton, NJ, p. 556-573.

Archer, N. J. and Yee, K. K. (1978). *Wear, 48*:237.

Armas, B. and Combescure, C. (1977). "Chemical Vapor Deposition
at Low Pressure in the System Silicon-Boron," Proceedings
of 6th International Conference on Chemical Vapor Deposition,
Atlanta, Georgia, The Electrochemical Society, Inc., Princeton,
NJ, pp. 181-182.

Armas, B., Combescure, C., and Trombe, F. (1976). *J. Electrochem. Soc., 123*:308.

Arnold, H., Biste, L., and Kaufmann, Th. (1978). *Kristall und
Technik, 13(8)*:929.

Aylett, B. J. and Tannahill, A. A. (1985). *Vacuum, 35(10-11)*:435.

Baik, D. S., Kim, M. S., and Chun, J. S. (1984). "Chemical Vapor
Deposition of TiC Using Propane," Proceedings of 9th International Conference on Chemical Vapor Deposition, Cincinnati,
Ohio, The Electrochemical Society, Inc., Pennington, NJ, pp. 745-
756.

Balog, M., Schieber, M., Michman, M., and Patai, S. (1977). *Thin
Solid Films, 47*:109.

Barin, I. and Knacke, O. (1973). *Thermochemical Properties of
Inorganic Substances*, Springer-Verlag, New York.

Barin, I., Knacke, O., and Kubaschewski, O. (1977). *Thermochemical Properties of Inorganic Substances - Supplement*, Springer-
Verlag, New York.

Bell, A. T. (1981). "Plasma-assisted CVD," Proceedings of 8th
International Conference on Chemical Vapour Deposition, Gouvieux,
France. The Electrochemical Society, Inc., Pennington, NJ,
p. 185.

Bergmann, E. and Vogel, J. (1986). *J. Vac. Sci. Technol., A4(6)*:
2867.

Bernard, C. (1981). "The Application of Thermodynamics to Chemical
Vapor Deposition Processes," Proceedings of 8th International
Conference on Chemical Vapor Deposition, Gouvieux, France.
The Electrochemical Society, Inc., Pennington, NJ, pp. 3-16.

Besmann, T. M. (1977). SOLGASMIX-PV, A Computer Program
to Calculate Equilibrium Relationships in Complex Chemical Systems, *ORNL/TM-5775*, Oak Ridge National Laboratory, Oak Ridge,
TN.

Bhat, D. G. (1980). Investigation of the CNTD Mechanism and
Its Effect on the Microstructure and Properties of Silicon Nitride,
*U.S. Naval Air Systems Command Contract N00019-78-C-0557*,
Final Report.

Bhat, D. G., Cho, T., and Woerner, P. F. (1986). *J. Vac. Sci.
Technol., A 4(6)*:2713.

Bhat, D. G. and DeKay, Y. R. (1987). Comparison Between Laboratory Characterization and Field Performance of Steel Mud Pump Liners Coated with CM 500L, A Tungsten-Carbon Alloy, *Slurry Erosion: Uses, Applications and Test Methods*, ASTM STP 946 (J. E. Miller and F. E. Schmidt, Jr., eds.). American Society for Testing and Materials, Philadelphia, pp. 103-117.

Bhat, D. G. and Holzl, R. A. (1982). *Thin Solid Films, 95*:105.

Bhat, D. G. and Panos, R. M. (1981). Investigation of the CNTD Mechanism and Its Effect on Microstructural Properties of SiC and AlN, *U.S. Air Force Office of Scientific Research Contract F49620-79-C-0041*, Final Report.

Bhat, D. G. and Roman, J. E. (1987). "Morphological Study of CVD $\alpha$-$Si_3N_4$ Deposited At One Atmosphere Pressure." Proceedings of 10th International Conference on Chemical Vapor Deposition, Honolulu, Hawaii. The Electrochemical Society, Inc., Pennington, NJ, p. 579.

Bhat, D. G., Shah, D. C., Kyle, J. R., and Woerner, P. F. (1987). Coated Silicon Nitride Cutting Tool and Process for Making, U.S. Patent, 4,640,643.

Bhat, D. G. and Woerner, P. F. (1986). *J. Metals, 38(2)*:68.

Bichler, R., Haubner, R., and Lux, B. (1987). "Low Pressure Diamond Deposition from Methane-Hydrogen Gas Mixture," Proceedings of 6th European Conference on Chemical Vapour Deposition, Jerusalem, Israel, pp. 413-422.

Bickler, D. B. (1986). *NASA Tech. Brief, 10(1)*: No. 5.

Blocher, J. M., Jr. (1966). Vapor Deposited Materials, *Vapor Deposition* (C. F. Powell, J. H. Oxley and J. M. Blocher, Jr., eds.), Wiley, New York, p. 3.

Bojarski, Zb., Wokulska, K., and Wokulski, Z. (1981). *J. Cryst. Growth, 52*:290.

Bonetti-Lang, M., Bonetti, R., and Hintermann, H. E. (1982). *Refractory and Hard Metals, 1*:161.

Bouteville, A., Royer, A., and Remy, J.-C. (1987). "LPCVD of Ta and Ti Silicides," Proceedings of 6th European Conference on Chemical Vapour Deposition, Jerusalem, Israel, pp. 264-271.

Boving, H., Hintermann, H. E., Begelinger, A., and De Gee, A. W. J. (1983). *Wear, 88*:13.

Bracuti, A. J., Lannon, J. A., Bottei, L. A., and Bhat, D. G. (1981). *J. Ballistics, 5(2)*:1083.

Brennfleck, K., Fitzer, E., and Schoch, G. (1985). "CVD of $ZrO_2$, $Y_2O_3$ and Y-stabilized $ZrO_2$ from Metalorganic Compounds," Proceedings of 5th European Conference on Chemical Vapour Deposition, Uppsala, Sweden, pp. 63-70.

Brutsch, R. (1985). *Thin Solid Films, 126*:313.

Bunshah, R. F. (1982). *Deposition Technologies for Films and Coatings - Developments and Applications*, Noyes Publications, Park Ridge, NJ.

Bunshah, R. F. and Deshpandey, C. V. (1985). "Plasma-assisted Vapor Deposition Processes and Some Applications," Proceedings of 11th International Plansee Seminar, vol. 2 (Bildstein, H. and Ortner, H. M., eds.), Metallwerk Plansee GmbH, Reutte, Tirol, Austria, pp. 931-959.

Caputo, A. J., Lackey, W. J., and Wright, I. G. (1985). *J. Electrochem. Soc.*, *132*:2274.

Chase, M. W., Curnutt, J. L., Hu, A. T., Prophet, H., Syverud, A. N., and Walker, L. C. (1971). *J. Phys. Chem. Ref. Data*, *3(2)*:311, Reprint # 50.

Chase, M. W., Curnutt, J. L., Prophet, H., McDonald, R. A., and Syverud, A. N. (1975). *J. Phys. Chem. Ref. Data*, *4(1)*:1, Reprint # 60.

Chase, M. W., Curnutt, J. L., Downey, Jr., J. R., McDonald, R. A., Syverud, A. N., and Valenzuela, E. A. (1982). *J. Phys. Chem. Ref. Data*, *11(3)*:695, Reprint # 205.

Cheng, D. J., Shyy, W. J., and Hon, M. H. (1987). *Scripta Met.*, *21*:637.

Cho, J. S., Nam, S. W., and Chun, J. S. (1982). *J. Mater. Sci.*, *17*:2495.

Cho, T., Bhat, D. G., and Woerner, P. F. (1986). *Surface and Coatings Technology*, *29*:239.

Clerc, G. and Gerlach, P. (1975). "Pyrolytic BN," Proceedings of 5th International Conference on Chemical Vapor Deposition, Fulmer Grange, England, The Electrochemical Society, Inc., Princeton, NJ, pp. 777-785.

Cochran, A. A. and Stephenson, J. B. (1970). *Met. Trans.*, *1*:2875.

Colombier, C. and Lux, B. (1986). *Refractory & Hard Metals*, *5*:222.

Constant, G. and Moranch, R. (1983). "A Systematic Approach of the Low Temperature CVD with Organometallic Compounds," Proceedings of 4th European Conference on Chemical Vapour Deposition, Eindhoven, The Netherlands, pp. 36-43.

Dapkus, D. P. (1982). *Ann. Rev. Mater. Sci.*, *12*:243.

Das, D. and Kumar, K. (1983). *Thin Solid Films*, *108*:181.

Demyashev, G. M., Krasovskii, A. I., Kuz'min, V. P., Maladin, M. B., and Chuzhko, R. K. (1985). *Neorganicheskie Materialy*, *21(7)*:1155.

Desmaison, J., Roche, J. L., Yoon, K. H., and Billy, M. (1987). "Protection Against Corrosion or Oxidation of Porous Non-oxide Ceramics by a CVD Coating of Same Nature," Proceedings of 6th European Conference on Chemical Vapour Deposition, Jerusalem, Israel, pp. 144-153.

Dirkx, R. R. and Spear, K. E. (1984). A Morphological Study of Si Borides Prepared by CVD, *Emergent Process Methods for High Technology Ceramics*, Materials Science Research, vol. 17 (R. F. Davis, H. Palmour III and R. L. Porter, eds.), Plenum Press, New York, pp. 359-369.

Dischler, B. and Brandt, G. (1985). *Industrial Diamond Review, 45 (508)* :131.

Dor, L. B., Elshtein, A., and Shappir, J. (1983). "Deposition and Characterization of MO-CVD $ZrO_2$," Proceedings of 4th European Conference on Chemical Vapour Deposition, Eindhoven, The Netherlands, pp. 444-450.

Driesner, A. R., Storms, E. K., Wagner, P., and Wallace, T. C. (1973). "High Temperature, Low-density ZrC Insulators made by Chemical Vapor Deposition," Proceedings of 4th International Conference on Chemical Vapor Deposition, Boston, Massachusetts, The Electrochemical Society, Inc., Princeton, NJ, pp. 473-487.

Egashira, M., Katsuki, H., Takahashi, S., and Iwanaga, H. (1987). *Yogyo-Kyokai-Shi, 95*:138.

Eriksson, G. (1971). *Acta Chem. Scand., 25*:2651.

Esrom, H. and Wahl, G. (1987). "Modeling of Laser CVD," Proceedings of 6th European Conference on Chemical Vapour Deposition, Jerusalem, Israel, pp. 367-380.

Fredriksson, E. and Carlsson, J.-O. (1985). *Thin Solid Films, 124*:109.

Fredriksson, E. and Carlsson, J.-O. (1986). *J. Vac. Sci. Technol., A4 (6)* :2706.

Fonzi, F. (1976). "Co-deposited Aluminum-Titanium Oxide Coatings on Cemented Carbide Inserts," Proceedings of International Conference on Hard Material Tool Technology, Carnegie Press, Pittsburgh, Pennsylvania, pp. 172-182.

Fox, R. W. and McDonald, A. T. (1978). *Introduction to Fluid Mechanics*, 2nd ed., Wiley, New York, p. 317.

Funke, V. F., Klemant'ev, A. A., Kosukhin, V. V., Tyutyunnikov, A. I., and Yamskov, N. S. (1969). *Poroshkovaya Metallurg., 84 (12)* :985.

Gass, H., Mantle, H., and Hintermann, H. E. (1975). "The Influence of Cobalt on the Nucleation and Preliminary Growth of TiC," Proceedings of 6th International Conference on Chemical Vapor Deposition, Atlanta, Georgia, The Electrochemical Society, Inc., Princeton, NJ, pp. 99-110.

Gates, A. S., Jr. (1986). *J. Vac. Sci. Technol., A4 (6)* :2707.

Gebhardt, J. J. (1973). "CVD Boron Nitride Infiltration of Fibrous Structures: Properties of Low Temperature Deposits," Proceedings of 4th International Conference on Chemical Vapor Deposition, Boston, Massachusetts, The Electrochemical Society, Inc., Princeton, NJ, pp. 460-472.

Gebhardt, J. J. and Cree, R. F. (1965). *J. Amer. Ceram. Soc.*,
    *48 (5)* :262.
Glaski, F. A. and Crowson, A. (1977). "Tantalum Alloy Chemical
    Vapor Plating of Gun Barrels," Proceedings of 6th International
    Conference on Chemical Vapor Deposition, Atlanta, Georgia,
    The Electrochemical Society, Inc., Princeton, NJ, p. 542.
Grishachev, V. F., Maslov, V. P., Vesna, V. T., and Shcherbakova,
    L. E. (1984). *Poroshkovaya Metallurg.*, *4 (256)* :40.
Gross, M. E. and Schnoes, K. J. (1987). "Chemical Vapor Deposition
    of Cobalt and Formation of Cobalt Silicide," Proceedings of 10th
    International Conference on Chemical Vapor Deposition, The
    Electrochemical Society, Inc., Pennington, NJ, p. 759.
Gyarmati, E., Mehner, A.-W., and Wallura, E. (1983). "Material
    Properties of SiC Deposited by CVD in a Fluidized Bed," Proceed-
    ings of 4th European Conference on Chemical Vapour Deposition,
    Eindhoven, The Netherlands, pp. 313-320.
Habig, K.-H. (1986). *J. Vac. Sci. Technol.*, *A4 (6)* :2832.
Hakim, M. J. (1975). "Chemical Vapour Deposition of HfN and HfC
    on Tungsten Wires," Proceedings of 5th International Conference
    on Chemical Vapor Deposition, Fulmer Grange, England, The
    Electrochemical Society, Inc., Princeton, NJ, pp. 634-649.
Hamamura, K., Yamagishi, H., and Nagakura, S. (1974). *J. Cryst.
    Growth*, *26*:255.
Hannache, H., Naslain, R., Bernard, C., and Heraud, L. (1983).
    "The CVD of Boron Nitride from $BF_3$-$NH_3$ Mixtures," Proceedings
    of 4th European Conference on Chemical Vapour Deposition,
    Eindhoven, The Netherlands, pp. 305-312.
Hannache, H. E., Langlais, F., and Naslain, R. (1985). "Kinetics
    of Boron Carbide Chemical Vapour Deposition and Infiltration,"
    Proceedings of 5th European Conference on Chemical Vapour
    Deposition, Uppsala, Sweden, pp. 219-233.
Haskell, R. W. and Imam, A. R. (1975). "Erosion and Heat Transfer
    Characteristics of Tungsten-coated Steel," Proceedings of 5th
    International Conference on Chemical Vapor Deposition, Fulmer
    Grange, England, The Electrochemical Society, Inc., Pennington,
    NJ, p. 829.
Haubner, R. and Lux, B. (1987). *Refrac. and Hard Metals*, *6 (4)* :210.
Hayashi, T., Mikoya, M., Yamai, I., Saito, H., and Hirano, S.
    (1987). *J. Mater. Sci.*, *22*:1305.
Heestand, R. L., Short, D. W., and Robinson, W. C. (1967).
    "Chemical Vapor Deposition of Ceramic Compounds," Proceedings
    of Conference on Chemical Vapor Deposition of Refractory Metals,
    Alloys and Compounds, Gatlinburg, Tennessee, The American
    Nuclear Society, Hinsdale, IL, pp. 175-191.
Hickey, G., Boone, D., Levy, A., and Stiglich, J. (1984). *Thin
    Solid Films*, *118*:321.

Hieber, K. (1974). *Thin Solid Films, 24:*157.

Hieber, K. and Stolz, M. (1975). "Preparation of CVD Molybdenum, Tantalum and Tantalum Oxide Layers," Proceedings of 5th International Conference on Chemical Vapour Deposition, Fulmer Grange, England, The Electrochemical Society, Inc., Princeton, nj, pp. 436-449.

Hintermann, H. E. (1981). *Thin Solid Films, 84:*215.

Hirai, T. and Hayashi, S. (1982). *J. Mater. Sci., 17:*1320.

Hirayama, M. and Shohno, K. (1975). *J. Electrochem. Soc., 122:*1671.

Hollabaugh, C. M., Wahman, L. A., Reiswig, R. D., White, R. W., and Wagner, P. (1977). *Nuclear Technol.* 35:527.

Holleck, H. (1986). *J. Vac. Sci. Technol., A4(6):*2661.

Holzl, R. A. (1968). Chemical Vapor Deposition Techniques, *Techniques of Materials Preparation and Handling, Part 3,* Techniques of Metals Research Series, vol. 1, Part 3 (R. F. Bunshah, ed.), Interscience Publishers, New York, p. 1377.

Holzl, R. A. (1978). Investigation of the CNTD Mechanism and Its Effect on Microstructure and Properties of Silicon Nitride, *U.S. Naval Air Systems Command Contract N00019-77-C-0395,* Final Summary Report.

Holzl, R. A., Benander, R. E. and Davis, R. D. (1984). Tungsten Alloys Containing A15 Structure and Method for Making, U.S. Patent, 4,427,445.

Hough, R. L. (1972). "Chemical Vapor Deposition of Metal Oxides from Organometallics," Proceedings of 3rd International Conference on Chemical Vapor Deposition, Salt Lake City, Utah, The American Nuclear Society, Hinsdale, IL, pp. 232-241.

Hunt, L. P. (1987). "Thermodynamic Equilibria in the Si-H-Cl and Si-H-Br Systems," Proceedings of 10th International Conference on Chemical Vapor Deposition, Honolulu, Hawaii, The Electrochemical Society, Inc., Pennington, NJ, p. 112.

Ikawa, K. (1972). *J. Less Comm. Metals, 29:*233.

Jackson, R. L., Spenser, J. E., McGuire, J. L., and Hoff, A. M. (1987). *Solid State Technol., 30:*107.

Jansson, U. and Carlsson, J.-O. (1985). *Thin Solid Films, 124:*101.

Jensen, K. F. and Graves, D. B. (1983). *J. Electrochem. Soc., 130(9):*1950.

Jensma, J. P., Jacobs, A. P. G., Hendriks, W. E., and Verspui, G. (1985). "Low Pressure CVD of SiC at Moderate Temperatures," Proceedings of 5th European Conference on Chemical Vapour Deposition, Uppsala, Sweden, pp. 405-412.

Kamata, K., Maeda, Y., and Moriyama, M. (1986). *J. Mater. Sci. Lett., 5:*1051.

Kamimori, T. and Mizuhashi, M. (1981). "CVD of $SnO_2$ Films From Organotin Compounds," Proceedings of 8th International Conference on Chemical Vapour Deposition, Gouvieux, France, The Electrochemical Society, Inc., Pennington, NJ, pp. 436-449.

Kaplan, R. B. (1972). "Chemical Vapor Deposition of Hafnium Carbide," Proceedings of 3rd International Conference on Chemical Vapor Deposition, Salt Lake City, Utah, The American Nuclear Society, Hinsdale, IL, pp. 176-182.

Kato, A. and Tamari, N. (1975). *J. Cryst. Growth*, 29:55.

Kato, A. and Tamari, N. (1980). *J. Cryst. Growth*, 49:199.

Kato, A., Yasunaga, M., and Tamari, N. (1977). *J. Cryst. Growth*, 37:293.

Kawahara, K., Fukase, K., Inoue, Y., Taguchi, E., and Yoneda, K. (1987). "CVD Spinel on Si," Proceedings of 10th International Conference on Chemical Vapor Deposition, Honolulu, Hawaii, The Electrochemical Society, Inc., Pennington, NJ, p. 588.

Kehr, D. E. R. (1977). "Chemical Vapor Deposition of Silicides of Molybdenum, Niobium and Tantalum," Proceedings of 6th International Conference on Chemical Vapor Deposition, Atlanta, Georgia, The Electrochemical Society, Inc., Princeton, NJ, p. 561.

Kieda, N., Mizutani, N., and Kato, M. (1987). "CVD of 5A Group Transition Metal Nitrides," Proceedings of 10th International Conference on Chemical Vapor Deposition, Honolulu, Hawaii, The Electrochemical Society, Inc., Pennington, NJ, p. 1203.

Kieffer, R., Fister, D., Schoof, H., and Mauer, K. (1973). *Powder Metall. Int.*, 5(4):188.

Kijima, M. and Konishi, M. (1985). *Yogyo Kyokai-Shi*, 93(9):35.

Kim, D. G., Yoo, J. S., and Chun, J. S. (1986). *J. Vac. Sci. Technol.*, A4(2):219.

Komiyama, H., Osawa, T., Shimogaki, Y., Wakita, N., Minamiyama, M., and Ueoka, T. (1987). "Particle Precipitation Aided Chemical Vapor Deposition For Rapid Growth of Ceramic Films - Preparation of 1mm Thick AlN, $TiO_2$ and $ZrO_2$ Films," Proceedings of 10th International Conference on Chemical Vapor Deposition, Honolulu, Hawaii, The Electrochemical Society, Inc., Pennington, NJ, p. 1119.

Kurtz, S. R. and Gordon, R. G. (1986). *Thin Solid Films*, 140:277.

Landingham, R. L. and Austin, J. H. (1969). *J. Less Common Met.*, 18:229.

Lartigue, J. F., Ducarroir, M., and Armas, B. (1984). "Vapor Deposition of $Si_3N_4$ From $Si(CH_3)_4/NH_3$ Mixtures Under Low Pressure," Proceedings of 9th International Conference on Chemical Vapor Deposition, Cincinnati, Ohio, The Electrochemical Society, Inc., Pennington, NJ, pp. 561-574.

Lee, S.-L. (1982). *Surface Treatment*, 15(3):138.

Lepie, M. P. (1964). *Trans. Brit. Ceram. Soc.*, 63(8):431.

Lewis, D. W. (1970). *J. Electrochem. Soc.*, 117:978.

Lindstrom, J. N. and Stjernberg, K. G. (1975). "Rate Determining Steps At CVD of $Al_2O_3$, TiC and TiN," Proceedings of 5th European Conference on Chemical Vapour Deposition, Uppsala, Sweden, pp. 169-182.

Lo, J.-S., Haskell, R. W., Byrne, J. G., and Sosin, A. (1973). "A CVD Study of Tungsten Silicide System," Proceedings of 4th International Conference on Chemical Vapor Deposition, Boston, Massachusetts, The Electrochemical Society, Inc., Princeton, NJ, p. 74.

Logar, R. E., Wauk, M. T., and Rosler, R. S. (1977). "Low Pressure Deposition of Phosphorus-doped Silicon Dioxide at 400 C in a Hot Wall Furnace," Proceedings of 6th International Conference on Chemical Vapor Deposition, Atlanta, Georgia, The Electrochemical Society, Inc., Princeton, NJ, pp. 195-202.

Mantle, H., Gass, H., and Hintermann, H. E. (1975). "Chemical Vapour Deposition of Tungsten Carbide (WC)," Proceedings of 5th International Conference on Chemical Vapor Deposition, Fulmer Grange, England, The Electrochemical Society, Inc., Princeton, NJ, pp. 540-555.

Mantyla, T., Telama, A., Vuoristo, P., and Kettunen, P. (1985). "Deposition of $Al_2O_3$ by the $AlCl_3$ + $H_2O$ Reaction," Proceedings of 5th European Conference on Chemical Vapour Deposition, Eindhoven, The Netherlands, pp. 347-353.

Matsuda, T., Uno, N., Nakae, H., and Hirai, T. (1986). *J. Mater. Sci.*, *21*:649.

Messier, R., Badzian, A. R., Badzian, T., Spear, K. E., Bachmann, P., and Roy, R. (1987). *Thin Solid Films*, *153*:1.

Michalski, A., Sokolowska, A., Wronikowski, M., and Wesolowski, L. (1987). "$B_7O$ Layers," Proceedings of 6th European Conference on Chemical Vapour Deposition, Jerusalem, Israel, pp. 305-310.

Million-Brodaz, J. F., Vahlas, C., Bernard, C., Torres, J., and Madar, R. (1987). "Thermodynamic and Experimental Studies of APCVD of $TiSi_2$," Proceedings of 6th European Conference on Chemical Vapour Deposition, Jerusalem, Israel, pp. 280-297.

Milewski, J. V., Gac, F. D., Petrovic, J. J., and Skaggs, S. R. (1985). *J. Mater. Sci.*, *20*:1160.

Minet, J., Langlais, F., Naslain, R., and Bernard, C. (1987). "Thermodynamic and Experimental Study of the CVD of Zirconia from $ZrCl_4$-$H_2$-$CO_2$ Gas Mixtures," Proceedings of 6th European Conference on Chemical Vapour Deposition, Jerusalem, Israel, pp. 68-75.

Monnig, K. A., Brors, D. L., Fair, J. A., and Saraswat, K. C. (1984). "Deposition Parameters and Properties of Low Pressure Chemical Vapor Deposited Tungsten Silicide for IC Applications," Proceedings of 9th International Conference on Chemical Vapor Deposition, Cincinnati, Ohio, The Electrochemical Society, Inc. Pennington, NJ, pp. 275-286.

Motojima, S., Iwamori, N., and Hattori, T. (1986). *J. Mater. Sci.*, *21*:3836.

Motojima, S., Iwamori, N., Hattori, T., and Kurosawa, K. (1986). *J. Mater. Sci.*, *21*:1363.

Motojima, S., Kohno, M., and Hattori, T. (1987). *J. Mater. Sci.*, *22(2)*:547.

Motojima, S. and Kuzuwa, S. (1985). *J. Cryst. Growth*, *71*:682.

Motojima, S., Yagi, H., and Iwamori, N. (1986). *J. Mater. Sci. Lett.*, *5*:13.

Maury, P., Oquab, D., Morancho, R., Nowak, J. F., and Gauthier, J. P. (1987). "Low Temperature Deposition of Chromium Carbide by LPCVD Process Using Bis Arene Chromium as a Single Source," Proceedings of 10th International Conference on Chemical Vapor Deposition, Honolulu, Hawaii, The Electrochemical Society, Inc., Pennington, NJ, p. 1213.

Moustakas, T. D., Dismukes, J. P., Ye, L., Walton, K. R., and Tiedje, J. T. (1987). "Polycrystalline Diamond Deposition From Methane-Hydrogen Mixtures," Proceedings of 10th International Conference on Chemical Vapor Deposition, Honolulu, Hawaii, The Electrochemical Society, Inc., Pennington, NJ, p. 1164.

Mullendore, A. W. and Pope, L. E. (1987). *Thin Solid Films*, *153*: 267.

Nakae, H., Matsunami, Y., Uno, N., Matsuda, T., and Hirai, T. (1985). "Preparation of BN-TiN Compound and BN-$Si_3N_4$ Compound by Chemical Vapor Deposition," Proceedings of 5th European Conference on Chemical Vapor Deposition, Uppsala, Sweden, pp. 242-249.

Nakamura, R. (1985). *J. Electrochem. Soc.*, *132*:1757.

Niihara, K. and Hirai, T. (1976). *J. Mater. Sci.*, *11*:593.

Oxley, J. H. (1966). Transport Processes, *Vapor Deposition* (C. F. Powell, J. H. Oxley and J. M. Blocher, Jr., eds.), Wiley, New York, p. 124.

Pauleau, Y., Bouteville, A., Hantzpergue, J. J., Remy, J. C., and Cachard, A. (1982). *J. Electrochemical Soc.*, *129*:1045.

Park, C.-S., Kim, J. G., and Chun, J. S. (1983). *J. Electrochem. Soc.*, *130*:1607.

Perry, A. J. and Archer, N. J. (1980). Wear Resistant Coatings Made by Chemical Vapour Deposition, *Materials Coating Techniques*, NATO-AGARD Lecture Series 106, Paper 10, p. 10-1.

Pierson, H. O. (ed.) (1980). *Chemically Vapor Deposited Coatings*, American Ceramic Society, Columbus, Ohio.

Pierson, H. and Mullendore, A. W. (1980). *Thin Solid Films*, *72*:511.

Piton, J. P., Ladouce, B., and Vandenbulcke, L. (1987). "Thermodynamic Approach of the CVD of Titanium Carbide at 750-850°C," Proceedings of 6th European Conference on Chemical Vapour Deposition, Jerusalem, Israel, pp. 120-130.

Ploog, K. (1974). *J. Cryst. Growth*, *24/25*:197.

Powell, C. F., Campbell, I. E., and Gonser, B. W. (1955). *Vapor Plating*, Wiley, New York.

Powell, C. F., Oxley, J. H., and Blocher, J. M., Jr. (eds.) (1966). *Vapor Deposition*, Wiley, New York.

Powell, C. F. (1966). Chemical Vapor Deposition, *Vapor Deposition* (C. F. Powell, J. H. Oxley, and J. M. Blocher, Jr., eds.), Wiley, New York, pp. 249-276.

Qureshi, J., Levy, A., and Wang, B. (1986). *J. Vac. Sci. Technol.*, *A4(6)*:2638.

Randich, E. (1980). *Thin Solid Films*, *72*:517.

Randich, E. and Gerlach, T. M. (1981). *Thin Solid Films*, *75*:271.

Rebenne, H. and Pollard, R. (1985). *J. Electrochem. Soc.*, *132(8)*: 1932.

Rode, E. Harshbarger, W., and Watson, L. (1987). "Investigation of Gas Reactions During CVD of $WSi_2$," Proceedings of 10th International Conference on Chemical Vapor Deposition, Honolulu, Hawaii, The Electrochemical Society, Inc., Pennington, NJ, p. 711.

Rosenberger, F. (1987). "Flow Dynamics and Modelling of CVD," Proceedings of 10th International Conference on Chemical Vapor Deposition, Honolulu, Hawaii, The Electrochemical Society, Inc., Pennington, NJ, p. 11.

Salles, P., Bernard, C., Ducarrior, M., and Nadal, M. (1987). "Deposition of Nonstoichiometric Zirconium Carbide," Proceedings of 10th International Conference on Chemical Vapor Deposition, Honolulu, Hawaii, The Electrochemical Society, Inc., Pennington, NJ, p. 1129.

Saraie, J., Kwon, J., and Yodogawa, Y. (1985). *J. Electrochem. Soc.*, *132*:890.

Sarin, V. K., Buljan, S.-T., and D'Angelo, C. (1983). Carbide-coated Composite Silicon Nitride Cutting Tools, U.S. Patent, 4,416,670.

Sarin, V. K. and Buljan, S.-T. (1983). Nitride-coated Silicon Nitride Cutting Tools, U.S. Patent, 4,406,668.

Sarin, V. K. and Buljan, S.-T. (1984). Coated Composite Modified Silicon Aluminum Oxynitride Cutting Tools, U.S. Patent, 4,469,489.

Schick, H. (1966). *Thermodynamics of Certain Refractory Compounds*, vol. I and II, Academic Press, New York.

Schilling, C. L., Wesson, J. P., and Williams, T. C. (1983). *Ceram. Bull.*, *62(8)*:912.

Setaka, N. (1987). "Vapor Deposition of Diamond," Proceedings of 10th International Conference on Chemical Vapor Deposition, Honolulu, Hawaii, The Electrochemical Society, Inc., Pennington, NJ, p. 1156.

Shlichta, P. J. and Holliday, R. J. (1985). Beta Silicon Carbide Whiskers, *NASA Tech. Briefs*, *9(4)*: No. 3.

Singh, R. N. (1987). "LPCVD of Boron Nitride From Trichloro-
    borazine," Proceedings of 10th International Conference on
    Chemical Vapor Deposition, Honolulu, Hawaii, The Electrochemical
    Society, Inc., Pennington, NJ, p. 543.
Sjostrand, M. E. (1979). "Deposition of Wear Resistant TiN on
    Cemented Carbides Using Mixtures of $NH_3/N_2$ and $TiCl_4/H_2$,
    Proceedings of 7th International Conference on Chemical Vapor
    Deposition, Los Angeles, California, The Electrochemical Society,
    Inc., Princeton, NJ, pp. 452-462.
Solanki, R., Moore, C. A., and Collins, G. J. (1985). *Solid State
    Technol.*, *28*:220.
Spear, K. E. (1979). "Applications of Phase Diagrams and Thermo-
    dynamics to CVD," Proceedings of 7th International Conference
    on Chemical Vapor Deposition, Los Angeles, California, The
    Electrochemical Society, Inc., Princeton, NJ, pp. 1-16.
Spear, K. E. (1984). "Thermochemical Modelling of Steady-state
    CVD Processes," Proceedings of 9th International Conference
    on Chemical Vapor Deposition, Cincinnati, Ohio, The Electro-
    chemical Society, Inc., Pennington, NJ, pp. 81-97.
Spear, K. E. (1987). *Earth and Mineral Sciences Newsletter*, *56(4)*:53.
Stauf, G. T., Driscoll, D. C., Dowben, P. A., Barfuss, S., and
    Grade, M. (1987). *Thin Solid Films*, *153*:421.
Steinmann, P. A., Tardy, Y., and Hintermann, H. E. (1987).
    *Thin Solid Films*, *153*:333.
Stjernberg, K. G., Gass, H., and Hintermann, H. E. (1977). *Thin
    Solid Films*, *40*:81.
Stiglich, J. J., Jr. and Bhat, D. G. (1980). *Thin Solid Films*,
    *72*:503.
Stiglich, J. J., Jr., Bhat, D. G., and Holzl, R. A. (1980).
    *Ceramurgia Int.*, *6(1)*:3.
Stolyarov, E. V., Bondarenko, V. P., Badrak, S. A., and Turov,
    V. P. (1983). *Poroshkovaya Metallurg.*, *9*:83.
Stull, D R. and Prophet, H. (1971). *JANAF Thermochemical Tables*,
    *Second Edition*, NSRDS-NBS 37, U.S. Government Printing
    Office, Washington, DC.
Sun, W.-P., Lin, H. J., and Hon, M.-H. (1987). *Thin Solid Films*,
    *146*:55.
Sundgren, J.-E. and Hentzell, H. T. G. (1986). *J. Vac. Sci. Tech-
    nol.*, *A4 (5)*:2259.
Suzuki, M. and Tanji, H. (1987). "CVD of Polycrystalline Aluminum
    Nitride," Proceedings of 10th International Conference on Chemi-
    cal Vapor Deposition, Honolulu, Hawaii, The Electrochemical
    Society, Inc., Pennington, NJ, p. 1089.
Tabata, O. (1975). "$SnO_2$ Film on Si Synthesized by CVD Method,"
    Proceedings of 5th International Conference on Chemical Vapor
    Deposition, Fulmer Grange, England, The Electrochemical Society,
    Inc., Princeton, NJ, pp. 681-694.

Takahashi, T., Sugiyama, K., and Hideaki, I. (1970). *J. Electrochem. Soc.*, *117(4)*:541.

Tanji, H., Monden, K., and Ide, M. (1987). "CVD Mechanism of Pyrolytic Boron Nitride," Proceedings of 10th International Conference on Chemical Vapor Deposition, Honolulu, Hawaii, The Electrochemical Society, Inc., Pennington, NJ, p. 562.

Taylor, H. L. and Trotter, J. D. (1972). "Cerium Oxide Films From a Chloride Source," Proceedings of 3rd International Conference on Chemical Vapor Deposition, Salt Lake City, Utah, The American Nuclear Society, Hinsdale, IL, pp. 475-480.

Thornton, J. A. (1980). "The Use of Low Pressure Plasmas in Materials Processing," Part II-Applications, AIAA 13th Fluid and Plasma Dynamics Conference, Snowmass, Colorado, Paper AIAA-80-1324, American Institute of Aeronautics and Astronautics, New York.

Thornton, J. A. (1982). Plasmas in Deposition Processes, *Deposition Technologies For Films and Coatings - Developments and Applications* (R. F. Bunshah, ed.), Noyes Publications, Park Ridge, pp. 19-62.

Thompson, D. G. (1984). Diamond-like Carbon Coatings, *Current Awareness Bulletin, No. 142*, Metals and Ceramics Information Center, Battelle Columbus Laboratories, Columbus, Ohio.

Tsui, P. and Spear, K. E. (1984). A Morphological Study of SiC Prepared by CVD, *Emergent Process Methods for High Technology Ceramics*, Materials Science Research, vol. 17 (R. F. Davis, H. Palmour III, and R. L. Porter, eds.), Plenum Press, New York, pp. 371-380.

van den Brekel, C. H. J. (1977). Part I, *Philips Res. Repts.* *32*:118.

van den Brekel, C. H. J. and Bloem, J. (1977). Part II, *Philips Res. Repts.*, *32*:134.

Veprek, S. (1985). *Thin Solid Films, 130*:135.

Wahl, G. and Hoffmann, R. (1980). *Rev. int. hautes Temper. Refract., Fr., 17*:7.

Wahl, G., Schmaderer, F., Huber, R., and Weber, R. (1987). "Simulation of Stagnation Flow Reactors," Proceedings of 10th International Conference on Chemical Vapor Deposition, Honolulu, Hawaii, The Electrochemical Society, Inc., Pennington, NJ, p. 42.

Wahl, G., Schmaderer, F., Metzger, M., and Nicoll, A. R. (1981). "The Chemical Vapour Deposition of Ti-Si containing Coatings on Ni-base Superalloys," Proceedings of 8th International Conference on Chemical Vapor Deposition, Gouvieux, France. The Electrochemical Society, Inc., Pennington, NJ, p. 685.

Wallace, T. C. (1973). "Chemical Vapor Deposition of ZrC in Small Bore Carbon Composite Tubes," Proceedings of 4th International Conference on Chemical Vapor Deposition, Boston, Massachusetts, The Electrochemical Society, Inc., Princeton, NJ, pp. 91-106.

Wang, J., Zhang, S., and Wang, Y. (1987). "Theoretical Model of LPCVD Thickness Distribution and Comparison Between Model and Experiments," Proceedings of 10th International Conference on Chemical Vapor Deposition, Honolulu, Hawaii, The Electrochemical Society, Inc., Pennington, NJ, p. 23.

Wang, M. S. and Spear, K. E. (1984). "Experimental and Thermodynamic Investigations of the V-Si-H-Cl CVD system," Proceedings of 9th International Conference on Chemical Vapor Deposition, Cincinnati, Ohio, The Electrochemical Society, Pennington, NJ, pp. 98-111.

West, G. and Beeson, K. (1987). "Chemical Vapor Deposition of Molybdenum Silicide," Proceedings of 10th International Conference on Chemical Vapor Deposition, Honolulu, Hawaii, The Electrochemical Society, Inc., Pennington, NJ, p. 720.

Wicks, C. E. and Block, F. E. (1963). *Thermodynamic Properties of 65 Elements - Their Oxides, Carbides and Nitrides*, Bureau of Mines Bulletin No. 605, U.S. Government Printing Office, Washington, DC.

Wieczorek, C. (1985). *Thin Solid Films, 126*:227.

Wokulski, Z. (1987). *J. Cryst. Growth, 82*:427.

Wong, P. and Robinson, McD. (1970). *J. Amer. Ceram. Soc., 53*:617.

Yamai, I. and Saito, H. (1978). *J. Cryst. Growth, 45*:511.

Yee, K. K. (1978). *International Metals Reviews*, Review No. 226.

# 3

# Ion Beam–Based Techniques for Surface Modification

HILLARY SOLNICK-LEGG and KEITH O. LEGG   *Ionic Atlanta, Inc.,*
*Atlanta, Georgia*

## 1   INTRODUCTION

Bulk materials are generally chosen to provide mechanical, chemical, or structural integrity. However, all materials (and thus, tools, components, and parts) react with their environment via their surfaces in ways which may ultimately determine their useful life. Even some properties, such as fatigue, which were once thought to be strictly bulk properties are markedly affected by surface finish and treatment.

This chapter describes in detail techniques which can modify surface characteristics of a wide range of materials at low temperatures and frequently without altering tolerances in any measurable manner. In most surface treatments, heat provides the energy that allows a reaction to proceed, and for many processes the heat required is quite high (up to 2000°F). In ion beam-based surface treatments, high-energy ions provide the energy for reactions to occur and thus form new compounds or alloys. Since the ions do not penetrate deeply into the treated surface, but remain in the upper 0.5 micron (or 20 microinches), ion implantation is truly a surface modification process. This shallow penetration depth might lead one to expect few if any long-term effects, particularly in environments where wear and friction are the major causes of tool or component degradation. This, however, is not the case with ion beam-modified materials, as we shall see in more detail.

Ion beams are now being used in a wide variety of processes. This chapter covers ion implantation and ion-assisted coating of nonelectronic materials. Other processes, such as sputtering, cluster beam deposition, and direct ion beam deposition of coatings are covered elsewhere in this book.

## 2   DEVELOPMENT OF ION BEAM TECHNOLOGY

For nearly 20 years ion implantation, the firing of ions into a surface has been a standard method for doping semiconductor materials (silicon wafers) to produce integrated circuit chips such as micro-processors and computer memories. The reliability, controllability, and reproducibility of ion implantation methods have made this technology the backbone of the semiconductor industry.

In semiconductor processing, trace amounts of dopants, which impart specific electrical characteristics to silicon to make it a semiconductor, are introduced through ion implantation of elements such as arsenic, phosphorus, and boron. Because of the high precision of the ion implantation method, combined with the ability to use it in conjunction with lithographic masks, and the possibility of utilizing highly focused ion beams for fine-line applications, this technology has become the backbone of the semiconductor industry and will become more important as device and circuit sizes decrease. Excellent reviews of ion implantation for semiconductor applications can found by Townsend et al. (1976) and for electronic and nonelectronic applications, by Williams (1986).

Current implanter technology for treatment of nonelectronic materials is based on equipment developed for and used by the semiconductor industry. There are still many similarities between implanters used for either application. A standard semiconductor ion implanter consists of an ion source to produce beams of the desired dopant species, a higher-resolution analyzing magnet to bend and separate the dopant elements from contaminant elements, a beam line, an end station for treatment of the wafers, which is a high-vacuum chamber, and a carousel within the end station for high-speed scanning of the wafers during implantation to ensure uniform doping. Chamber pressures, prior to implantation, are typically about $10^{-6}$ to $10^{-7}$ torr. Doping uniformity and accuracy is required to be better than 1%.

Ion doses for doping of silicon wafers are generally at least two to three orders of magnitude *less* than those needed to induce chemical, optical, or physical property changes in nonelectronic materials such as metals or ceramics. Electronic materials generally require additions of approximately 1% or less of a trace element to fabricate the desired circuit. Thus ion fluences (or doses) for implantation of electronic materials are in the range of $10^{14}$ to a few times $10^{15}$ ions/cm$^2$. Wafer processing also requires implantation of only a few different elements (B, P, As, and occasionally Sb in silicon; S, Si, Se, Te, Be, Mg, Zn, or Cd in gallium arsenide); thus semiconductor implanters are usually dedicated machines that are set up to produce one, or several, ion species. After ion implantation, thermal annealing is required to remove implantation damage from the near-surface regions of semiconductor materials.

For nonelectronic materials, the new alloying or compounding element must be added in amounts of at least several atomic percent, sometimes as much as 20-50 at.%. For metals implant doses of most elements are from $10^{16}$ to $10^{18}$ ions $cm^{-2}$. Annealing is generally not required when metals are implanted, since it is this very damage that forms the basis of much of the improvement seen, particularly for chemical and mechanical property changes. The ion implanters used for nonelectronic materials are different from standard semiconductor implanters primarily in the need for beam scanning to reduce local heating effects (the silicon wafers themselves are usually rapidly scanned during implantation, eliminating any need for beam scanning) and in the design of the treatment chambers. The treatment chamber, or end station, is generally a stainless steel high-vacuum system with specialized internal fixturing capable of handling a variety of tools or parts to be implanted. For industrial production processing, a dedicated implanter with a specialized chamber is the most economical means of using ion implantation as a commercial process. Ion implanters for nonelectronic materials are usually required to generate high-current (0.5 to 10 mA) beams of a large variety of ion species, depending upon the specific property to be altered and the substrate material. Ion species which are commonly required for nonelectronic materials modification include N, C, CO, Ti, Al, Si, Ni, B, Cr, Ar, Y, P, O, Pt, and many others.

In the late 1960s and early 1970s British scientists began to explore the effects of ion beams on nonsemiconductor materials, particularly metals (Dearnaley, 1969; Hartley, 1976). These early efforts concentrated on the effect of ion implantation on friction and microhardness in metals, but soon turned to modifying sliding wear characteristics. The result has been a relatively straightforward implantation technique, that of implanting nitrogen ions, which has been very successful in reducing wear in metal and carbide tooling and components in a wide number of industrial applications. Tools and parts successfully treated by this technique include punches, steel and carbide drawing and extrusion dies, stamping and pressing tools, mill rolls, and molds, bearings, turbine blades, barrel nozzles, and injection screws used in the manufacture of abrasive plastics, and many others. Over the past 10 years, ion implantation technology for nonsemiconductor materials has been extended to include implantation of many other implant species for improvement of surface properties.

Today ion beam technology is being used to treat semiconductors, metals, ceramics, glasses, composites, polymers, and minerals. It is used industrially to lengthen the life of bearings, Ti alloy knee and hip prostheses, to provide high-temperature corrosion resistance in metals, such as those used for nuclear fuel rod cladding, and to create buried insulating layers in silicon chips. Most recently, ion implantation has been teamed with various deposition techniques to produce surface treatments known as ion-assisted coatings (IAC), which combine the best of two useful technologies. It is now quite possible to produce coatings that can be tailored in such a way that their thickness can be chosen to be anywhere from 0.01 micron (0.4 microinch) to several microns, be strongly adherent to nearly any type of substrate, and have a coating composition itself selected from a wide range of materials. No longer does the substrate material need to be chosen only from a limited group of materials (e.g., to nitride a steel one must choose from a small group of nitridable steels); nearly any coating can be grown on nearly any surface.

## 3   ION IMPLANTATION

### 3.1   Basics of Ion Implantation

Ion implantation processes are generally done in a vacuum, at approximately $10^{-6}$ torr, using an ion implanter. In its simplest form the implanter contains an ion source, in which the ions are created from a plasma, and a chamber containing the items to be treated. The ions are fired as a beam from the source to the object in the chamber through a beam line held under high vacuum. Biasing the source at a voltage from a few thousand electron volts up to 200 keV (and in some cases up to 4 MeV) allows the ions to be accelerated to high speeds and aimed into the chamber (see Fig. 1). Since the ions enter the chamber as a beam the process is line-of-sight. Because many ions are produced from compounds, if a single ion species is to be used the beam must be filtered by bending it through a magnetic field. This is usually essential for all but gas ion implants.

When an ion enters the surface, it collides with the atoms of the solid, forcing its way in, blazing a trail ahead, behind, and to the side of its path as it knocks atoms away (see Fig. 2). These atoms in turn collide with other atoms, and those with yet others, creating, for a time of about $10^{-11}$ s, a region of the material containing hundreds of interstitials and vacancies. This is known as a collision cascade and is often visualized as a very localized spike of thermal energy, although it cannot truly be thought of as a thermal process. An ion with 100 keV of energy will generally

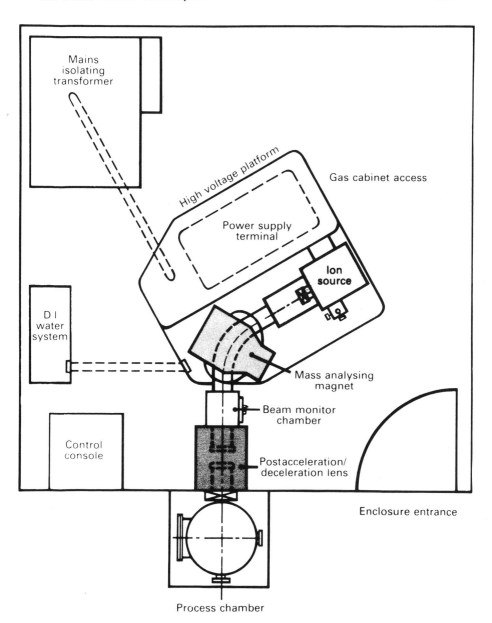

FIGURE 1  Schematic of commercially available mass-analyzed ion implanter for treatment of materials. (Courtesy of Whickham Ion Beam Systems, Ltd., UK)

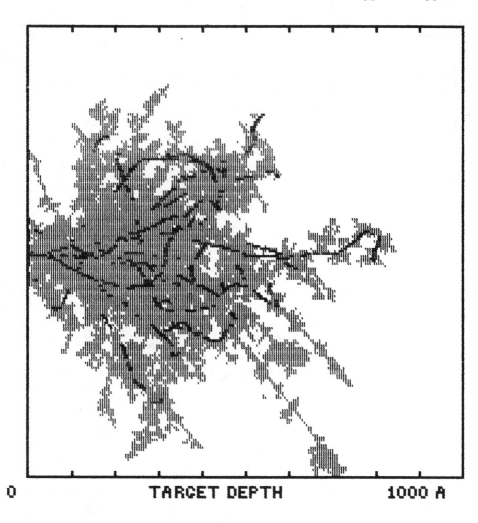

FIGURE 2   TRIM calculation of collision cascades (light shading)
and ion tracks (dark lines) of 100-keV Ti$^{+}$ ions implanted in iron.
Long shaded "fingers" are caused by fast recoils. (Program courtesy
of IBM Corporation)

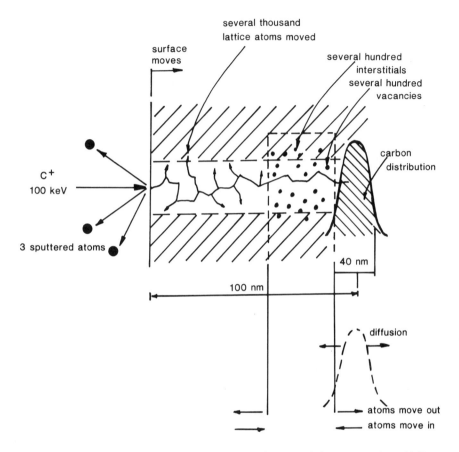

FIGURE 3  Processes occurring in a typical collision cascade. (After Colligon, 1986)

penetrate from a few hundred to a thousand atomic layers (depending on its mass and the substrate material) before its energy is exhausted and it comes to rest (Fig. 3). As the material settles back to equilibrium, most of the atoms return to normal lattice sites, leaving some vacancies and interstitials "frozen in." This process creates a layer rich in the implanted material buried beneath the surface, associated with a damage layer. The shapes of both the ion and damage distributions are roughly gaussian (see Fig. 4). Ion implantation for surface modification is usually done at energies ranging from 35 to 200 keV, so this implanted layer lies at a depth between 0.01 and 0.5 micron (0.4 to 20 microinches).

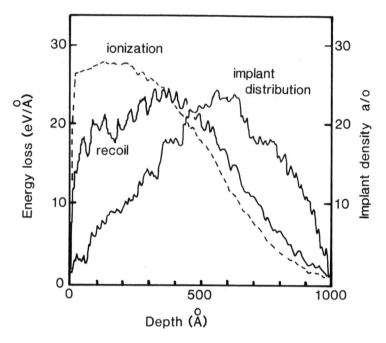

FIGURE 4  Ion distribution and energy lost to recoils and ionization
for 100-keV $N^+$ in iron. Recoils give rise to vacancies and interstitials

The entire stopping process takes only $10^{-11}$ s, and the displaced atoms come to rest in a similar time, so the whole process
is like a very fast heat and quench taking place in a cylinder of
material perhaps 0.1 micron long and 0.02 micron in diameter. This
is a great deal more localized and 100 times faster than the shortest
readily available laser pulse. The effects of this rapid heat-quench
cycle together with the new atoms lodged in the original material
give the ion implantation process some of its unique properties.

The depth of penetration of the ions is a function of their energy
and mass and the mass of the substrate atoms: the higher the energy,
the deeper the ions penetrate. In general, the lighter the ion or
the lighter the substrate the greater the depth of penetration,
with ions penetrating deeper if they have more energy (see Fig. 5).

The ions themselves, once in the surface, are neutralized and
become an integral part of the material, so the implanted layer
cannot flake or peel off as can a standard coating. Implanted ions
can combine with the atoms of the solid, or with each other, or
even with residual gases in the vacuum chamber to produce conven-

tional alloys or compounds. Researchers at the Naval Research Lab (Singer and Barlak, 1983) have found that Ti ion beams will react with hydrocarbons present in the treatment chamber, cracking them to form Ti-C compounds on the surface of implanted steels.

Since high-energy ion beams provide the driving force behind reactions during implantation, it is also possible to form metastable or "nonequilibrium" compounds in implanted materials that cannot be formed by normal, thermodynamic means. This makes it possible to put in far more of an element than one could normally dissolve thermally. Thus, some of the properties of implanted materials are what would be expected from experience, while others are far less familiar.

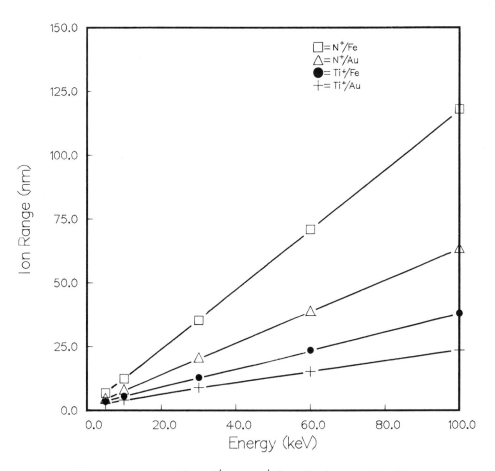

FIGURE 5   Ion ranges for Ti$^+$ and N$^+$ ions in iron and gold.

The general advantages of ion beam processes for surface modification, many of which are also applicable to semiconductors, can
be summarized as follows (see, for example, Townsend, 1986):

Implanted species need not be in thermodynamic equilibrium.
Low processing temperatures.
Controllable addition of different ions for desired surface character-
    istics.
No measurable change in surface dimensions after implantation,
    but surface finish may be improved.
Implanted material is buried beneath surface and cannot come off.
High-vacuum cleanliness, low toxicity of process and materials used.
Use of multienergy implants can offer tailored depth profile.
Implantation through existing layers or surface barrier oxide.
Bulk properties of treated material unaffected.
Since only the surface region is implanted, expensive or strategic
    materials are conserved.

## 3.2   Ion Implantation Equipment and Methods

Ion implantation processes are done in a *vacuum* of approximately
$10^{-6}$ torr by using an *ion implanter*. In its simplest form an implanter
consists of an *ion source* (Fig. 6), in which the ions are created,
and a chamber containing the items to be treated. The ion source
usually creates positive ions by stripping an electron off the atoms
of its feed material. Applying a positive voltage to the source fires
these ions out into the processing chamber as a beam. If the ions
are derived from a simple gas, such as nitrogen, the gas is ionized
in the source and the ions can be simply fired directly into the
part to be treated. Some ions (such as boron) are derived from
a compound ($BF_3$), while metal ions are produced by evaporating
the metal or one of its compounds (usually a chloride or fluoride)
into the ion source. Since all the atoms put into the source will
become ionized, ions produced in this way usually have to be filtered
(or *mass analyzed*) to remove the unwanted ions and allow only the
desired species to reach the sample. This is usually done by passing
the beam through a magnetic field and an aperture to separate the
different ions by mass. If this is not done, the sample will be
implanted with all the ions produced in the source. For example,
if it was to be implanted with boron derived from $BF_3$ gas, the
implant without mass separation would actually be $B^+$, $BF^+$, $BF_2$,
$BF_3^+$, $F_2^+$, $F^+$ and small amounts of any other ions produced in
the source, including water vapor, hydrogen, and carbon monoxide.
On the other hand, implantation with a simple gas, such as nitrogen,
can generally be done without mass analysis, since nitrogen gas in
the source produces almost exclusively $N_2^+$ and $N^+$ ions.

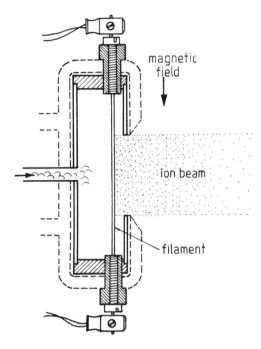

FIGURE 6  Schematic of Freeman ion source, typically used in ion implanters. (From Byers, 1985)

The depth of penetration of the ions is determined by their *energy* (i.e., by the accelerating voltage applied to the source). This energy is usually measured in kiloelectron volts (keV or kV). For most implanters it is in the range of 30,000 to 200,000 V (30 to 200 keV), although some machines can go as high as 4,000,000 V (4 MeV). The amount of material implanted into the sample is measured by the number of ions which hit each square centimeter of its surface, and so is measured in ions per square centimeter (ion/$cm^2$ or ions $cm^{-2}$). This is known as the *ion dose* or *ion fluence*. For most applications outside the semiconductor industry this dose is usually between $10^{15}$ and $10^{18}$ ion/$cm^2$. Since a surface contains about $10^{15}$ atoms $cm^{-2}$, these doses represent 1 to 1,000 monolayers worth of material. The rate at which implantation occurs is determined by the *beam current*. This may be anything from 1 μA to 100 mA (6 × $10^{12}$ to 6 × $10^{17}$ ions $s^{-1}$). However, since this current must usually be spread over the whole surface of the item being implanted (by defocusing the beam, scanning the beam, or scanning the item itself), the rate at which implantation occurs is determined by the *beam current density*, measured in mA cm$^{-2}$. In general, very high voltage machines are capable of only rather low currents (up to a few hundred microamps), while at very low voltages currents of many tenths of an amp are possible. Beams of simple gas ions, such as nitrogen and argon, are relatively easy and inexpensive to produce, and their beam currents are high. Metal ions, on the other hand, are more difficult to produce and give lower beam currents, except in specialized machines.

Because the ions arrive at the object as a beam, the process is line-of-sight. Only those parts of the surface which the beam can hit directly can be implanted, necessitating sample manipulation to treat complex shapes. Furthermore, in order for the implant to be effective the beam must strike the surface well away from grazing incidence. If the beam strikes the surface at right angles to it (90° or normal incidence), most of the ions will lodge beneath the surface, but as the beam arrives at a shallower angle the ions come to rest closer to the surface and the surface itself begins to erode by sputtering, removing most of the implanted ions. Therefore ion implantation is best done 45-90° from the surface, while for some ions the angle must be at least 60°. Consequently, the interiors of deep holes cannot be reached. As a general rule, a hole can only be effectively implanted over a depth about equal to its diameter.

Currently there are three main types of ion implanters:

1.  *Mass analyzed implanters*, basically similar to those used in the semiconductor industry, which can implant any species.
2.  *Nitrogen implanters*, which can produce beams only of gases (almost exclusively nitrogen) primarily used for implantation of tools.
3.  *Plasma source ion implanters*, which at the time of writing are only in the development stage. These machines generate the ion beam from a plasma within the implantation chamber.

We shall describe and evaluate each type of machine.

## Mass Analyzed Implanters

The mass analyzed machine is the most common type of implanter (Byers, 1985; see Fig. 1) and is essentially the same as that used in the semiconductor industry. A narrow beam of ions is used, and either the beam is scanned across the samples or the samples are mechanically scanned in front of the beam. Since it can produce pure beams of any ion species it is the most useful for R&D activities, and is currently the only type of machine capable of producing beams of metal ions or other nongaseous materials. Its advantages are as follows:

1. Any ion species can be generated.
2. Pure, single-energy ion beams are produced, which is especially good for well-defined research and development purposes.
3. The process parameters, such as beam current, energy, and uniformity, are very well defined.
4. Because the implantation chamber is separated from the ion source, the pressure in the chamber is low, thus limiting contamination.
5. Ion beam energy can be very high (up to several million volts for some specialty machines) or very low (down to a few kilovolts), although most machines operate from 30 to 200 keV.

Its disadvantages are

1. Beam currents are generally limited (a few tens or hundreds of microamps for research machines, and up to 10 mA for production machines).
2. The machines are expensive and complex and so generally must be operated and maintained by specialists.
3. Sample manipulation is required to process complex shapes. Either beam or sample scanning (or both) must be used to implant the entire sample surface.

*Nitrogen Ion Implanters*

This type of machine does not employ any mass analysis, so all the ions produced in the ion source are fired into the sample. For this reason it is almost exclusively used for nitrogen implantation. Machines of this design can be very small and simple (see Fig. 7) or very large, to accommodate large and heavy components (see Fig. 8). Because this type of implanter is intended for use in metallurgical processing, it is generally simpler than the mass analyzed type of machine. For this reason sample or beam scanning, while sometimes used, can often be discarded in favor of simply using a broad ion beam of reasonable uniformity. Because of its simplicity this type of machine can be operated and maintained by less specialized personnel than the mass analyzed machine requires. The advantages of this type of machine are

1. Simplicity of operation and maintenance.
2. Large beam currents (5-10 mA for most machines, and 0.1-1 A for some high-current models). Because the beam is not mass separated and there is little loss of beam during transport to the sample, the arrangement is efficient.
3. This type of system can be made very large.

FIGURE 7  Prototype ion implanter for treatment of small engineering components. (From Hartley, 1979)

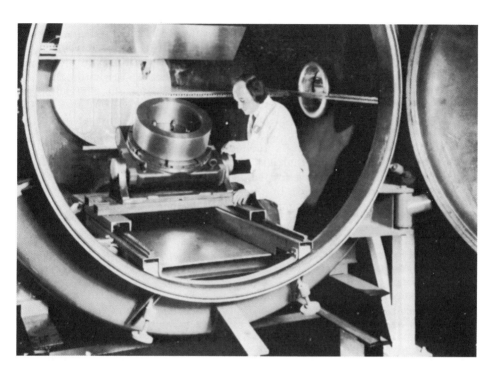

FIGURE 8 Large-scale ion implanter for gaseous ion implantation of industrial components up to 1 m$^2$ in size at AERE Harwell, UK (European Plastic News, 1982).

Its disadvantages are

1. Beam uniformity is often relatively poor (although usually sufficient for tool processing).
2. Because the beam is a mixture of atomic and molecular ions ($N^+$ and $N_2^+$) and the relative proportions of the constituents are not always stable or well known, the ion beam energy (and thus penetration depth) and dose are not well defined.
3. Because the ion source is generally close to the chamber, the chamber pressure is generally high during processing, which can lead to oxidation of the treated surface.
4. Sample manipulation is required to process complex shapes.

*Plasma Source Ion Implantation (PSII)*

At present this type of machine is in the early development stage (Conrad et al., 1988). Rather than generating the ion beam in an ion source and then firing it at the samples in a separate chamber, the PSII arrangement surrounds the samples by the ion source. This is done by generating a plasma (a mixture of ions and electrons) within the implantation chamber so that the plasma surrounds the items to be implanted, thus eliminating the line-of-sight restriction (see Fig. 9). A train of short, high-voltage, negative pulses is then applied to the samples. On each pulse, ions are sucked out of the plasma directly into the sample surface from all directions simultaneously. The pulse is turned off after a few microseconds or milliseconds so that the plasma does not lose all of its ions, and is turned on again a few milliseconds later after the plasma has become reestablished. In this way the implant dose is built up by a series of very short pulsed ion implants. The beam current in each pulse can be very high (many amperes), but the dose rate is determined by the average current. This is limited by sample

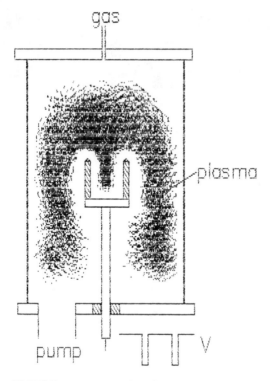

FIGURE 9  Schematic of PSII type treatment chamber. Note that ion source is the plasma within the chamber itself.

heating considerations and by the capacities of the power supplies. Although the method is new and not yet fully tested, it does appear to have many advantages which might gain it wide currency, especially for the implantation of tools. The advantages of PSII are

1. Simplicity and low cost. There is no ion beam to generate or manipulate and only one vacuum system to run.
2. Sample rotation and scanning are usually unnecessary since the beam comes from all directions at once.
3. Normal incidence implantation. The ions travel along the electric field lines to the sample surface and so, in general, impinge on it at right angles. This eliminates problems of sputtering at low beam incidence angles.
4. Very high beam currents spread over the entire surface, thus eliminating local beam heating caused by the scanning of intense ion beams.

The disadvantages of PSII are as follows:

1. Any plasma nonuniformities will lead to uneven implantation. For example, extra plasma intensity (and hence higher implant dose) at sharp points will lead to variations in local implant dose.
2. Only gaseous ions can be implanted, such as N and Ar.
3. Ion energy is likely to be limited to the region of 100 keV.
4. All ion species present in the chamber are implanted, so dose and energy are not well defined.
5. Any gas released from the samples or the chamber walls and fixtures will also be ionized and implanted.
6. The effect of current pulses is not yet documented. High-current pulses may produce local annealing and quenching effects, which may be either advantageous or deleterious, depending upon the material.

*Other Ion Implanters*

The types of equipment described above are those most generally in use at the time of this writing. However, the field is changing quite rapidly, and it is probable that new types of equipment will become available. One promising new development, which is currently in the R&D stage, is the metal vapor, vacuum arc ion source. This source, generally referred to as the MEVVA source, generates an ion beam by creating electric arcs (or sparks) on the surface of an electrode composed of the implant species (Brown et al., 1986). The ions are generated as a series of pulses rather than as a continuous beam. Therefore, the same considerations concerning annealing and quenching apply to the MEVVA source as for the PSII source. In contrast to the PSII source, the MEVVA source can only produce beams of solid ions species, such as Ti, C, U, Pt, etc. Because of the way in which the ion beam is generated, it does not require any supporting gas and thus the ions can be implanted directly without the need for mass analysis.

### 3.3 Factors to Consider in Choosing Ion Implantation

A decision on whether ion implantation would be a reasonable approach to a problem should take into account the following:

*Size*—At the time of this writing most implanters are limited to objects less than 2 to 3 ft long. (The current exception is the very large chamber implanter at UKAERE Harwell, England, as shown in Fig. 8.)

*Line-of-sight access*—The insides of deep holes or tubes or reentrant geometries cannot be implanted. (Very large holes and tubes could in principle be implanted by specialized implantation sources.)

*Vacuum compatibility*—It must be possible to put the object in a vacuum. (Porous oil-impregnated bearings, for example, would be a problem.)

*Surface condition*—Since implantation modifies the surface, only finished components or tools should be treated. No further finishing treatment is required.

*Cost*—Costs are difficult to gauge since the technology is new, but cost is comparable with high-quality physical vapor deposition (PVD) coatings. Cost goes up roughly linearly with the area to be treated, the dose needed, and the beam current available. Simple nitrogen implants on, say, 3/8-in taps and 1-in.-diameter wire drawing dies are likely to be about $10 per item. Large items with complex shapes can cost several thousand dollars. Costs are generally higher for metal ion beams than for gases. Prices are likely to drop as larger machines become available.

The circumstances in which ion implantation would probably provide a distinct advantage over other surface modification processes would include

*Low temperatures*: when the object may be damaged or warped by temperatures greater than 200°C (e.g., high-precision tools, plastics).

*Integrity*: when severe problems would result if a coating were to flake off (e.g., biomedical applications, bearings).

*Tolerance and finish*: when no degradation can be permitted in tolerance or surface finish (e.g., high-precision bearings where surface finish can actually be improved).

*Partial coverage*: when only part of the object should be processed (this is one reason for its use in the semiconductor industry where spatial tolerances of a micron are required).

*Repeatability*: when precise reproducibility of conditions from run to run is critical.

*Refurbishing*: unlike many hard coatings, it is not necessary to grind off the implanted layer when refurbishing a tool.

*Wear rate*: unlike many coated tools and parts, which may fail catastrophically when the coating wears off, implanted components usually degrade slowly.

Situations where ion implantation is probably not a good candidate include

*Cutting tools*: currently, simple nitrogen implantation is generally ineffective in cutting tools because the cutting edge becomes too hot. Other implant species may be effective in this application, but none have been sufficiently proven.

*Deep holes and reentrant geometries*: line-of-sight is needed.

*Inexpensive tools and parts* (such as simple drills and cutters): except where downtime and scrap generation are costly, or the part is difficult or expensive to replace.

## 3.4 Characteristics of Implanted Materials

As the ions penetrate the surface of the material, they produce localized damage regions. For semiconductor materials the damage so produced must be removed for the device to function effectively. In other materials, however, this "damage" is an asset, producing effects not obtainable otherwise.

In metals, ion beams produce extensive dislocation networks. In steels, the dislocations caused by the ion beam damage harden the surface, rather like a miniature shot-peening treatment. In softer materials, such as copper, they can actually make the material softer and its deformation more homogeneous and reversible. These opposite effects lead to improved fatigue behavior in both nitrogen-implanted steels and aluminum-implanted copper (Fig. 10, Kujore et al., 1981).

As ion dose increases, more ions are loaded into the near-surface region, and more and more compressive stresses are created. In metals this improves fatigue and wear resistance properties. In ceramics the surface becomes tougher and less prone to cracking

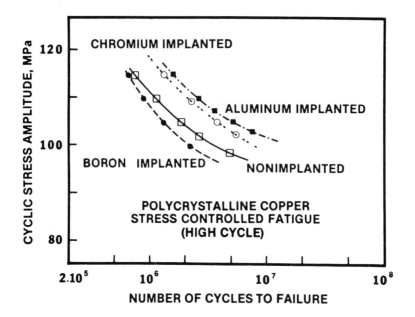

FIGURE 10   Effect of ion implantation on cyclic stress-life relationship of polycrystalline copper (Kujore et al., 1981).

(Roberts and Page, 1982). The effect can even be enhanced by
choosing larger ions for implantation. In some cases small ions,
such as B, can actually lead to the creation of tensile stress, which
worsens fatigue properties (Kujore et al., 1981). For some materials,
such as steels, there are optimum dose levels beyond which little
or no improvement in mechanical properties occur. The optimal
nitrogen dose to reduce the volumetric wear rate in a nitriding
steel occurs at about $2 \times 10^{17}$ ions $cm^{-2}$ (see Fig. 11). There is
some evidence that ion dose is different for different metals; hence
in the literature one will note that most doses are given as a range.
Nitrogen implants for most ferrous and nonferrous metal tooling
range from 2 to $8 \times 10^{17}$ ions $cm^{-2}$.

If enough ions are implanted at sufficiently low temperatures,
the crystallinity of the surface can be eliminated and made amorphous.
In metals this removes grain boundaries and reduces the variations
in surface chemistry which can lead to galvanic corrosion and pitting
(Ashworth et al., 1980; Clayton, 1981). In some ceramics the use
of high ion doses can create a softer surface which deforms plastically

under impact or scratching, as a metal does, rather than cracking, as shown in Fig. 12 (Roberts and Page, 1982; see also McHargue, 1987). In other ceramics high ion doses produce so much stress that the surface can become separated from the underlying material or even spall off completely (Fig. 13; Legg et al., 1985).

The changes we observe in surface properties are caused by the combination of these effects (see, for example, the review by Dearnaley, 1986). For example, when Ti is implanted into 52100 bearing steel or tool steel, it can combine with carbon from cracked pump oils in the vacuum chamber to create an amorphous Ti-Fe-C surface resistant to wear (Smidt and Sartwell, 1985). The surface properties therefore result from direct combination of surface atoms with the ion beam, cracking of gas molecules onto the surface by the ion beam, and diffusion of these materials into the surface by ion-enhanced diffusion. At the same time the surface is made amorphous and put under compressive stress. Greater control of this process is possible through deliberate introduction of methane during implantation (Sartwell, 1987). Other methods of achieving the same results include sequential implantation of Ti and C for bearing applications (Sioshansi, 1987).

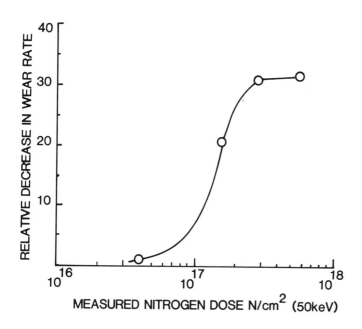

FIGURE 11  Wear rate as a function of $N^+$ dose for an implanted nitriding steel disk wearing against an unimplanted stainless steel pin. (After Dearnaley and Hartley, 1978.)

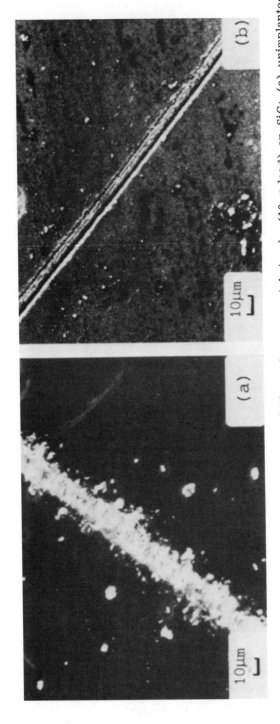

FIGURE 12 Scanning electron micrographs of diamond cone scratch tracks (10-g load) on SiC: (a) unimplanted and (b) implanted to $6 \times 10^{17}$ $N_2^+$ $cm^{-2}$. (From Roberts and Page, 1982)

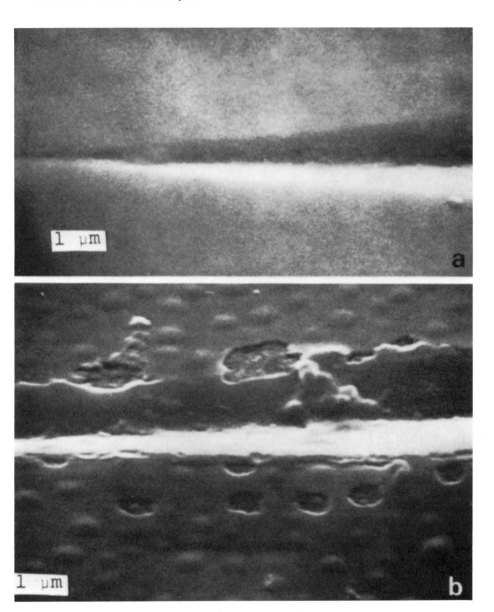

FIGURE 13   Knoop microhardness indentation of (a) unimplanted yttria stabilized zirconia (YSZ) and (b) as-implanted YSZ ($6 \times 10^{16}$ Al + $cm^{-2}$) showing blistered and spalled implanted region. (From Legg et al., 1985)

The ion beam thus creates changes in the surface region, introduces foreign atoms to form compounds or alloys (such as nitride
precipitates), produces compressive stresses and dislocations, and
pins existing dislocations. In these ways the most common implantation treatment, implantation of nitrogen ions, can harden the surface
of many metals by well-known hardening mechanisms. This means
that, while ion implantation can be used to harden many materials,
if all the hardening mechanisms have already been used (as in fully
hardened 52100 bearing steel), no further benefit will be derived
from nitrogen ion implantation. In these circumstances further improvements can only be achieved by implanting other ions. For
example, implantation of both yttrium and nitrogen together creates
yttrium-nitrogen substitutional/interstitial pairs which are highly
effective in pinning dislocations (Dearnaley et al., 1985; Dearnaley
et al., 1986).

In bearings, and perhaps in other tribological applications,
the ion beam can also greatly improve the surface properties by
ion beam smoothing [i.e., by sputtering away microscopic asperities
(Dearnaley, 1987)]. This not only immediately reduces friction but
also reduces wear by removing a primary source of particles which
tend to oxidize and scratch the bearing surfaces. Ion beam smoothing
is also observed in implanted ceramics. Ion implantation of aluminum
coated with a thin film of plasma-assisted hydroxylapatite (Fig. 14)
smoothed the surface out when compared with unimplanted film
(Stevenson et al., 1989).

In Co-cemented WC (tungsten carbide) tools the mechanisms
of wear improvement are uncertain, but probably include hardening
of the WC particles by compressive stress and preventing adhesion
of the Co to the workpiece and its subsequent loss by dissolution
in the material being machined (Dearnaley and Hartley, 1978a).

### 3.5  Laboratory Test Results

*Corrosion*

An important application of ion implantation technology is that of
providing improved corrosion protection for many different materials
in a variety of aqueous corrosion and oxidative environments. This
is often done by using implants to deliberately create stable oxides
on the surface that are resistant to an oxidizing environment at
high temperatures. These types of conditions are found in such
applications as aircraft engines and nuclear reactors. In one experiment, researchers implanted a Fe-23Cr-1.45Al-0.1Y alloy with aluminum
ions and then oxidized the samples in dry $O_2$ at 1100°C for periods
up to 24 h (Bernabai et al., 1980). As Fig. 15 shows, the aluminum-
implanted samples showed a very high degree of oxidation resistance.

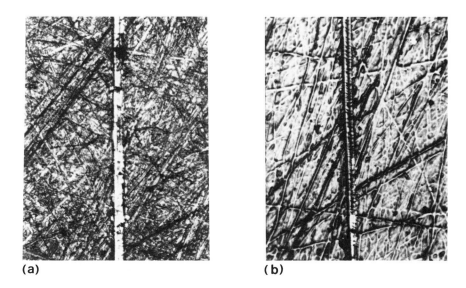

**(a)**          **(b)**

FIGURE 14 Optical micrograph of plasma-enhanced coating of hydroxylapatite on mechanically rough Al substrate: (a) unimplanted and (b) ion implanted. Magnification 50×. (From Stevenson et al., 1989)

However, the high potential of ion implantation techniques to improve corrosion properties of surfaces resides in the ability of this technique to form surface alloys of otherwise immiscible elements. The importance of this is evident when one considers that multiphase conventional alloys are susceptible to local galvanic attack at phase boundaries (Clayton, 1981). Combining the capability of ion implantation to produce the metastable alloys with another engineering feature of the process, that is, the ability to render a surface layer amorphous and thus eliminate all grain boundaries, we now have a process that is quite versatile and adaptable to numerous industrial applications, as the selected examples below will demonstrate.

1. In Great Britain it was found that corrosion-resistant coatings on stainless steel, nuclear reactor fuel rods worked well initially but flaked off due to radiation-induced swelling, uncovering fresh surfaces to oxidize further. Yttrium ion implantation prevented spalling of the oxide and so reduced oxidation, thus permitting the plant to be operated at sustained high temperatures (Dearnaley et al., 1986).

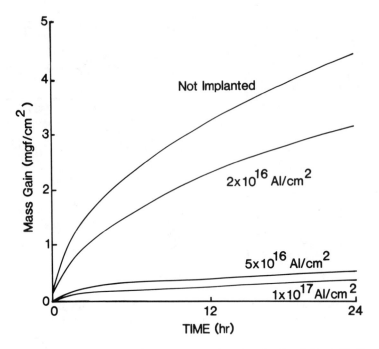

FIGURE 15   Mass gain (degree of oxidation) of Fe-24% Cr-1.45%
Al-0.1%Y in oxygen at 1100°C. (After Bernabai et al., 1980)

2. Pure titanium metal begins corroding in 20% sulfuric acid after
   immersion for between 6 and 10 h; after implantation of only
   3.5% Pt$^+$ ions (at a depth of 100 nm), the onset of active behavior was delayed for approximately 100 days (Munn and Wolf,
   1985).
3. Implantation of superpure Fe with low doses of Pt$^+$ (Kasten
   and Wolf, 1980) or Au$^+$ ions (Ferber et al., 1980) strongly
   increased catalytic activity. Platinum has only limited solubility
   in iron, while gold is insoluble but forms an intermetallic phase.
   The effect of implanting either element was to increase activity
   compared with that for smooth or sputter coated surfaces of
   either Pt or Au (Wolf, 1981).
4. After implantation of 304 stainless steel samples with high doses
   of P, the treated surfaces were found to be an amorphous alloy
   of chromium and iron phosphates which had superior resistance
   to Cl$^-$ ion attack in acidic (0.1 N NaCl) solution (Clayton et al.,
   1982).

5. Burner tips used in a British oil-fired power station were implanted with $Ti^+$ and $B^+$ (Dearnaley, 1980) and lasted for over
   8000 h of operation under conditions where fuel oil and air
   are injected into the power generating plant at 550°C (Fig. 16).
   For conditions such as these the primary problems are corrosion
   and wear of the tip orifice. At the end of 8000 the orifice diameter
   of the tips had increased 30 to 50 microns; untreated tips have
   an operating life of only 3000 h, after which the orifice diameter
   has enlarged by 100 microns.
6. Moderate doses of $Pt^+$ ions were found to protect stainless steel
   from attack by a 20% solution of sulfuric acid when immersed
   for 80 days (Picraux and Peercy, 1985).
7. Ion implantation can also be used to produce "reverse" chemical
   effects. For example, ion implantation has been used to form
   optical waveguides in $LiNbO_3$ by increasing the reactivity of
   the complex oxide in hydrofluoric acid, which enhances its
   etch rate (Colligon, 1986).
8. Titanium, although generally corrosion resistant in most environments, suffers from crevice corrosion. Ion implantation of $Pd^+$
   prevented start of crevice corrosion in Ti samples that were
   boiled in $MgCl_2$ solution during long-time corrosion studies.
   Untreated Ti shows pitting attack in five days; the Pd-
   implanted Ti coupons remained passive for four weeks with
   total absence of pits inside crevices (Ferber and Wolf, 1987).
9. Generally, implantation of B or P ions into iron or steel produces
   amorphous surface layers which are effective in inhibiting anodic
   corrosion in acid solutions (Chen et al., 1983).
10. Implantation of O or P ions has been found to substantially
    decrease permeation of hydrogen in iron foils (see Fig. 17).
    Oxygen implantation produced an "internal oxide layer," while
    phosphorus implantation produced an amorphous layer. Both
    served as barriers to hydrogen migration (Ferber and Wolf,
    1987).

Ion implantation alone is sometimes the only viable technique,
as in the nuclear fuel rod corrosion problem described above. However, for many corrosive environments superior protection can be
found when a physical barrier to corrosion is produced. For this
reason coatings are more frequently used for corrosion protection.
When the coating must also survive adverse tribological conditions,
ion-assisted coatings might well be considered, as described in
Section 4.

*Wear and Fatigue*

Of primary importance for any process that aims to reduce wear
in metals is the effects it may have on the major components of

FIGURE 16  Burner tip for oil-fired power station; ion implantation
reduced erosion of the orifice. (From Dearnaley, 1980)


tribological systems [e.g., adhesion, friction, oxidation, and deformation (Singer, 1984)]. These factors are quite sensitive to surface
chemistry and microstructure and are therefore affected by ion
implantation. For many metals, wear is improved with gaseous implants
alone through at least several different mechanisms. The deformation
mode of wear in metals is affected by such factors as microhardness
of the surface, stacking fault energy, coefficient of friction, and
residual compressive stresses, all of which can be modified to suit
user needs through various ion implantation techniques. Adhesion
and friction are particularly sensitive to surface chemistry: for
many metals the formation of a surface oxide not only eliminates
severe adhesion but reduces friction as well.

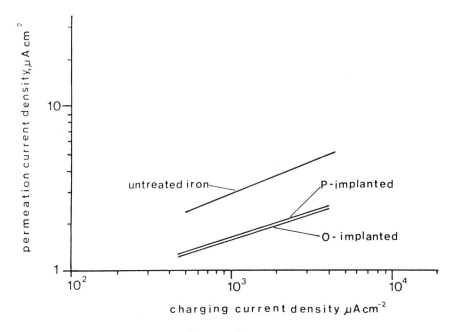

FIGURE 17   Influence of $O_2^+$ and $P^+$ implantation on hydrogen permeation through an iron foil. (From Moine et al., 1987)

Some properties, such as fatigue, which were once thought to be strictly bulk properties are markedly affected by surface finish and treatment. Work by Vardiman and Kant (1982) and Heydari et al. (1982) has shown that ion implantation can significantly improve the fatigue lives of various metals. Vardiman (1986) has also published a review of the principal studies of the effects of ion implantation on fatigue life in pure metals and alloys.

Many tribosystems benefit from the implantation of nitrogen ions alone. Some representative examples of the benefits of nitrogen implantation are given:

Friction and wear in crankshaft bearings (Fig. 18) has been reduced by N implantation, perhaps largely by the sputter removal of sharp asperities, resulting in ion beam smoothing (Dearnaley, 1987).

Ductile fracture during the run-in stages of 304 and 1018 steels can be largely eliminated. This delayed the onset of wear even though the friction was as high as unimplanted material (Singer, 1984).

FIGURE 18   Three crankshafts undergoing nitrogen ion implantation
in 800-mm-diameter beam within large-scale ion implanter shown in
Fig. 8. (From Dearnaley, 1986)

Increases of 20% in microhardness have been observed for implanted cobalt-cemented tungsten carbide (Townsend, 1986).

Wear is reduced by a factor of 10 during standard lubricated pin-on-disk tests on implanted manganese steel; unlubricated pin-on-disk tests yielded improvements of about four to five times (Dienel, Kreissig, and Richter, 1986).

The onset of cavitation erosion is delayed 50% in nitrogen implanted 1018 steel. After aging the implanted material at 100°C, the improvement is increased to a factor of 3 (Hu et al., 1978).

The coefficient of friction is decreased while the hardness is increased for nitrogen implanted Ti-6Al-4V (Singer, 1984).

Industrial NiTi shape memory alloys were implanted with nitrogen ions, which produced a shallow amorphous layer. Friction testing of the implanted alloys showed a significant drop in the coefficient of friction compared with untreated NiTi specimens [Moine et al. (1987); see Fig. 19].

Cyclic fatigue tests of nitrogen implanted low-carbon steel samples showed a 100-fold increase in the number of cycles to failure over unimplanted samples (Free, 1984).

Cyclic fatigue tests of the industrially significant titanium alloy Ti-6Al-4V have been conducted for both carbon and nitrogen implants (Vardiman and Kant, 1982). At low stresses (less than 100 ksi), both types of implants increase the fatigue life by approximately seven to eight times, with a concomitant increase in the endurance limit of 10% for nitrogen implants and 20% for the carbon implants. At higher stresses (100 to 115 ksi) only the carbon implanted samples exhibited improved fatigue life (four to five times), while the effect of the nitrogen implants had largely disappeared.

Nitrogen implanted 1018 steel shows little improvement over unimplanted material until aged for 6 h at 100°C when fatigue life increases by a factor of 3 and the endurance limit by about 15% [Fig. 20 (Hu et al., 1978)].

Although nitrogen ion implantation has become an industry standard for extending the useful industrial life of a wide variety of tooling and components (see Table 1), this particular technique is not suitable for all metals and allows or for all problems. For some materials, different implants may be more effective.

The wear of phosphor bronze-bearing material has been improved significantly by implantation of $B^+$ or $P^+$ and less so when implanted with $N^+$ (Saritas et al., 1982).

Implantation of Stoody 3 alloy (50 Co,31Cr,12.5W) with titanium ions greatly reduced friction and surface damage by forming

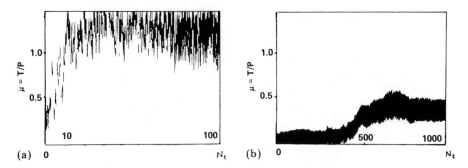

FIGURE 19   Coefficient of friction as a function of the number of
turns under 0.5-N applied load for a NiTi disk: (a) unimplanted
and (b) after $3 \times 10^{17}$ $N^+$ $cm^{-2}$ implantation. (From Moine et al., 1987)

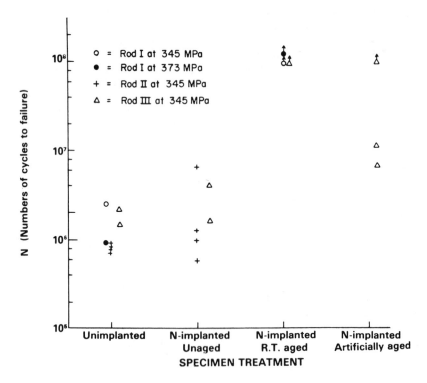

FIGURE 20   Fatigue life of four different treatments for AISI 1018
steel specimens. (From Hu et al., 1978)

TABLE 1   Performance of Ion Implanted Tools and Parts

| Item | Ion | Life | Comment |
|------|-----|------|---------|
| Metal-forming tools | | | |
| WC wire drawing dies | N,C,CO | 3–5× | |
| WC deep drawing dies | N | 2× | |
| WC swaging dies | N | 2× | |
| Stellite 4-wire drawing dies for copper | C | 5× | |
| WC punches and dies | N | 5× | |
| Cr-C, HSS, and Cr-plated HSS punches | N | 4× | Overall life can be up to 100× by implanting around end and re-grinding several times |
| HSS punches for electric motor laminations | N | 5× | |
| Drills for brass-covered steel strips | N | 3× | |
| Ring steel press tool | N | 10× | |
| 12Cr-2C steel forming tools | N | 2.5× | |
| M2 thread cutting dies | N | 4× | |
| HSS gear cutting tools used at low temperature | N | 2× | |
| H-13 steel mill rolls | N | 4× | |
| HSS progression tools | N | 6× | |
| Tools for plastics production | | | |
| Injection molds, sprue bushes, feed wear pads | N | up to 20% | Generally reduces friction and corrosion |
| Tool taps for thermo-setting resins | N | 5× | |
| M2 taps for phenolics | N | 10× | |
| Tool steel injection molding screws | N | 10× | |
| WC printed circuit board drills | N | 2× | Cleaner hole, less adhesion to drill, cooler running |
| General mold tools used with abrasive fillers | N | 10× | |
| Tool steel injection molding nozzles | N | 2–5× | |
| Aluminum dies and tools for injection molding | N | 3× | Aluminum prototype molds have been used in production |

(continued)

Table 1 (continued)

| Item | Ion | Life | Comment |
|------|-----|------|---------|
| [Tools for plastics production] | | | |
| Chrome-plated brass extrusion tools | N | 3× | |
| WC slitters for rubber | N | 2× | |
| Other items | | | |
| Cr-C steel paper slitters | N | 2× | |
| Cr-plated steel acetate punches | N | 3-10× | |
| Steel bread knives | N | 6× | |
| Burner tips for oil-fired power stations | Ti, B | 2-4× | |
| Nuclear fuel rod casings (stainless steel) | Y | | Prevents corrosion, does not flake off under irradiation |
| Permalloy recording heads | B | 1.5× | Improved abrasion resistance |
| 52100 steel bearings | N | 2× | |
| 440C stainless steel bearings | Ti+C | | Reduces rolling contact fatigue |
| Ti alloy turbine blades for jet engines | Pt | 100× | Fatigue life increase; wear and abrasion not measured |
| Ti alloy prostheses (hip and knee joints) | N | 1000× | Prevents corrosive wear; greatly reduces wear of polyethylene cup |
| Sn-coated steel hip joints | N | 1000× | |
| Co-Cr orthopaedic prostheses | N | | Reduced corrosion and ion release |

an abrasion-resistant layer at the surface (Dillich, Bolster, and Singer, 1984).

Aluminum ion implantation into copper forms a low-stacking-fault-energy surface layer leading to greatly improved fatigue life for both strain- and stress-controlled tests (Kujore et al., 1981). In strain-controlled fatigue studies of polycrystalline copper, ion implantation of $Al^+$ was found to increase fatigue life by 50 to 70% due to slip homogenization (Fig. 21).

Platinum ion implantation of titanium alloy turbine blades, tested under simulated high-temperature engine operating conditions showed fatigue life increased by a factor of more than 100 (Eckler, 1979).

Dual implants of Ti and C ions into 440 C stainless steel resulted in a 40% reduction in coefficient of friction and an 80% reduction in wear rate, even at hertzian stresses much larger than the bulk yield strength (Pope et al., 1984).

Implants of both Fe and Ce into 2124 Al alloy samples with subsequent anneals produced stable Al-Fe and Al-Fe-Ce particles which imparted both precipitate and subgrain boundary strengthening to the alloy with a resultant decrease in creep strain of about 30% (Jata and Hubler, 1985).

Detailed, careful, instrumented lathe tests have been conducted on Ti implanted M2 high-speed steel inserts, used to machine unlubricated annealed 4140 steel, and compared with nonimplanted inserts run under the same conditions. Test results showed a reduction in both the cutting force and flank wear rate for the ion implanted inserts. The decrease in cutting force was measured by a reduction in power consumption (Smidt, Hirvonen, and Ramalingam, 1983).

The fracture toughness of ceramics such as zirconia has been improved 70% by using a combination of aluminum ion implantation and heat treatments to produce alumina-rich, nucleation-controlled microstructures (see Fig. 22), while Ni implantation into alumina also increases the fracture toughness (Hioki et al., 1986).

The effect of ion implantation on the microhardness of ceramics has been studied by several groups (see, for example, McHargue et al., 1986; Burnett and Page, 1984; and Cochran et al., 1984). Microhardness was found to be a function of both implant fluence (or dose) and implant temperature, illustrated in Fig. 23 for implantation of $Cr^+$ ions into alumina (McHargue et al., 1986). The same study also showed that the hardness of a material such as titanium diboride, already quite hard, can be increased by 70 to 110% through implantation of high-energy $Ni^+$ ions.

*Polymers*

Since ion implantation is a low-temperature process, it can be used to modify properties of materials such as polymers. Weber et al. (1982) demonstrated that implantation of $F^+$ stabilized the electrically conducting polymer polyacetylene against oxidation. Structure and electrical conductivity of implanted films remained unchanged even after one year in air, whereas unimplanted films, in that time, oxidized and became brittle. Ion implantation of poly(p-phenylene

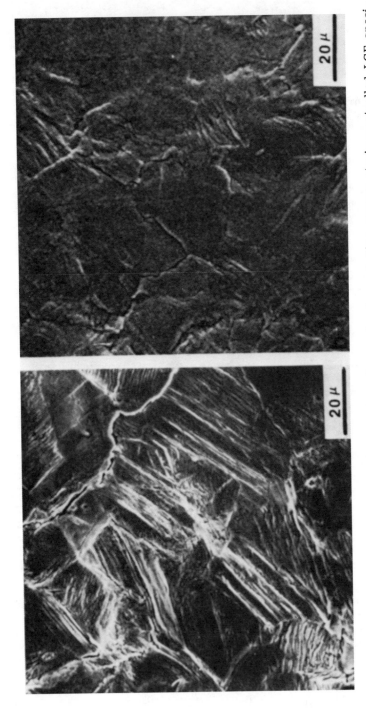

FIGURE 21  Scanning electron micrographs of surfaces of polycrystalline copper strain-controlled LCF specimens: (a) unimplanted and (b) aluminum implanted (Kujore et al., 1981).

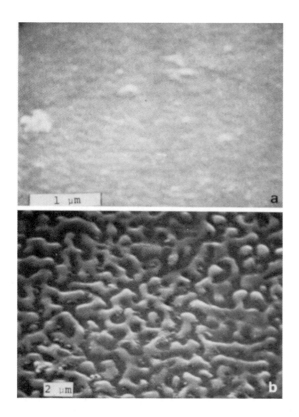

FIGURE 22   $ZrO_2$ implanted with $4 \times 10^{17}$ ion $cm^{-2}$ $Al^+$: (a) annealed 6 h at 1200°C; (b) annealed 6 h at 1150°C and 6 h at 1400°C, showing nucleation-controlled microstructure of interlocking ribbons of $ZrO_2$ and $Al_2O_3$. (From Legg et al., 1985)

sulfide) polymer increased electrical conductivity by approximately 14 orders of magnitude (Abel et al., 1982). Implantation of a heavy ion such as $Ar^+$ at high energies (2 MeV) will induce property changes in polymers, such as increased chain scission (or cross-linkage) or production of very hard surfaces that cannot be scratched by quartz or garnet (Venkatesan, 1984). Venkatesan (1984) has reviewed the modification of polymer films using ion beams.

*Biomedical Applications*

A fairly recent but exciting application of ion implantation technology is for biomedical prostheses (bone pins, knee and hip joints). Since

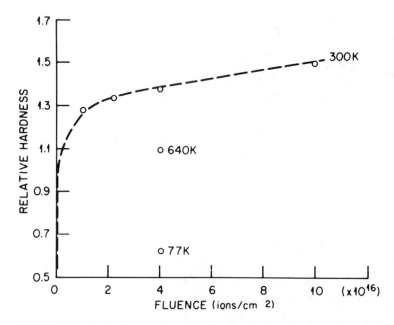

FIGURE 23  Relative hardness (implanted/unimplanted) of $Al_2O_3$ implanted with Cr at different temperatures. (From McHargue, 1986)

the testing of a new process for surgical implants in humans is of necessity more complex and time-consuming than for nonbiomedical applications, there is not yet any published data on actual in vivo tests. Most of the data in the literature describes research on Ti-6Al-4V coupons in order to understand the general effects of ion implantation on this surgical alloy (see for example, Williams et al., 1984). All results to date are from laboratory testing which, even so, indicates some very encouraging improvements that should ensure that the lifetime of the prostheses greatly exceeds that of the patient. Some of these test results follow:

1.  A factor of 1000 improvement in nitrogen implanted tin-coated steel hip joints has been reported (Dearnaley, 1985a,b), as well as 300-fold improvement in Ti-6Al-4V alloy hip joints (Fig. 24, Dearnaley, 1986).
2.  Ion implantation of Co-Cr alloy orthopaedic implants has significantly reduced corrosion attack and thus decreased ion release of toxic elements (Co, Cr, Ni) (Sioshansi, 1987).
3.  Implantation of the Ti-6Al-4V prosthesis reduces wear of the metal and of the high-density polyethylene cup in which it moves (Mathews, Greer, and Armstrong, 1986).

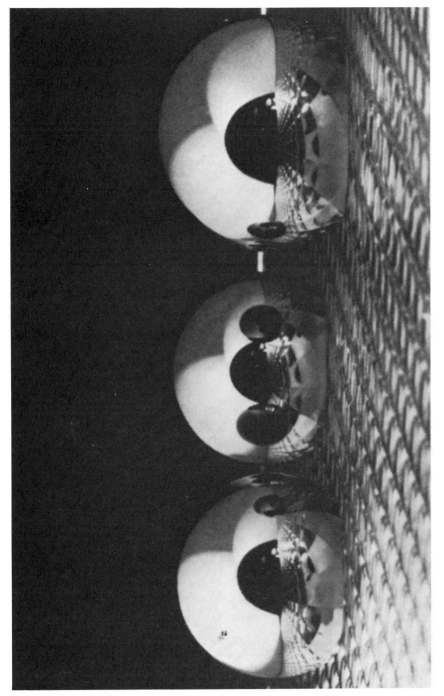

FIGURE 24  Three nitrogen ion implanted and ion beam-smoothed Ti-6Al-4V alloy hip joint balls.  (From Dearnaley, 1986)

4.  Nitrogen implantation of 316LVM surgical-grade stainless steel improved fatigue life by nearly two orders of magnitude, while B and Ta implantation significantly reduced pitting corrosion (Higham, 1986).

### 3.6 Industrial Tool Laboratory Tests

Extensive tests of ion implanted industrial tooling and parts have been carried out since the mid-1970s, when the pioneering group at AERE Harwell (Great Britain) approached industry with the new technology. This group has worked with over 300 British firms in conducting careful on-site tests on both implanted and nonimplanted tooling. Tools and components treated included case carburized steel press tools (Fig. 25), hot rolling mills for nonferrous rods (Fig. 26), tool steel gear cutters (Fig. 27), and forming tools for

FIGURE 25  Nitrogen ion implanted, case carburized steel forming tool after an increase in production life of 2.5×. (From Dearnaley, 1980)

FIGURE 26  Nitrogen ion implanted hot rolling mills for nonferrous rod after an increase in production life of 5×. (From Dearnaley, 1980)

daisy wheel printers (Fig. 28). By the late 1970s to early 1980s such tests had begun in the United States. There now exists a fairly extensive body of literature documenting the successes of various types of ion implantation methods in a wide range of indus-trial applications. Indeed, there are now several commercial companies in the United States whose primary source of income is nitrogen ion implantation of industrial tooling.

Table 1 is a compilation of the principal types of industrial tool (and components) tests reported to date from industries all over the world. Since wear is such a complex process and tool use conditions in different plants are variable, results may differ significantly from plant to plant and tool to tool. Furthermore, the existence of good laboratory data does not guarantee correspond-ing production use improvements. Nor does poor test performance necessarily imply unsuitability of the process in production. This is because the complexities of the wear process and the variability of tool use conditions makes it very difficult to design any laboratory

FIGURE 27  Nitrogen ion implantation increased the working life of tool steel gear cutter by 200%. (From Dearnaley, 1980)

FIGURE 28 Nitrogen ion implanted forming tool for daisy wheel printers. (From Harwell publication 2192/86)

test for ion implanted surfaces which truly predicts tool performance
in the wide variety of in-service conditions which exist in practice.
Therefore, while laboratory tests are very useful and informative,
especially concerning underlying wear mechanisms, their predictive
power is far from perfect, which should always be borne in mind
when considering the results.

## 4  ION-ASSISTED COATING PROCESSES

### 4.1  Introduction

The search for better materials, with more stringent demands on
their surface properties, has led over the years to many improvements in coating techniques. The applications for such surface
treatments include thin films for isolation and protection of electronic
devices, optical coatings, laser mirror coatings, diffusion and corrosion barriers, and wear-resistant coatings for tools and bearings.
Requirements for purity and controllability have frequently mandated
the use of various vacuum deposition and high-temperature chemical
reaction methods, primarily physical and chemical vapor deposition
(CVD). Many of these techniques now use plasmas to enhance coating
quality and reduce process temperatures. For example, plasma-
assisted coating techniques have evolved that combine plasmas (as
generated, for example, in ion plating or sputter deposition) with
gas-phase reactants to produce a process known as plasma-assisted
chemical vapor deposition (PACVD). The effect of such a combination
is to blend the best of the original coating technologies by using
the energy deposited from the plasma to enhance the properties
of the coating formed by CVD. These techniques are described
in detail elsewhere in this book. Ion-assisted coatings are often
designed to accomplish a similar effect by using an ion beam rather
than a plasma. However, the beam is not necessarily employed during
coating deposition, as a plasma is, but can be used subsequently.
Furthermore, the beam can also be used to mix the coating into
the substrate to improve coating adhesion.

In general, an ion beam is easier to define, control, and use,
and it is more readily adaptable to different substrate materials
and geometries than a plasma is. With a plasma, control and reproducibility can be difficult, since plasma parameters frequently depend
upon factors such as the geometry of the set of parts being treated.
An ion beam also has a well-defined current, ion species, and energy,
which can be anywhere from 100 eV to several hundred keV and
can be directed at will to any part of the object being treated.
A plasma is made up of whatever gas is in the process chamber,
while the particle energies and current densities bombarding the

sample are poorly defined. On the other hand, an ion beam generally carries a much lower current than a plasma does and can require complex manipulation of treated objects to reach all the necessary areas. Therefore ion beam treatment often takes longer and is more expensive than equivalent plasma treatment. For these reasons in our laboratory we use a combination of ion- and plasma-assisted treatments to meet the requirements for reproducibility and efficiency.

Ion-assisted coatings are generally used for high-precision surface treatments and for surfaces which cannot easily be treated by more common techniques. Because of their adaptability, reproducibility, and low temperatures, they also offer a relatively rapid and controllable method of developing new coatings or treating new types of materials.

As the demands on thin films increase and as high precision, in terms of film thickness and homogeneity as well as general process reliability, becomes crucial, the demand for

> . . . thin films of precisely defined composition, stoichiometry, crystalline structure and stability . . . has led to the development of deposition techniques based on the use of ion beams to assist in the deposition process by adding activation energy to the growing film. . . . (Armour, Bailey, and Sharples, 1986)

Because it injects material beneath the surface rather than placing it on the surface, ion implantation is very different from coating and has its own advantages and problems, which have already been detailed. However, the combination of these two technologies, now broadly known as *ion-assisted coating (IAC)*, promises to combine the best of both methods as a singularly effective means of tailoring the surface properties of a very wide variety of materials. Improvements arise partly for the same reasons that plasmas aid in the deposition of coatings (activation of atoms to improve bonding and densifying the coating), but with the advantage that the ion beam is well-controlled and highly efficient in imparting energy to the substrate surface and the film during deposition, without necessarily heating it. The ion species can even be chosen to provide direct chemical modification of the coating and substrate before, during, and after deposition, if required. Yet, because of the low ion doses which can often be used, the costs can be considerably less than those of ion implantation alone.

## 4.2  Mechanisms of IAC

We have outlined the general interaction mechanisms between an ion beam and a solid. All these mechanisms also take part in IAC,

but the ion energies used in IAC can be as low as a few tens or
hundreds of volts for depositing the impinging ions and their energy
very close to the surface, or as high as several hundred kilovolts,
for intermixing the film with the substrate. The precise role of
the ions in aiding formation of the coating is uncertain and depends
upon the specific process used, but it includes several mechanisms.
First, the ion beam sputter-cleans the surface of the substrate,
which promotes better coating adhesion. It is known that ion bombardment during deposition of a coating knocks in and sputters
off atoms in the growing coating, filling the microscopic pores which
would normally form and serving to densify the coating (Muller,
1986a,b). Higher density and a smoother surface can also be achieved
by implantation after deposition. Atoms arriving at the surface can
be ionized or electronically excited by the ion beam, making them
much more chemically active and so better able to bond properly
with the other atoms of the coating. Gas molecules arriving on the
surface can be split apart by the beam, thus depositing atoms such
as carbon and species such as oxygen or nitrogen.

## 4.3  Factors to Consider

Since IAC involves ion implantation, most of the factors which should
be considered in selecting ion implantation processes (such as vacuum
compatibility and line-of-sight requirements) also apply to IAC.
A major difference is that IAC usually produces a true coating,
whereas ion implantation does not.

The circumstances under which IAC would provide an advantage
over other coating processes are similar to those for ion implantation
and include the following:

*Low temperatures*: when the object may be damaged or warped by
     high temperatures (e.g., high-precision tools, plastics)
*Integrity*: when coating adhesion is a particularly difficult problem
     or the results of its decohesion are severe (e.g., polymer or
     composite substrates, bearings, materials subjected to extreme
     temperature cycling or to radiation, or biomedical components)
*Continuity*: when a dense, pinhole-free coating is essential
*Tolerance and finish*: when thin, well-controlled coatings are required
     to prevent degradation of tolerance or surface finish (e.g.,
     high-precision bearings)
*Partial coverage*: when only part of the object should be coated,
     or when one part of the object requires a more strongly bonded,
     chemically modified, or otherwise different coating
*Repeatability*: when precise repetition of conditions from run to
     run is critical

*Avoidance of catastrophic failure*: can be done if the coating is
   highly intermixed at the interface, such that the underlying
   material is implanted
*Coating development*: when a new coating or a coating for a difficult
   substrate is to be developed with a minimum of research effort
*Chemistry*: when unusual chemistry, stoichiometry, or crystallo-
   graphic phases are needed

In contrast to nitrogen ion implantation, IAC can be used for
cutting tools and other high-temperature applications. Some of the
coating methods used for IAC can access difficult geometries, but
these areas cannot be reached by the ion beam. In general, IAC
is comparable in cost with other high-technology coatings and is
therefore not generally recommended for inexpensive objects (unless
replacement or downtime costs are high). Since IAC encompasses
a broad range of different techniques, no estimate of current costs
is possible.

## 4.4 Ion-Assisted Coating Methods, Uses, and Results

The simplest way in which an ion beam can be used to obtain better
coating adhesion is by sputter-cleaning the surface prior to deposition
to remove adsorbed gas atoms which would impair the nucleation and
adhesion of the intended film. Although this type of cleaning is
standard in many coating systems, the use of a beam rather than
a simple discharge lends itself to much better defined and more
accurate surface preparation. For example, a controlled dose of
ion irradiation on sapphire prior to film deposition can be used
to adjust the surface stoichiometry to give much improved adhesion
(Baglin and Clark, 1985).
   Numerous arrangements for ion-assisted coating have been
developed. These are illustrated in Fig. 29. There is currently
no standard arrangement since the equipment employed is varied
to suit the processes to be used. The techniques used and results
so far obtained are described below.
   The most basic form of a combined ion beam/coating process
is *ion beam mixing*. This method consists of depositing a thin film
of coating material (up to about 50 nm) on a substrate and firing
an ion beam such as $Ar^+$ through the film into the substrate. The
recoil of ions from nuclear stopping collisions mixes the two materials
together. This method is used to integrate a thin layer into the
surface or to implant a metal ion without the necessity of creating
a beam of it. A brief review of this field is given by Paine and
Averback (1985). For example, Pd can be ion beam-mixed into Ti
to form an excellent corrosion barrier (Munn and Wolf, 1985). Depth

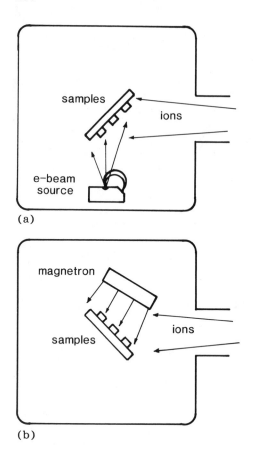

(a)

(b)

FIGURE 29 Various arrangements used for ion-assisted coating:
(a) electron beam deposition; (b) sputter deposition; (c) dual ion
beam sputter deposition (low energy ions); (d) ion beam-enhanced
gas phase deposition.

profiles for this process are shown in Fig. 30. Ion beam mixing
has been used to form molybdenum disilicide on silicon wafers by
depositing a 30-nm film of Mo and mixing it into the silicon substrate
by using $Ar^+$ ion beams (Van Ommen et al., 1987). The interest
in $MoSi_2$ stems from its ability to reduce resistance of interconnections
and contact resistance to the silicon substrate in electronic applications. The technique of ion beam mixing has also been used extensively to intermix multiple layers of different materials on a surface
(see, for example, King et al., 1985). Unusual phases and structures
have been produced by this technique (Hung and Mayer, 1985).

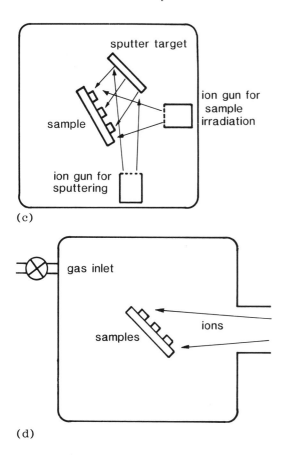

(c)

(d)

When the substrate is simultaneously heated and implanted, atoms from a deposited film can be made to diffuse along the depth profile of the beam and so penetrate to quite large depths, a process known as *radiation-enhanced diffusion*. A thin film of titanium deposited on the surface of M2 tool steel and then implanted with nitrogen follows the implant profile (see Fig. 31) and remains in the worn region even after material has been removed to a depth greater than the original implant depth (Solnick-Legg and Legg, 1983). During ion beam mixing some ion-enhanced diffusion almost always occurs, making the mixing much more efficient and thorough.

It is not always necessary to physically ion beam-mix the interface atoms to achieve improved film/substrate bonding, especially on ceramic and polymer surfaces. A thin coating can be bonded to a substrate by lightly implanting through the interface, a process called *ion beam stitching*. The interface remains sharp, with no

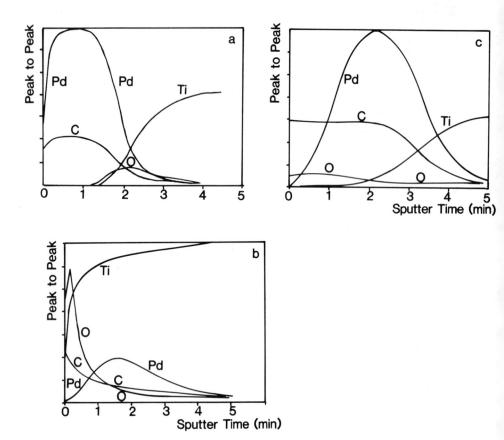

FIGURE 30   Auger electron spectroscopy depth profiles: (a) Ti
with a 50-nm evaporated Pd layer; (b) Ti implanted with Pd$^+$ (400
keV, $2 \times 10^{16}$ ion cm$^{-2}$); (c) Ti with 50-nm Pd layer, ion beam-mixed
with 200-keV Kr$^+$ ions. (After Munn and Wolf, 1985)

intermixing. There is some evidence to suggest that for thin films
deposited on insulators or metals covered by a thin oxide layer
that the generation of secondary electrons within the oxide may
play a key role in the efficacy of ion beam stitching (Sai-Halasz
and Gazecki, 1984). Impurities that exist on the substrate surface
prior to deposition (hydrocarbons), as well as the chemical reactivity
of the film and substrate atoms, all play important roles in the ultimate adhesive strength of ion beam-stitched films (J. S. Williams,
1986). Electron spectroscopy for chemical analysis (ESCA) often
shows evidence of changes in the atomic bonds across the interface.

Baglin and Clark (1985) report that ion irradiation of copper thin films on sapphire, quartz, and glass-ceramic substrates increases the strength of the film/substrate bond eightfold (see Fig. 32). While heating an unimplanted copper film on such substrates causes it to ball up, the interface bond on the implanted material is actually improved. The strength of the bond rises with dose, with no effect below $3 \times 10^{14}$ ions cm$^{-2}$, but apparently saturating at about $1$-$3 \times 10^{16}$. Similar methods can be used on polymers such as Teflon, polyamide, and Lexan.

Improvements in bonding of metal films to carbon fiber composites have also been found (Legg, 1987). Again, low doses of ions improved bonding as measured by the tape pull test. However, we have found that similar techniques applied to films deposited on parylene-C (a polymer used in biomedicine) were deleterious and caused the parylene to embrittle.

Ion beam stitching requires only the passage of ions near the interface atoms rather than actual atomic displacement. Thus, low doses are adequate for this technique, in the range of $10^{15}$ to $10^{16}$

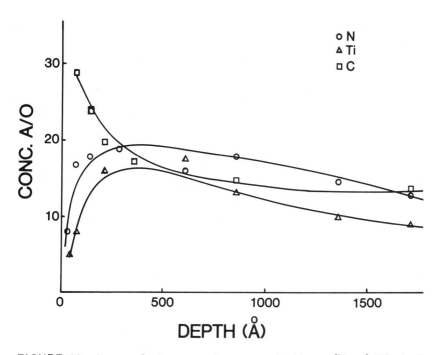

FIGURE 31  Auger electron spectroscopy depth profile of M2 steel coated with titanium, driven in by ion beam-enhanced diffusion using $N_2^+$ ions. (From Solnick-Legg and Legg, 1983)

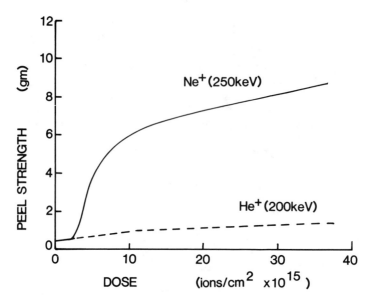

FIGURE 32   Peel strength as a function of ion dose for copper
on polished alumina. (After Baglin and Clark, 1985)

ions cm$^{-2}$ (that is, 1 to 10 ions for every interface atom). The
low ion doses result in substantial cost savings over direct higher-
dose implantation.

   *Ion beam-assisted deposition* (IBAD) [sometimes also called ion
beam-enhanced deposition (IBED)], in which deposition and ion irradiation are simultaneous, is useful in forming compound films, since
both structural and chemical properties can be modified. Film adhesion to nearly any substrate is improved by concurrent ion bombardment during early film growth stages or immediately prior to deposition.
Low-energy IBAD (50 eV to 5 keV) is used to produce optical dielectric films through carefully controlled structural modification of the
growing film. This process has been successfully used to tailor
the refractive index of dielectric optical films (see Fig. 33) and
reduce adsorption of water vapor into the films (Martin, 1986) as
well as to modify properties such as electrical resistivity, optical
density, surface topography, film density, and porosity (Cuomo
and Rossnagel, 1987).

   If the bombarding ions are of a reactive species (nitrogen,
oxygen, or carbon, for example), then a compound thin film will
be formed [*reactive ion beam-assisted deposition (RIBAD)*]. Control
over parameters such as deposition rates, ion species, dose and
energy, substrate temperature, composition of the ambient gas

mixture, and chamber pressures would permit chemistry and stoichiometry to be altered. According to Cuomo and Rossnagel (1987),
current methods of producing high-precision optical coatings generally
lack reproducibility of the refractive indices of the optical films
and are unstable; these films are used for mirrors in laser gyro-
scopes and for antireflection coatings. RIBAD produces high-quality
optical films using oxygen ion beams which alter film microstructure
and stoichiometry. Dielectric films produced by RIBAD include $TiO_2$,
$SiO_2$, $HfO_2$, BeO, $Ta_2O_5$, and $ZrO_2$. These films are characterized
by low extinction coefficients and are generally hard and stable.

The techniques of IBAD and RIBAD are now being used to
produce many other types of coatings. For example, TiN coatings

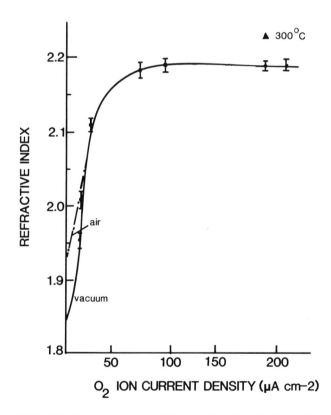

FIGURE 33   Variation of the refractive index of a $ZrO_2$ film with
the number of ions implanted during growth, measured by ion current
density. Changes result from increased packing density. (After
Martin, 1986)

are grown by simultaneous electron beam deposition and ion irradiation at low temperatures (Kant et al., 1985; Sartwell, 1986). Using an in situ electron beam evaporator, researchers grew about 100- to 200-nm films while simultaneously implanting them with nitrogen ions. The microstructure of films that were simply deposited and not implanted was found to consist of 10-nm equiaxed grains. The IBAD film microstructure was typical of that usually found in standard films deposited at high temperatures with a bimodal size distribution. When adhesion testing was used on N implanted and nonimplanted films, the as-deposited films cracked and spalled off along the scratch track while the IBAD film deformed with the substrate and remained bonded to it (Fig. 34, Sartwell, 1986).

Alternatively, a very thin film can be first thoroughly intermixed into the substrate using an ion beam, with or without prior ion beam cleaning, and then given sequential depositions and ion implants to produce a final coating on top of a highly intermixed interface. This type of ion-assisted coating, usually done with reactive ion beams such as nitrogen, carbon, or oxygen, can be called *reactive ion-assisted coating* (RIAC). The thin films are deposited by a variety of methods, including sputtering and electron beam evaporation, both of which can be done in situ, and ion plating. CVD, chrome plating, etc., outside the implantation chamber. The earliest example of ion-assisted coating is from Weissmantel's work (Weissmantel et al., 1972), in which hard-carbon thin films were deposited by bombarding surfaces with low-energy ions (about 1 keV) in a hydrocarbon ambient. More recently, radiofrequency (rf) sputter deposition of a 35-nm film of Ti on M43 tool steel in a mixed Ar-N plasma followed by high-dose N ion implantation, produces a very thin film (10 to 20 nm thick) with excellent friction, wear, and corrosion properties (Solnick-Legg et al., 1986). The gas ion dose required for this is in the mid-$10^{17}$ ions cm$^{-2}$ range, similar to that used in direct nitrogen implantation. Although a Ti-O-N surface is formed rather than a pure titanium nitride coating, this surface shows greatly reduced adhesion, friction, and wear. In unlubricated wear and friction tests of an uncoated pin against a treated substrate, the average coefficient of friction was 0.35. This can be compared with a coefficient of friction of 0.75 obtained by activated reactive evaporation (ARE) techniques (Jamal, Nimmagadda and Bunshah, 1980) in dry tests of an uncoated pin on stainless steel substrates coated with 4 to 8 microns of TiN. Furthermore, the ion-assisted coating had excellent corrosion resistance and wear characteristics and continued to show reduced corrosion, friction, and wear rates on the implanted substrate even after the coating had worn off.

Ti-Pd layers produced by ion-assisted coating have been corrosion tested by boiling in $MgCl_2$ solution (Ferber and Wolf, 1987). The

FIGURE 34 Optical micrographs of deformations produced during adhesion testing using a Rockwell C indenter loaded to 30 N for (a) unimplanted and (b) $N_2^+$ implanted (IBAD) films. (From Sartwell, 1986)

(a)

(b)

$\dfrac{100 \ \mu m}{}$

specimens remained passive even after four months in boiling $MgCl_2$. Whether implanted or ion-assisted-coated, the presence of the catalyst (palladium) on the surface of the titanium permitted a hydrogen reduction reaction to occur, which in turn prevented crevice pH from dropping, and the titanium remained passive. The thickness and uniformity of the ion-assisted coatings were not important; only enough Pd was needed to remain on the surface to catalyze hydrogen evolution at the titanium surface.

Coatings of $Fe_6Al_4$ were deposited on Z2CN1809 steel samples and then implanted with argon to produce a highly adherent ion-assisted coating. The FeAl alloy was chosen because of its good high-temperature oxidation resistance. Figure 35 shows treated and untreated alloys after pin-on-disk wear testing. Wear scars on the untreated steel are deep, well-marked, and pitted, indicating strong adhesion between the pin and sample surface. Even after eight times longer wear testing, the ion-assisted-coated steel samples showed only shallow, smooth wear scars.

In another application of RIAC technology, thin, hard films of boron nitride have been deposited by evaporation of B under $N^+$ ion irradiation at 25 to 40 keV (Satou et al., 1985). The crystal structure of the films varied, depending upon the B/N ratios. For a B/N ratio of 0.9 to 1.1 the films were smooth with a substantial cubic phase. At lower B/N ratios the films became more granular and hexagonal. For a B/N ratio of about 1.2, films contained both cubic and hexagonal phases of BN but had a hardness of 3000 to 5000 $H_V$ as measured with a 10-g Vickers indenter (Andoh et al., 1987). Since boron nitride is second only to diamond in hardness, RIAC may make it possible to create very hard surfaces without the brittleness of diamond. Cubic BN is not only extremely hard, but is also chemically stable with high thermal conductivity and high resistivity.

Solid lubricant coatings have been made by sputter-coating a surface with a thin $MoS_2$ layer and then implanting with Ar to bond the coating to the surface (Kobs et al., 1987). The result is a lubricating layer with a very low friction coefficient which adheres well to the surface even under prolonged wear. Without implantation to intermix coating atoms with substrate atoms, the coating would have rapidly worn off.

In contrast to more conventional coating techniques, which usually require careful control, RIAC or RIBAD films can often be deposited over a broad range of deposition and implantation conditions, provided that the ion dose and energy suffice to supply a certain minimum amount of energy to each atom in the film (Kant, 1986). This means that new coatings or coatings for novel applications can be more readily developed. In addition, with the creation

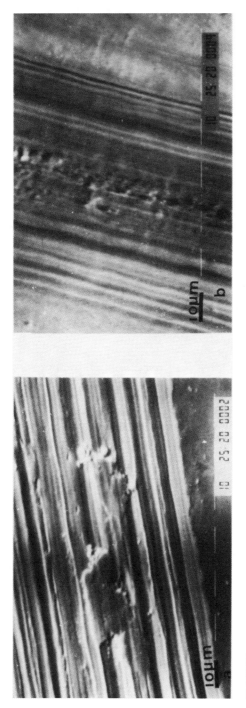

FIGURE 35 Wear scars on Z2CN1809 steel: (a) untreated surface after 50 passes and (b) FeAl ion-assisted coating after 400 passes. (From Moine et al., 1987)

of very thin films, such as the ion-assisted deposition of TiN described above, which possess the same excellent low friction and
wear properties as the much thicker PVD/CVD TiN coatings, the
applications of RIAC techniques to devising coatings for high-
precision, high-tolerance tooling and components, such as precision
bearings and laser mirrors, is now possible.

The ion beam can also be used to modify a deposited coating
as well as to improve its adhesion. For example, in our laboratory,
we produce a Ni-B coating for forming tools by first depositing
a nickel film and then implanting it with $B^+$. This process produces
a presumably amorphous nickel which has been found to be very
resistant to galling under high load.

In an unusual application of ion-assisted coatings to biomedicine,
a very complex and highly adherent thin coating of hydroxylapatite
has been developed and applied to prostheses, metals, ceramics,
and polymers (Stevenson et al., 1989). This coating is a synthetic
bone mineral which should be either biocompatible or bioactive,
depending on the manner in which it is produced or ion beam-
modified. In this application it is important that the material be
produced at low temperature and yet be crystalline.

The coatings so far reported in the literature are only a small
fraction of the possibilities presented by ion-assisted coating techniques. New ion-assisted coatings and coatings for specialized applications are continually being developed. In our own laboratory,
for example, we have readily produced well-adhering coatings of
compounds such as ZrN, ZrC, BN, and $TiO_2$ by RIAC methods.
These compound coatings (which would by other methods require
high temperatures) have been deposited on metals, ceramics, and
even polymers. Similar techniques can be used to deposit adherent
metal coatings on glasses, ceramics, polymers, and composites.

## 5 SUMMARY

Coating processes have developed over the years from such relatively
simple processes as electroplating to the more sophisticated PVD
and CVD techniques, where strongly adherent coatings are produced.
More recently, it has become increasingly difficult to distinguish
between CVD and PVD processes as plasma enhancement is developed
for both methods, resulting in vastly improved coatings. Ion implantation, however, is more intimately involved with alteration of the
substrate surface and near-surface regions. It is a low-temperature
process capable of creating nonequilibrium alloys and compounds
in the surface. Smoothness, dimensional tolerances, and temper
are totally unaffected; since the implant is an integral part of the

substrate material, there is no sharp interface and the implanted layer cannot easily come off. Disadvantages include the relatively high cost of the process and its inability to provide a protective layer to completely isolate the underlying material from its environment.

The emergence of ion-assisted coating is the result of the convergence of two surface modification technologies: conventional coating methods (e.g., PVD, PACVD, CVD) and ion implantation. The ion beam not only provides a controlled method for depositing energy over a depth of 100 nm or more to improve film growth and adhesion, but is also available to modify the initial substrate surface, the coating itself, or the final coating surface. The ion beam permits many forms of process control that are not possible with plasma techniques.

Ion-assisted coatings are seldom used for volume production at this time but are being developed and used to obtain coatings with properties and under conditions which defy normal coating techniques (for example, difficult or low-temperature substrates, pinhole-free films, coatings of unusual chemical composition or physical properties). It should also be regarded as a means to develop new coatings rapidly with a minimum of R&D effort. With the use of ion beams we expect the development of new types of films which can be deposited on substrates that, until recently, were difficult to coat.

Both ion implantation and ion-assisted coating exploit the properties of ion beams to create materials often outside the bounds of normal phase equilibria and to produce surface modifications in a controlled and reproducible manner. While ion implantation is being done on a commercial scale, ion-assisted coating is still largely under development. It is to be expected that in both cases the next few years will see developments in equipment which will lower processing costs and increase the size of objects which can be treated. This will lead to much wider use of these technologies as standard treatments for surface modification.

## REFERENCES

Abel, J. S., Mazurek, H., Day, D. R., Maby, E. W., Senturia, S. D., Dresselhaus, G., and Dresselhaus, M. S. (1982). *Metastable Materials Formation by Ion Implantation*, 173.

Andoh, Y., Ogata, K., Suzuki, Y., Kamijo, E., Satou, M., and Fujimoto, F. (1987). *Nucl. Instr. Meth.*, B19/20:787.

Armour, D. G., Bailey, P., and Sharples, G. (1986). *Vacuum, 36* (11/12):769.

Ashworth, V., Procter, R. P. M., and Grant, W. A. (1980). In
    *Treatise on Materials Science and Technology, 18, Ion Implanta-*
    *tion* (J. K. Kirvonen, ed.), Academic Press, New York, p.
    175.
Baglin, J. E. E. and Clark, G. J. (1985). *Nucl. Instr. Meth.*,
    *B7/8*:881.
Baglin, J. E. E., Schrott, A. G., Thompson, R. D., Tu, K. N.,
    and Segmuller, A. (1987). *Nucl. Instr. Meth.*, *B19/20*:782.
Bernabai, U., Cavallini, M., Bombara, G., Dearnaley, G., and
    Wilkins, M. A. (1980). *Corr. Sci.*, *20*:19.
Brown, J. G., Galvin, J. E., Gavin, B. F., and MacGill, R. A.
    (1986). *Rev. Sci. Instrum.*, *57*:1069.
Brown, W. L., Venkatesan, T., and Wagner, A. (1981). *Nucl.*
    *Instr. Meth.*, *191*:157.
Burnett, P. D. and Page, T. F. (1984). *J. Mater. Sci.*, *19*:3524.
Byers, P. (1985). Proc. Conf. on Application of Ion Plating and
    Implantation to Materials, Atlanta, Georgia, June.
Chen, Q. M., Chen, H. M., Bai, X. D., Zhang, J. Z., and Wang,
    H. H. (1983). *Nucl. Instr. Meth.*, *209/210*:867.
Clayton, C. R. (1981). *Nucl. Instr. Meth.*, *182/183*:865.
Clayton, C. R., Doss, K. G. K., Wang, Y.-F., Warren, J. B.,
    and Hubler, G. K. (1982). In *Ion Implantation into Metals*
    (V. Ashworth, W. A. Grant, and R. P. M. Procter, eds.),
    Pergamon Press, Oxford, p. 67.
Cochran, J. K., Legg, K. O., and Baldau, G. R. (1984). In
    *Emergent Process Methods for High Technology Ceramics* (R. F.
    Davis, H. Palmour III, and R. L. Porter, eds.), Plenum, New
    York.
Colligon, J. S. (1986). *Vacuum, 36* (7-9):413.
Conrad, J. R., Radtke, J. L., Dodd, R. A., Worzala, F. J., and
    Tran, N. C. (1988). *J. Appl. Phys.* (To be published.)
Cuomo, J. J. and Rossnagel, S. M. (1987). *Nucl. Instr. Meth.*,
    *B19/20*:963.
Dearnaley, G. (1969). *Rep. Prog. Phys.*, *32*:405.
Dearnaley, G. (1980). In *Ion Implantation Metallurgy* (C. Preece
    and J. K. Hirvonen, eds.), AIME Pub., p. 1.
Dearnaley, G. (1985a). *Nucl. Instr. Meth.*, *B7/8*:158.
Dearnaley, G. (1985b). *Mater. Sci. Eng.*, *69*:139.
Dearnaley, G. (1986). *Surface Eng.*, *2*:213.
Dearnaley, G. (1987a). Personal communication.
Dearnaley, G. (1987b). Private communication.
Dearnaley, G., Goode, P. D., Minter, F. J., Peacock, A. T.,
    Hughes, W., and Proctor, G. W. (1986). *Vacuum, 36*:807.
Dearnaley, G., Goode, P. D., Minter, F. J., Peacock, A. T.,
    and Waddell, C. N. (1985). *J. Vac. Sci. Tech.*, *A3* (6), p. 2684.

Dearnaley, G. and Hartley, N. E. W. (1978a). Metal Forming Dies, U.S. Patent, 4,105,443.

Dearnaley, G. and Hartley, N. E. W. (1978b). *Thin Solid Films*, 54:215.

Dienel, G., Kreissig, U., and Richter, E. (1986). *Vacuum*, 36 (11/12):813.

Dillich, S. A., Bolster, R. N., and Singer, I. L. (1984). In *Ion Implantation and Ion Beam Processing of Materials*, North-Holland, Amsterdam, p. 637.

Eckler, T. A. (1979). Ion Plated Coating on Titanium Alloy, *Technical Report AFML-TR-79-4109*.

Ferber, H., Kasten, H., Wolf, G. K., Lorenz, W. J., Schweickert, H., and Folger, H. (1980). *Corrosion Sci.*, 20:117.

Ferber, H. and Wolf, G. K. (1987). *Mater. Sci. Eng.*, 90:213.

Free, J. (1984). "Beam Magic," *Popular Sci.*, December, p. 80.

Hartley, N. E. W. (1976). *Inst. Phys. Conf. Series*, 28, p. 210.

Hartley, N. E. W. (1979a). *Radiation Effects*, 44:19.

Hartley, N. E. W. (1979b). *Surfacing J.*, 10:1.

Heydari, P., Starke, E. A., Chakrabortty, S. B., and Legg, K. O. (1982). *Ion Implantation into Metals* (V. Ashworth, W. A. Grant, and R. P. M. Proctor, eds.), Pergamon Press, Oxford, p. 172.

Higham, P. A. (1986). *Biomedical Materials*, vol. 55, MRS Symposia Proceedings, p. 253.

Hioki, T., Itoh, A., Ohkubo, M., Noda, S., Doi, H., Kawamoto, J., and Kamigaito, O. (1986). *J. Mater. Sci.* (In press.)

Hu, W. W., Clayton, C.R., Herman, H., and Hirvonen, J. K. (1978). *Scr. Metall.*, 12:697.

Hung, L. S. and Mayer, J. M. (1985). *Nucl. Instr. Meth.*, B7/8:676.

Jamal, T., Nimmagadda, R., and Bunshah, R. F. (1980). *Thin Solid Films*, 73:245.

Jata, K. V. and Hubler, G. K. (1985). *J. Vac. Sci. Tech. A 3* (6):2677.

Kant, R. (1986). Paper given at Conf. on Appl. of Accelerators in Research & Industry, Denton, Texas, 1986. *Nucl. Instr. Meth.* (To be published.)

Kant, R. A. and Sartwell, B. D. (1986). In *The Use of Ion Implantation for Materials Processing* (F. A. Smidt, ed.). NRL Memorandum Report 5898, p. 83.

Kasten, H. and Wolf, G. K. (1980). *Electrochim. Acta*, 25:1581.

King, B. V., Tonn, D. G., and Tsong, I. S. T. (1985). *Nucl. Instr. Meth.*, B7/8:607.

Kobs, K., Dimigen, H., Leutenecker, R., and Ryssel, H. (1987). Paper given at 14th International Conference on Metallurgical Coatings, San Diego, California.

Kujore, A., Chakrabortty, S. B., Starke, E. A., and Legg, K. O. (1981). *Nucl. Instr. Meth.*, 182/183:949.

Legg, K. O. (1987). *Nucl. Instr. Meth.*, *B24/25*:565-567.

Legg, K. O., Cochran, J. K., Solnick-Legg, H., and Mann, X. L. (1985). *Nucl. Instr. Meth.*, *B7/8*:535.

Martin, P. J. (1986). *Vacuum*, *36*(10):585.

Mathews, F. D., Greer, K. W., and Armstrong, D. L. (1986). *Biomedical Materials*, Vol. 55. MRS Symposia Proceedings, p. 243.

McHargue, C. J. (1987). *Nucl. Instr. Meth.*, *B19/20*:797.

McHargue, C. J., Farlow, G. C., White, C. W., Appleton, B. R., Williams, J. M., Sklad, P. S., Angelini, P., and Yust, C. S. (1986), *J. Mater. Energy Sys.*, *8*:255.

Moine, P., Popoola, O., Villain, J. P., Junqua, N., Pimbert, A., Delafond, J., and Grilhe, J. (1987). *Surf. Coating Tech.*, *33*:479.

Muller, K.-H. (1986a). *J. Vac. Sci. Tech.*, *A4*:461.

Muller, K.-H. (1986b). Paper given at AVS Topical Symposium on Sputtering.

Munn, P. and Wolf, G. K. (1985). *Nucl. Instr. Meth.*, *B7/8*:205.

Paine, B. M. and Averback, R. S. (1985). *Nucl. Instr. Meth.*, *B7/8*:666.

Picraux, S. T. and Peercy, P. (1985). *Sci. Amer.*, p. 102.

Pope, L. E., Yost, F. G., Follstaedt, D. M., Picraux, S. T., and Knapp, J. A. (1984). In *Ion Implantation and Ion Beam Processing of Materials* (G. K. Hubler, D. W. Holland, C. R. Clayton, and C. W. White, eds.), North-Holland, Amsterdam, p. 661.

Roberts, S. G. and Page, T. F. (1982). In *Ion Implantation into Metals* (V. Ashworth, W. A. Grant, and R. P. M. Procter, eds.), Pergamon Press, Oxford, p. 235).

Sai-Halasz, G. A. and Gazecki, J. (1984). *Appl. Phys. Lett.*, *45*: 1067.

Saritas, S., Procter, R. P. M., Ashworth, V., and Grant, W. A. (1982). *Wear*, *82*:233.

Sartwell, B. (1986). *J. Mater. Energy Sys.*, *8*:246.

Sartwell, B. (1987). Conf. on the Applications of Accelerators to Research and Industry, Denton, Texas, *Nucl. Instr. Meth.* (To be published.)

Satou, M., Yamaguchi, K., Andoh, Y., Suzuki, Y., Matsuda, K., and Fujimoto, F. (1985). *Nucl. Instr. Meth.*, *B7/8*:910.

Singer, I. L. (1984). In *Ion Implantation and Ion Beam Processing of Materials*, North-Holland, Amsterdam, p. 585.

Singer, I. L. and Barlak, T. M. (1983). *Appl. Phys. Lett.*, *43* (5):457.

Sioshansi, P. (1987a). International Conference on Metallurgical Coatings, San Diego, California, *J. Vac. Sci. Tech.* (To be published.)

Sioshansi, P. (1987b). *Materials Eng.*, 19.

Smidt, F. A., Hirvonen, J. K., and Ramalingam, S. (1983). In *Ion Implantation for Materials Processing*, Noyes Data Corp., p. 109.

Smidt, F. A. and Sartwell, B. D. (1985). *Nucl. Instr. Meth.*, *B6*:70.

Solnick-Legg, H. and Legg, K. O., Phase 1 Final Report, NSF Grant No. PHY-8260465.

Solnick-Legg, H., Legg, K. O., Rinker, J. G., and Freeman, G. (1986). *J. Vac. Sci. Tech.* *A4*(6):2844.

Spooner, S. and Legg, K. O. (1980). In *Ion Implantation Metallurgy* (C. M. Preece and J. K. Hirvonen, eds.), AIME Pub., p. 162.

Stevenson, J. R., Solnick-Legg, H., and Legg, K. O. (1989). In *Biomedical Materials and Devices* (J. Hanker and B. Giammara, eds.). Materials Research Society, V. 110 (to be published 1989).

Townsend, P. D. (1986). *Contemp. Phys.*, *27*:241.

Townsend, P. D., Kelly, J. C., and Hartley, N. E. W. (1976). In *Ion Implantation, Sputtering, and Their Applications*, Academic Press, New York.

Van Ommen, A., Willemsen, M., and Wolters, R. (1987). *Nucl. Instr. Meth.*, *B19/20*:742.

Vardiman, R. G. (1986). In *Ion Plating and Implantation* (R. F. Hochman, ed.), ASM Publishers, 107.

Vardiman, R. G. and Kant, R. (1982). *J. Appl. Phys.*, *53*:690.

Venkatesan, T. (1984). *Nucl. Instr. Meth.*, *B7/8*:461.

Weber, D. C., Brant, P., and Carosella, C. (1982). In *Metastable Materials Formation by Ion Implantation* (S. T. Picraux and W. J. Choyke, eds.) Elsevier, New York.

Weissmantel, C., Fiedler, O., Hecht, G., and Reisse, G. (1972). *Thin Solid Films*, *13*:359.

Williams, J. S. (1986). *Rep. Prog. Phys.*, *49*:491.

Williams, J. M., Beardsley, G. M., Buchanan, R. A., and Bacon, R. K. (1984). *Mat. Res. Soc. Symp. Proc.*, *27*:735.

Wolf, G. K. (1981). *Nucl. Instr. Meth.*, *182/183*:875.

# 4
# Sputtering Techniques

JOHN KEEM  *Ovonic Synthetic Materials Company,*
*Troy, Michigan*

## 1  INTRODUCTION

The objective of this chapter is to compare and understand aspects
of surface modifications obtained by three different thin-film vapor
phase deposition techniques. The three surface modification techniques
are glow discharge deposition, planar magnetron sputtering, and ion
beam sputtering.

Because most practical applications of these surface modification
techniques use commercially designed equipment, we will not discuss
details of equipment design. The details of designs and techniques
will be examined only when it is necessary to understand them
to obtain the desired surface modification. The discussion will be
oriented toward crucial issues associated with obtaining desired
surface modifications.

Excellent up-to-date detailed technical descriptions of these
three techniques and general discussions of thin-film technology
and glow discharge processes are contained in Bunshah et al. (1982),
Vossen and Kern (1978), Chapman (1981), Maissel and Glang (1970),
and Westwood and Cuomo (1985). References 1-4 and 6 in particular
contain extensive citations to review articles and the relevant original
literature.

## 2  SUMMARY

By examining the thin-film process steps common to these techniques, we can understand and compare the processes. The common features are

1. Control of the composition, flux, and degree of excitation of feedstocks in the gaseous state
2. Control of the condensation rate of components of the feedstock onto the substrates
3. Control of film growth conditions on the substrates

The three deposition processes all incorporate glow discharges to dissociate and excite the feedstocks. The important aspects of glow discharges are described in relation to their influence on the feedstocks. It is shown how control of the composition, flux, and degree of excitation of the feedstocks in the gaseous state for these processes is obtained through control of the glow discharge used in all three processes. Control of the condensation rate of components of the feedstock onto the substrates for these processes is shown to primarily involve the location of the substrates in relation to the glow discharge and the feedstock source or sources and the gas pressure in the reactor. Finally, the control of film growth conditions on the substrates is found to be coupled, to varying degrees depending on the process, to the factors which influence the two previous control issues as well as the electrical potential of the substrates.

The successful practical examples of surface modification serve to illustrate the unique interplay of the factors influencing the three basic process steps in each of the techniques and to illustrate aspects of the unique strengths of each of the techniques. The examples are surface modification by glow discharge deposition to increase corrosion resistance of stainless steel, surface modification by planar magnetron sputtering to obtain enhanced wear-resistant surfaces on machine tools, and surface modification by ion beam sputtering to increase the reflectivity of an optical surface to x-rays.

## 3  SURFACE MODIFICATION BY THIN-FILM VAPOR PHASE DEPOSITION

There are a wide variety of thin-film vapor phase deposition techniques by which surface properties of substrates may be modified. Though the details of the operation of these deposition techniques vary greatly, they all may be viewed as versions of a single basic process.

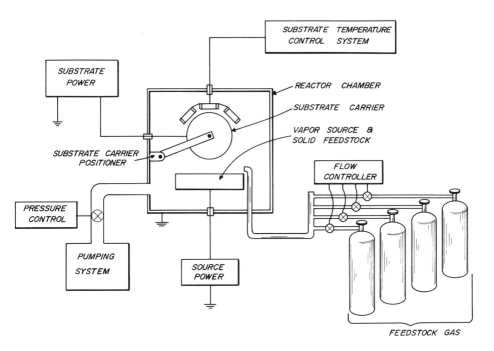

FIGURE 1  Schematic vapor phase deposition system.

A schematic vapor phase deposition system is shown in Fig. 1. There are eight subsystems:

1. A reactor chamber
2. A chamber pressure control system
3. A source for the vapor
4. A power supply for the vapor source
5. A feedstock supply for the vapor source
6. A carrier for the substrates
7. A substrate carrier power supply
8. A substrate temperature control system

In modern commercial systems these supply control packages and attendant factor-level monitors (like pressure, flow, and power-level sensors) are linked to a central microcomputer which closes the process feedback loop and is programed to control and monitor the factor levels during the entire process.

In this schematic system, the source is fed by feedstocks and the power supply to convert the feedstocks into vapor. The feedstock supply rate and the power level are used in conjunction with the chamber pressure control system to vary the composition, flux, and level of excitation of the vapor during the process. The substrate carrier transports the substrates through the vapor flux. The carrier is also electrically excited and temperature controlled so that the desired film growth conditions may be established. The chamber pressure control system controls the overall deposition system pressure.

Though all eight of these subsystems influence the quality of the surface modification, we can group them into broader categories, including effective generation of the vapor flux, effective capture of the vapor flux, and effective establishment of surface conditions to control film growth from the vapor flux. In glow discharge, planar magnetron sputtering, and ion beam-sputtering deposition techniques, we find that these broad categories translate to the following common aspects which have important influences on the quality of the surface modification:

1.  Control of the composition, flux, and degree of excitation of the vapor.
2.  Positioning of the substrates to intercept the correct vapor flux composition so that the required condensation rate is achieved even on complicated surface shapes.
3.  Control of the substrate surface conditions which influence the growth of the coating.

## 4   VAPOR FLUX FORMATION PROCESSES

In glow discharge deposition technique the vapor fluxes are created by dissociation and excitation of feedstock gases by glow discharges. The planar magnetron sputtering and ion beam-sputtering deposition techniques use glow discharges as a source of high-energy noble gas (generally argon) ions, which in turn excite the vapor fluxes by sputtering atoms from feedstock targets.

Theoretical descriptions of sputtering and glow discharges are outside the scope of this chapter. We highlight their important features, but we omit rigorous examination of the concepts or arguments. An essentially self-contained intermediate discussion of these phenomena is found in Ref. 3.

### 4.1   Sputtering

Sputtering excites material into the gas phase by momentum transfer. Sputtering is characterized by two essential features (Westwood and

Cuomo, 1985): first, high-energy ions or neutral atoms supplied by the glow discharge collide with the surface of the feedstock target and transfer their momentum to the target material; second, some of the feedstock material is ejected from the surface of the target as a result of the momentum transfer.

A useful measure of the strength of the transfer process for each feedstock material is the sputtering yield S, defined as the number of target atoms ejected per incident ion (see Figs. 2, 3). For each feedstock material, the sputtering yield depends on the energy of the sputtering ion, the depth of penetration of the ion into the target, the angle of incidence of the sputtering ion, and the total system pressure.

The depth of penetration enters the discussion because only the surface atoms can escape from the target. If more of the sputtering atom's energy can be deposited nearer to the target's surface, then more sputtered atoms will result and the yield will increase. For this reason, the yield increases as the angle of incidence of the sputtering ion becomes more grazing. At grazing incidence (Maissel and Glang, 1970) the ion deposits more of its energy in the first few layers of the surface of the target. As shown in

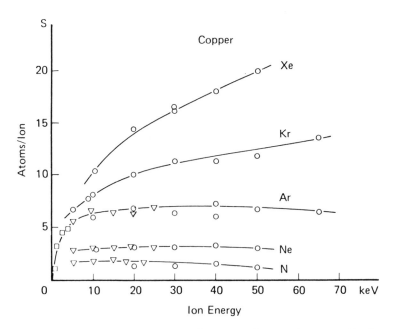

FIGURE 2   Sputtering yield of copper for various sputtering gases (from Ref. 3).

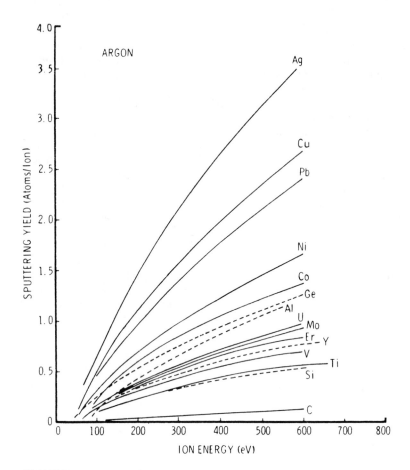

FIGURE 3  Sputtering yield per ion for various elements sputtered
by argon (from Ref. 3).

Fig. 4 this effect has a maximum at 50° to 70° from the normal,
depending on feedstock atom and sputtering ion mass. At larger
angles the probability of reflection for the incident ion increases.
At 90° the yield drops to zero (Westwood and Cuomo, 1985).

The number of excited vapor atoms produced is related to the
sputtering rate. This quantity decreases as the cosine of the angle
measured from the normal of the target surface. This dependence results because an atom will escape most easily by traveling straight
away from the surface of the target. However, with low-energy ions
(< 100-200 eV), because the ion does not penetrate into the target,

the angle of ejection is more likely to be equal to the angle of incidence of the sputtering ion.

In addition to the momentum transfer processes in the feedstock
target the density of gas above the feedstock target influences the
effective sputtering rate. As the system background pressure is
increased, the probability of collision of a vapor atom with a neutral
argon increases and thus the probability of that sputtered ion being
scattered back to the feedstock target increases, thereby reducing
the sputtering rate.

Because the sputtering process is based on momentum transfer
to a cooled feedstock target, the composition of the excited vapor
produced by sputtering is the same as that of the feedstock target,

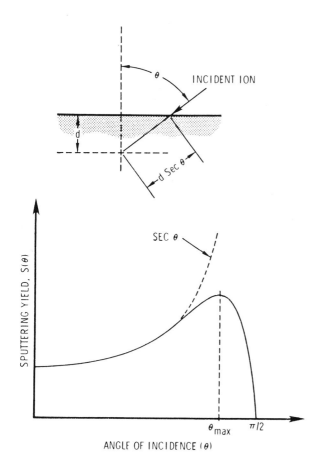

FIGURE 4  Influence of angle on sputtering yield (from Ref. 3).

even when the sputtering rates of the components of the feedstock target are different. The details of this effect are not well understood (Coburn, 1971, 1976), but it is known that the composition at the surface of the feedstock target changes, so the resulting vaporization rates yield a vapor flux composition equal to the feedstock target composition.

In summary, the control of composition, flux, and degree of excitation of the vapor in the sputtering process occurs in the following ways.

## Composition

The composition of the excited vapor is simply controlled by the composition of the feedstock target.

## Flux

The flux is controlled by controlling the vaporization rate of the feedstocks and the area of the feedstock target. Controlling the magnitude of the sputtering ion current, the energy of the sputtering ions, the grazing angle of incidence of the sputtering ions onto the feedstock target, and the system background pressure controls the vaporization rate.

By controlling the shape and the area of the feedstock target, we can shape the total flux density in the space between the source and the substrate for particular deposition profiles (Westwood and Cuomo, 1985).

## Degree of Excitation

The degree of excitation of the vapor is controlled by the sputtering ion's incident energy and the energy lost by collisions with the background gas as the sputtered atom moves to the substrate (Thornton, 1982).

### 4.2  Glow Discharge

The glow discharges we deal with serve two purposes: they are sources of energetic ions for the sputtering process by which we produce vapor fluxes from feedstock targets, and they provide a means of exciting molecules and gases so that they become more chemically reactive.

Any electrical discharge in a low-pressure gas is called a glow discharge. The glow discharges used in sputtering and glow discharge deposition are primarily supported by ionization of noble gases like argon or helium (frequently called the process gas), with significant partial pressures of reactant gases. Nonreactive noble gases are used to support the glow because they are not permanently lost from the glow by reaction with any part of the system.

The electrical current in the glow is carried by both ions and electrons. In the schematic dc glow discharge shown in Fig. 5, the electrons move from the negatively charged cathode to the positively charged anode while the ions move in the opposite direction. The ions in the discharge are formed by electrons which impact ionize the low-pressure gases.

Electrons leave the system when they reach the anode or diffuse to the walls of the chamber. The positive noble gas ions are lost from the glow when they are neutralized on impact with the cathode surface. As neutral atoms they are easily desorbed and rejoin the background gas by ion impact on the cathode.

In each process we are studying, the glow becomes self-sustaining in slightly different ways. In all the systems the ion losses are compensated by electron impact ionization. The electron losses are compensated for in different ways. In sputtering systems the electron losses are made up by electrons ejected from the cathode when it is struck by enough positive ions. The electron yield from the sputtering cathode for typical voltages is about 1 electron for 10 ions. In the ion beam systems a hot filament is the source of electrons sustaining the glow. In the rf glow discharge deposition system operating at 13.56 MHz, the high-frequency electric fields ionize enough atoms to overcome the electron losses.

The glow discharges we deal with have three distinct regions (see Fig. 5), each of which supports different phenomena which are important to the use of glow discharges in deposition processes:

1. The region where the glow is seen.
2. The narrow region between the cathode and the glow from which little or no light is emitted is called the cathode dark space.
3. The sheath regions. These are the regions between the glow discharge and all other components of the system including the anode, the walls of the chamber, and any substrates.

A useful image for the glow region is of an electrically neutral, uniform fluid of fast electrons moving to the anode and slow ions drifting to the cathode against a background of thermally vibrating atoms. (A typical ratio of atoms to ions is 10,000:1.) It is surprising that in the glow region the ion and electron thermal motion is *much faster* than the drift toward either electrode. The thermal velocities of atoms and ions are about the same.

In the region where the glow is seen, the number of moving electrons is greatest and exactly equal to the number of ions. In

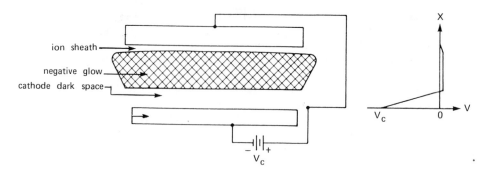

FIGURE 5   Schematic glow discharge (from Ref. 3).

this region inelastic collisions between electrons and atoms produce
ions and more electrons. Inelastic collisions between electrons and
ions can excite the ions to a few electron volts. Some excited ions
relax back to their unexcited state by the emission of light, which
is where the glow in the glow discharge comes from.

The electrical potential of the glow region has been named the
plasma potential. In the glow region the plasma potential is constant
(see Fig. 5) and more positive than any other potential in the system.

A useful image of the cathode dark space region is of ions
accelerating toward impact, with the cathode and electrons emitted
from the cathode being accelerated toward the glow region. Some
of the high-speed ions undergo collisions with slow-moving neutral
atoms of the *same type*. By a quantum mechanical charge exchange
collision, the high-speed ion may become a high-speed neutral,
and the low-speed neutral a low-speed ion. This interaction increases
the ion flux on the cathode, but also reduces the average ion impact
energy.

A useful image of the sheath regions is of the periphery of
the glow discharge where ions and electron diffuse from the neutral
glow. Because the electrons are moving faster than the ions, they
initially leave the glow at a higher rate. To compensate for this
loss, the plasma potential rises to slightly above the value of these
surfaces until the net flow of charge is reduced to zero. The potential rise is of the order of tens of volts. It is the mechanism of
sheath formation that maintains the glow at a slightly higher potential
than any other part of the system.

The glow discharges that we consider take place in argon plus
reactant gases at pressures from 0.1 mtorr to 1 torr. The magnitudes
of the rf and dc voltages necessary to maintain these discharges
are from a few hundred to 1000 or 2000 V. Typical current densities
are from 1 to 100 mA/cm$^2$.

## 5  SUMMARY OF THE TECHNIQUES

This section reviews the salient characteristics of the three deposition techniques we are studying and lists their attributes in the three functional areas of vapor flux production, vapor condensation rate control, and film growth control. Table 1 summarizes comparisons of important characteristics of each process.

### 5.1  Glow Discharge Deposition

Glow discharge deposition has been described as chemical vapor deposition with free radicals formed in the discharge. One of the primary strengths of the glow discharge deposition process is the relatively low operating temperature, generally less than 400°C. Similar gas-phase reactions without the glow discharge assistance require thermal energies near 1000°C. Lower operating temperatures are made possible by the highly excited state of the feedstock gases in the glow. Thermal energy required to dissociate outer electrons and promote reactions is normally supplied to the entire atomic system. In the glow discharge system the required energy is supplied to the electrons by the electric fields directly or by impact of free electrons with the outer electrons of the feedstock gases.

Figure 6 shows a schematic glow discharge deposition system. The gas feedstocks are injected at the center of the chamber and are exhausted at the perimeter of the deposition zone. The system pressure is about 200 mtorr. A 13.56-MHz rf potential ignites the glow discharge between the two electrodes. In this case one of the electrodes is also the substrate. Every half-cycle the electrodes will exchange roles as cathode and anode. As the feedstock gas enters the glow, it is dissociated and ionized. As mentioned earlier, at these pressures and frequencies the discharge is completely sustained by the coupling of the rf field to the electrons in the glow region. Ion bombardment of each electrode still takes place on alternate half cycles, but because of the increased scattering probability at higher pressures they arrive at the electrodes with much lower energies than in the lower-pressure glow discharges. Because of the low ion energies, very little sputtering occurs at either electrode. Sputtering rates in this process are also reduced because of the high system pressures. With a higher ambient pressure, collisions with the background gas and subsequent backscattering onto the electrodes of sputtered or desorbed species is very high.

This type of apparatus can produce coatings by both decomposition and compound formation reactions. For a reaction to work in

TABLE 1    Process Comparison

| Process factor | Glow discharge deposition | Planar magnetron sputtering | Ion beam sputtering |
|---|---|---|---|
| Reactor material | Nonmagnetic SS aluminum, quartz | Nonmagnetic SS aluminum | Nonmagnetic SS aluminum |
| Vacuum | 30-500 mtorr | 1-10 mtorr | Gun >0.1 mtorr Chamber <0.1 mtorr |
| Flow/cathode area | 0.5-3 sccm/cm$^2$ | 0.5-3 sccm/cm$^2$ | 0.1-0.5 sccm/cm$^2$ |
| Process gas | Reactive gas (i.e., $SiH_4$, $N_2O$, $N_2$, $O_2$, $C_2H_2$, $AlCl_3$) | $N_2$, $O_2$, $H_2$ | |
| Working gas | Ar, He | Ar | Ar |
| Exhaust gas purification | Extensive scrubbing and pump purge | | Pump purge |
| Power sources | 13.56 MHz | Dc & 13.56 MHz | Dc |
| Power densities (W/cathode area) | 0.03-1 W/cm$^2$ | 4-100 W/cm$^2$ | 1-10 W/cm$^2$ |
| Bias capabilities | Self, rf, dc | Self, rf, dc | |
| Utilization of material | 1-10% | 50% | 50% depositing |
| Deposition rate | To 1000 Å/min typ. 300 Å/min | To 10,000 Å/min typ. 5,000 Å/min | To 5,000 Å/min typ. 2,000 Å/min |
| Substrate geometry | Web, planar | Web, planar, 3-D | Planar, 3-D |
| Substrate temperatures | Ambient-500°C | Ambient-200°C | Subambient-500°C |
| Substrate location | Immersed in glow as anode or cathode, biased or floating | Isolated from the glow and/or biased | Completely removed from the glow could be biased |

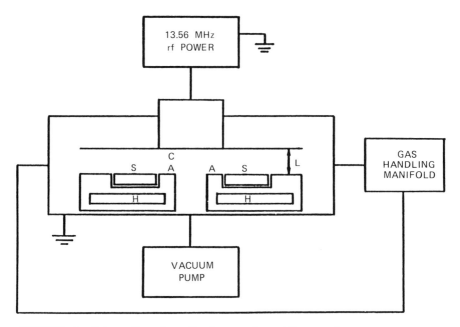

FIGURE 6   Schematic glow discharge deposition system.

this process the reaction products must contain the nonvolatile
surface-modifying material and volatile compounds or gases.

We define the process by the way it handles the three general
categories of deposition processing that we defined earlier. They
are the preparation of the vapor, control of the condensation rate,
and control of the film growth conditions on the substrate. To make
the process more familiar, we examine a specific reaction. For an
example, if the coating desired is simply a decomposition product,
such as amorphous silicon, the feedstock gas can be argon and
a suitable silicon-containing gas such as silane ($SiH_4$). The silane
is dissociated in the glow and deposited on both electrodes. In
principle, the complete reaction is

$$SiH_4 \rightarrow Si + 2H_2$$

In practice, the material deposited is termed a silicon-hydrogen
alloy with 10 to 35 mol.% hydrogen incorporated in the films. The
desired composition is achieved by careful control of the feedstocks
and dilution gas flow rates, rf power levels, and the temperature
of the substrate. Figure 7 shows the dependence of deposition rate
of amorphous silicon alloy on rf power for various concentrations

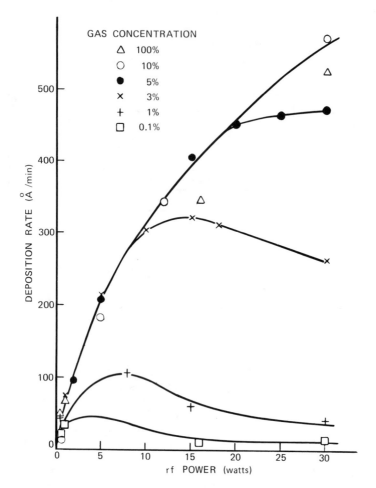

FIGURE 7   Deposition rate on rf power for amorphous silicon (from Ref. 3).

of silane in argon. It is also believed that the influence of ion bombardment may be important in determining the film quality.

Because of the dependence of sheath potentials and local gas flow patterns on the geometry of the reactor, knowledge of precise settings of these factors is not easily transferred from one reactor to similar settings on another reactor.

Composition of the vapor is controlled by the reactant gas composition and the reaction products. These quantities vary as a function of position in the reactor. Because of depletion effects as the gases flow through the reactor, the reactions take place constantly, thus changing the composition of the gas stream. The composition is also coupled to the flow pattern in the reactor. Because of the high pressures necessary to get good reaction rates, motion from the gas to the surface by the excited species is by diffusion. Thus the local composition is also influenced by local temperature fluctuations. The degree of excitation of the reactants is related to their residence time in the glow and the power in the glow. Because of the influence of the electric field distribution on vapor production, the electrode systems are generally kept symmetric and simple, with one of the electrodes being the substrate or having the substrates inset in pockets in the electrode so that there is as little influence as possible of substrate shape on the gas flow patterns.

Condensation rate is controlled by diffusion of the reactants to the surface of the substrate and the probability that they will react to form the desired product before they are excited off the substrate. Condensation rate is also controlled by proximity of the substrate to the other active electrode. Condensation is frequently confined to the two electrodes with similar condensation rates on each.

The film growth conditions are almost completely coupled to the vapor production processes. It is possible to independently heat the substrates, and, as in planar magnetron sputtering, a bias can be applied to the substrate electrode. Gas composition gradients and flow gradients which impact the rate at which reactants reach the substrate heavily influence the growth conditions.

## 5.2 Planar Magnetron Sputtering

The glow discharge in this system is an ion source and a means of dissociation and excitation of reactive gases. The glow is ignited by applying an rf or dc potential between the cathode and the anode ring surrounding it. A cathode dark space and an anode sheath are formed. The source of ionizing electrons is emission from the cathode surface by multiple ion impacts. The ionization efficiency of the electrons is improved manyfold because they are confined to a "race track" (see Fig. 8) around the cathode and cannot escape directly to the anode. The electron confinement is achieved with a magnetic field parallel to the surface of the cathode bridging the "race track." The use of magnets to confine the electron motion is the origin of the term *magnetron*. When an electron emitted from the surface of the cathode begins to accelerate through the cathode dark space, it finds itself deflected by the bridging magnetic field, according to the Lorentz force law, in a direction perpendicular

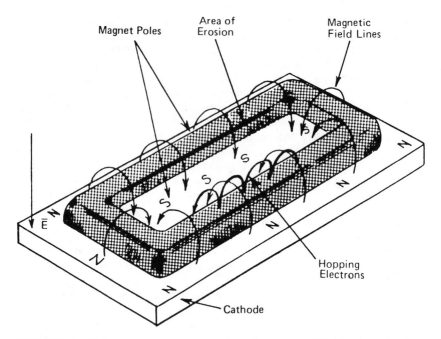

FIGURE 8   Planar magnetron showing "race track" (from Ref. 9).

to both the electric field and the bridging field. As illustrated
in Fig. 8, the electron is forced to move along the "race track"
and back down toward the cathode surface. As the electron approaches
the cathode surface, the cathode voltage slows it and begins to
force it back up toward the glow, and the orbit repeats until the
electron can escape from the magnetic confinement. During each
of the orbits that the confined electron makes, it collides with and
ionizes more atoms, which in turn impact the cathode and provide
electrons to sustain the discharge. Collisions finally allow the electron
to escape confinement.

Because magnetic confinement increases the electron path length
before it is captured at the anode, a lower density of atoms are
needed to produce the same number of ions. The optimum operating
pressure for a planar magnetron source is 1 to 10 mtorr, 10 to 30
times lower than for an unconfined glow.

The cathode erosion rate is proportional to the ion current
and sputtering yield and inversely proportional to the pressure
(since all collision probabilities increase with increasing pressure).
Because sputtering yields level off at higher ion energies (600 to
1000 V), increasing voltages above these levels does not significantly

increase the cathode erosion rate. The highest erosion rates are thus obtained at the highest currents and voltages and at the lowest pressures.

The sputtering deposition rate is the product of the cathode erosion rate and the substrate collection efficiency. Figure 9 shows a system in which the magnetrons are placed on the outside of a cylindrical reactor sputtering toward the center of the cylinder. The substrates are placed on a carousel which rotates around the axis of the reactor. In these systems it is possible to inject reactant gas into the glow near the periphery of the sputtering cathode. The sputtered material and the excited gas mix in the atmosphere of the reactor, but, because of conservation of energy and momentum (Chapman, 1981), it is almost impossible for them to interact in the gas state. If the partial pressure is not too high, the sputtered metal and the gas will react on the substrate and cathode (where momentum can be absorbed). As the partial pressure of the reactant gas increases, presence in the film will increase. At a critical value of partial pressure (and flow rate), the reaction rate between the reactant gas and the cathode material will begin to dominate the process. At this stage the feedstock target becomes poisoned with reaction layers. Frequently there is a decrease in sputtering rate. It is in this region where the coating becomes closest to stoichiometry, and unfortunately the process is somewhat destabilized.

The composition, concentration, and degree of excitation of the vapor in the planar magnetron sputtering process is controlled by the power level on the feedstock cathode and by the nature of the cathode. In reactive sputtering two extra degrees of freedom are brought into play: the partial pressure and the flow of the reactant gas.

The condensation rate is controlled by the position and motion of the substrate with respect to the feedstock cathode. There is also an effect on condensation rate with changes in system pressure and reactant gas pressure. In general, the higher the pressure the more loss channels there are and the lower the condensation rate will be.

The film growth conditions are controlled by substrate heaters and the position of the substrate with respect to the glow discharge. It is possible with a planar magnetron system to almost completely decouple the electrons and the high-energy ions and atoms generated in the glow from influencing the film growth on the substrate. It is also possible to bias the substrates with either an rf or dc potential. In effect, this makes the substrate another cathode in the system. Since the magnetron glow is confined to the surface of the cathode, the glow that is ignited by the bias can be very low energy and highly controlled. This technique can be used to cause

FIGURE 9   Photo of development magnetron sputtering system.

poorly adhered atoms in the coating to be sputtered away, thereby improving the overall quality of the film. The biasing technique can also help coverage by sputtering atoms off the line-of-site surfaces of the substrate and onto surfaces of the substrate that were partially hidden from the cathode.

## 5.3 Ion Beam Sputtering

In glow discharge deposition and magnetron sputtering, the substrate is always exposed to the energetic glow discharge. It is difficult to strictly control the influences of the glow discharge on the coating growing on the substrate. The ion beam deposition system separates the substrate from the glow discharge completely. With ion beam sputter deposition, the environment of the substrate can be independently controlled separately from the sputtering target.

Figure 10 shows a schematic ion beam system. The glow discharge in this situation is used as the source of ions for the sputtering process. The dc glow is ignited between the hot filament cathode and the anode in about 1 mtorr of argon. The glow then expands to fill the chamber separating itself from the anode, walls, and screen grid by the plasma sheath. The plasma potential, always the most positive potential in the system, is determined by the electron and ion loss rate from the glow, and is about 5 V higher than the anode voltage. Impact from high-energy electrons maintains the ionization of argon atoms. The source of electrons for the glow discharge is the hot filament cathode. The ionization efficiency of these electrons is further increased by application of an axial magnetic field near the surface of the anode. The Lorentz force on the moving electrons causes them to spiral back into the glow to ionize more atoms. No electrons need to be emitted from the cathode by ion impact. Typically, the discharge voltage is about 40 V, a few times the ionization potential of argon (15.8 eV). The cathode, chamber, and screen grid are all kept at the same potential.

Ions are extracted from the glow through the holes in the screen grid and accelerator grid by negatively biasing the accelerator with respect to the screen grid. This voltage difference determines the energy of the extracted ions.

Outside of the ion beam source the extracted ions are neutralized to keep coulomb repulsion forces between the ions from causing the beam to spread out. A thermal electron source is frequently used as a neutralizer. Once neutralized, another glow discharge has been formed. The net momentum of the ions and electrons in this glow is directed toward the sputtering targets. Since the targets are kept at ground potential, the plasma potential of the extracted glow is a few volts positive.

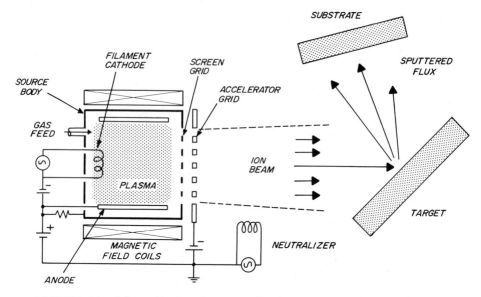

FIGURE 10   Schematic ion beam sputtering system.

Extraction is accomplished by making the accelerator grid slightly
negative, about -100 V, and raising the anode to the ion voltage
desired. The accelerator potential is slightly negative because if
it were left at ground potential, electrons from the neutralizer would
flow back up into the beam chamber through the screen grid circuit.
Application of this isolation voltage causes slight defocusing of the
extracted glow.

In an operating ion beam system, the ion beam deposition reactor
contains the ion beam source, the sputtering targets, and the substrate holder. The angle of incidence of the neutralized beam on
the targets is variable. This allows the sputter yield to be optimized
for a particular material and ion energy. It is also possible to vary
the orientation of the substrate holder in relation to the target
position, to heat or cool the holder, or to make the substrate rotate
on an axis perpendicular to the holder's surface.

The deposition rate on the substrate is controlled by the cathode
erosion rate and the substrate collection efficiency. The target
erosion rate is controlled by the ion current voltage and the orienta-
tions of the target and ion beam. The composition of the vapor, as
in planar magnetron sputtering, is essentially the same as the feedstock target composition. The degree of excitation of the vapor
flux can be controlled by controlling the energy of the ions in the
ion beam. Higher impact energies will shift the energy distribution

of the vapor flux to higher energies. The angle of inclination between the target and the ion beam influences the rate and degree
of excitation of the vapor flux.

The condensation rate is controlled by the orientation and location
of the substrate with respect to the vapor flux from the feedstock
target.

The film growth conditions are to the greatest extent possible
decoupled from the vapor production processes. This technique
offers the most flexibility in tailoring the growth conditions. Since
there is no bombardment by high-energy species from the vapor
production region, the surface mobility of the depositing species
can be controlled by the substrate temperature and the energy
spectrum of the sputtered atoms. In these systems the substrates
can be cooled to low temperatures or heated in a controlled way.

## 6 TECHNIQUE CHARACTERISTICS

Technical issues that come into play when choosing the appropriate
surface modification technique can be divided into two categories:
substrate-related requirements and coating-performance-related
requirements. The characteristics of the substrate which influence
the type of modification technique that can be used include size,
shape, electrical conductivity, temperature sensitivity, and chemical
composition. The coating-performance-related issues are composition
tolerance, uniformity requirements, coverage (throwing power),
deposition rate, thickness precision, and perfection (freedom from
impurities and defects).

Table 2 lists the process constraints imposed by each of the
three surface modification techniques. The substrate limitations
are based on proven equipment commercial designs. It is possible
to conceive of special designs which would remove some of these
constraints.

The cost of a surface modification technique is always a factor
in the choice between technical solutions. All of these techniques
require large capital investments for production scale equipment.
Amortization of these costs over a large number of parts always
has beneficial impact on the economic analysis of the total product
costs but is rarely the dominant factor. Facilities expenses (humidity
control and other environment control costs), utilities costs, feedstock costs and utilization rates, cleaning system costs, parts
handling costs, and rework costs all must be carefully considered
in the context of being integrated with the rest of the production
system and its economic batch sizes and process throughput for
each product specification. A meaningful analysis of the interplay

TABLE 2　Substrate and Coating Characteristics Comparison

| | Glow discharge | Planar magnetron sputtering | Ion beam sputtering |
|---|---|---|---|
| *Substrate* | | | |
| Size[a] | 2-D web × 40 cm × 0.5 mm<br>3-D | Web × 3 m × 10 cm<br>35 cm dia. | 10 cm × 10 cm × (1-4 cm)<br>4 cm dia. |
| Shape | Flat, web | Flat, web cylind. outside diameters | Flat, cylind. inside and outside diameters |
| Process temp. | Up to 400-500°C | 50-200°C | -196-200°C |
| *Coating* | | | |
| Composition | B,C,O,N,Al,Si (including compounds) incorporation of reaction products likely | "Unlimited"b; incorporation of sputter gas likely | "Unlimited"b; little or no incorporation of sputter gas |
| Uniformity | 2% across 80% of web | 1% across web or sheet | 10% without planetary |
| Throwing pwr. | Poor | Line of sight | Line of sight |
| Dep. rate | <100 nm/min | <1200 nm/min | <200 nm/min |
| Precision | 2-D 10 nm<br>3-D NA | 1 nm<br>100% of coating thickness | 1 nm<br>100% of coating thickness |
| Perfection[b] | Almost perfect | Dust and cathode debris limited | Dust and cathode debris limited |

a2-D shapes: dimensions are approximate maximums or minimums for length × width × thickness. 3-D shapes: dimensions are approximate cylinder diameters. Web means a continuous coil of substrate.
bThis includes reactively deposited coatings of nitrides, oxides, and carbides.

between the economic factors which influence the choice of a modification technique depend sensitively on the elements of the production process that are specifically associated with each product. This sort of analysis needs to be done on a case-by-case basis and is beyond the scope of this chapter, even for the successful applications we describe.

# 7  APPLICATIONS

## 7.1  Stainless Steel Corrosion-Resistant Coating Application

Failure of 400 series architectural quality stainless steel by salt spray corrosion significantly restricts its commercial use. The market for this material could be expanded by an increase of the corrosion resistance of the highly polished surface. The modification techniques we have discussed have the potential for solving this problem.

Salt spray corrosion of stainless is known to be related to establishment of small electrochemical discharges between certain defect sites and the steel substrate. The battery action of the discharge causes chlorine ions to move in the salt spray from the defect to the metal atoms and to react with the metal atoms to form a soluble compound which is washed away from the defect area by the natural flux of the liquid. To be effective as a corrosion inhibitor, the coating must reduce both electrical currents and chlorine ion transport rates from the defects to the steel. Stoichiometric, defect-free silicon dioxide is known to be a very good barrier to electron and ion transport. Silicon dioxide can adhere to adequately cleaned stainless steel. Though this coating is brittle, many architectural applications do not require severe deformation of the stainless steel, so coating cracking is minimized. Finally, because of its hardness, silicon dioxide has excellent abrasion resistance.

Basic research into the influence of thin stoichiometric $SiO_2$ coatings on 400 series stainless steel showed significant improvements in the corrosion resistance when the coating thickness was greater than 100 nm. The coatings were prepared by glow discharge deposition and reactive sputtering. On a research basis either of these processes could be made to produce the correct stoichiometry and thickness layer. The decision between the two processes was made by comparing the strengths of the processes using the substrate- and coating-related criteria we developed above.

Substrate-related criteria for the choice of the modification technique are

Size:  Large coils of stainless steel 30 cm wide and 100 m or more long (called a web)

Shape:   flat substrates about 0.1 cm thick and 30 cm wide
Conductivity:   adequate electrical conductivity to be used as an
   electrode
Temperature:   no significant surface or material degradation at
   temperatures below 500°C
Composition:   the stainless steel composition will neither degrade
   nor contaminate any of the vacuum systems

Coating performance specifications for this applications are

Composition:   Highly stoichiometric $SiO_2$
Uniformity:   ten percent thickness variation across the central 80%
   of the web to minimize variation in coating conductivities and
   optical interference effects
Coverage:   coverage of corrosion nucleation sites
Deposition rate:   the faster the better
Precision:   minimum thickness of 1000 Å; no interference fringe
   effects
Perfection:   The coating must have fewer imperfections per unit
   area leading to corrosion nucleation sites than the bare substrate
   had

According to Table 2, the size and the shape of the substrate
dictate either glow discharge or planar magnetron sputtering deposition. An initial strike against the sputtering process is the difficulty
in dealing with quartz cathodes. They are fragile, do not conduct
heat well, and have very low deposition rates. The requirement
for very low defect density and for high deposition rates swings
the decision toward glow discharge. The need for high temperatures
to drive off impurities and the requirement for good stoichiometry
control for a silica coating make the choice of glow discharge over
sputtering clear.
   As always with glow discharge deposition, the coating is immersed
in the glow and in the flowing dissociated, excited feedstock gas.
This close coupling and strong interaction between vapor formation
processes and coating growth factors is one of the chief engineering
challenges. The following list gives a capsule view of the interactions between the important process factors and their influences:

Power:   increases ionization and the volume that the glow occupies
Pressure:   increases reactant density, diffusion rate, and glow size
Electrode spacing:   decreases power density, decreases flow velocity
Temperature:   enhances desorbtion of undesirable reaction products,
   activates surface reactions, improves adhesion

Reactant dilution:   improves discharge uniformity and stability,
    reduces reactant density and residence time, changes excitation
    in glow
Reactant rations:   controls film composition and related properties

Though there are many coupled factors, this process can be
engineered so that it does not contain pathological regions in parameter space where the sensitivity of the film properties makes operation
impossible. The key to successful operation of a glow discharge
process is a good statistical design for the process optimization.

Figure 11 shows a glow discharge coating machine designed
and built to produce corrosion-resistant coatings on stainless steel.
The substrate is a 100-m coil high-quality stainless steel. Figure 11
shows the substrate control chambers (for web take-up and payout)
and the reactor deposition chamber. The control panel and the gas-
handling system are not in view.

FIGURE 11   Roll to roll stainless steel glow discharge deposition
system.

## 7.2 Planar Magnetron Sputtering of Machine Tool Coatings

One direction of machine tool development focuses on maximizing metal removal rate while maintaining or improving cutting precision. The cutting environment of a machining operation with a high metal removal rate combines high-temperature and high stress levels in an oxygen-containing ambient. Cutting temperatures typically reach 900°C (Kramer and Judd, 1985), and stress approach the ultimate strengths of the machine tool itself. Under these conditions the cutting tool must retain dimensional stability to produce a consistent metal removal rate. Tool dimensional tolerances are from 1 to 25 microns. Thus, in a certain sense, control of the final 1 to 25 microns of the tool surface wear properties is crucial to optimizing tool performance.

The coatings used to reduce wear need to have high hardness at high temperatures and low chemical activity. Wear mechanisms for machine tools have at least two components: abrasion and chemical reaction. Material systems with high hardness and high chemical stability have proven themselves as wear-resistant coatings (DuMond and Youtz, 1935). The class of materials of interest for wear-resistant tool coatings is the refractory carbides, nitrides oxides, and borides. One of the successful examples of this type of coating is TiN.

Any wear-resistant coating must have good adhesion to the tool. The effect of the deposition technique on adhesion, hardness, and chemical stability significantly influences the choice of a coating process. Experience has shown these influences to be comparable to the importance of choosing the correct materials system. Looking at our Table 2, we see immediately that the glow discharge deposition technique is not generally applicable because of its poor throwing power and limited substrate shape tolerance. Ion beam sputter deposition has potential but is limited by the substrate size and number. Planar magnetron sputtering is the process of choice.

Figure 12 shows a small planar magnetron development reactor used for the successful application of a HfC/HfN multilayer wear-resistant tool coating. This material system has the potential for surpassing the performance of TiN as a wear-resistant coating for tools cutting steel.

The multilayer coating system is instructive from the point of view of some of the pitfalls that can occur when combining regular and reactive deposition. Figure 13 shows fracture cross sections of two samples of multilayer. The stepped multilayer was deposited at a low rotation rate. The drill moved slowly from the deposition field of one cathode to the deposition field of the other. In the second part of this figure, we see an almost perfectly dense coating which resulted from a higher rotation speed. The key to understand-

FIGURE 12   Development planar magnetron sputtering unit and tool coating carousel.

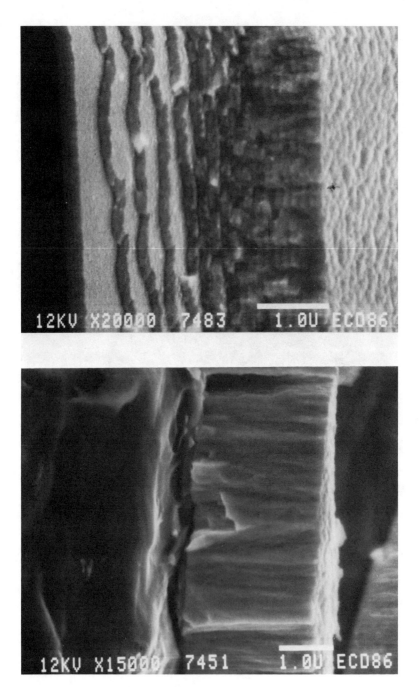

FIGURE 13  Fracture cross sections of two samples of multilayer coating.

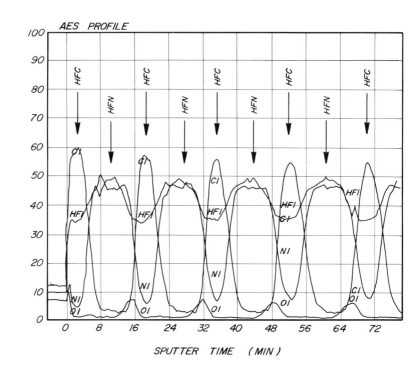

**FIGURE 14**   Auger depth profile of multilayer coating.

ing the results is shown in Fig. 14. This is an Auger depth profile
of the stepped coating. The oxygen between the layers is what
was causing the brittle fracture. The oxygen is an impurity in
the reactor. When the rotation rate was increased, the brittle fracture
was eliminated. The cause of the oxygen contamination was too long
a time between targets.

Though there are some problems in optimization of multifactor
systems, Fig. 15(a) shows a response surface analysis of two of
the factors influencing drill life, nitrogen flow, and power on the
HfC target. The response surface was obtained from a fractional
factorial experimental design. Figure 15(b) shows the agreement
between the predicted drill life and the measured drill life. The
R factor for this fit was 89%. Finally in Fig. 16 we show a small
sample of drill life data comparing uncoated, TiN-coated, and two
multilayer runs (773) and 787), indicating that doubling the life
of a TiN-coated drill is possible with this multilayer system.

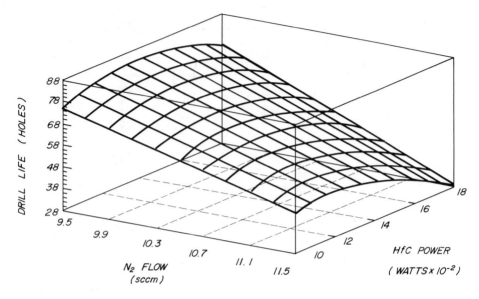

FIGURE 15a   Drill life response surface.

## 7.3   Ion Beam Sputtering of X-Ray Reflectors

A long-sought goal has been to extend the range of applicability
of mirrors, microscopes, telescopes, and diffraction gratings from
the visible and near-visible portions of the electromagnetic spectrum
to the x-ray region of the spectrum. Except for a few special grazing
incidence applications, it has not been possible to effectively use
the insights of geometrical optics in the x-ray wavelength range.
This failure in part is due to the low reflectivity of most surfaces
to x-rays. It has long been known (Kadin and Keem, 1986) that
x-ray reflectivity of surfaces can be enhanced by using coatings
in a fashion stimilar to the way they are used in optical multilayer
interference mirrors and filters. For x-ray mirrors, a stack of
two alternating materials (tungsten and silicon, for example) with
differing optical properties in the x-ray range are made so that
the layer spacing is of the same order as the wavelength of the
x-rays striking the surface. For certain angles of incidence such
a structure will produce reflectiveness or transmissions approaching
100% (13). The difficulty has been in synthesis of multilayer stacks
of the correct materials with repeat spacing of 1 to 10 nm.
    For success in this application it is crucial to have maximum
control over the film growth parameters. The deposition technique

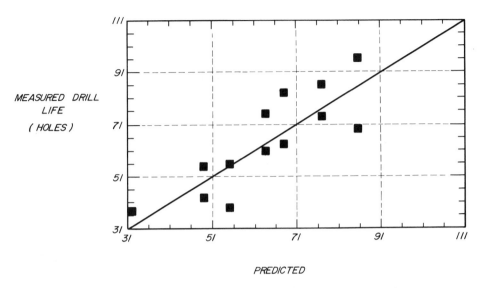

FIGURE 15b   Comparison of predictions from response surface analysis and actual drill life.

which operates at the lowest pressure and most completely decouples the substrate from the energetic region associated with the production and excitation of the vapor is the ion beam deposition system. This technique gives the greatest degree of engineering control over the processes which determine the quality of the multilayer system. Because of the physical separation of the source from the substrate, no uncontrolled flux of electrons or ions or fast-moving neutral particles reach the substrate. The only vapor species that reach the substrate are the sputtered atoms and the background gas. Further, by control of the sputtering ion beam energy and current, the energy spectrum and flux of the sputtered atoms can be controlled independently. Control of the energies and flux gives some control of the surface mobilities of the depositing atoms, thereby influencing the film growth morphology.

Figure 17 shows a high-resolution scanning transmission electron micrograph of an amorphous tungsten-amorphous carbon multilayer stack with a repeat spacing of 6.3 nm. The multilayer was deposited on a silicon wafer (single atoms can just be resolved in the micrograph) of the type used in the semiconductor industry. The extremely abrupt interfaces and the exact replication of these interfaces testify to the control of the deposition process that can be achieved.

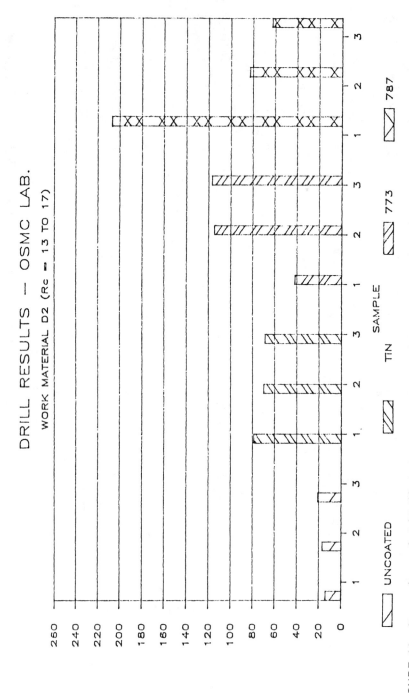

FIGURE 16  Comparison of drill life between uncoated, TiN-coated, and multilayer HfN/HfC coated. Drill size, 7 mm; material SKH9; cutting speed, 25 m/min; feed rate, 0.15 mm/rev; type of hole-through, 20 mm; coolant, emulsion.

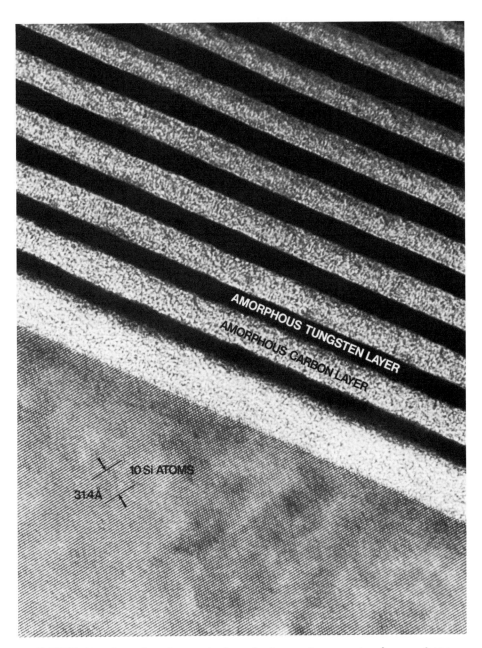

FIGURE 17  Scanning transmission electron micrograph of amorphous tungsten-amorphous carbon multilayer stack.

## 8  CONCLUSION

This chapter has emphasized the general understanding of the factors
which influence the quality of surface modification as implemented
by glow discharge deposition, planar magnetron sputtering, and
ion beam sputtering. We have separated the general process of
surface modification into three functional categories: control of
the production of the flux, control of the condensation rate of the
flux, and control of the film growth parameters. Each surface modification process handles these functions very differently, and each
has its special strengths and weaknesses. The value of this separation
is that it allows a consistent evaluation of the strengths of each
process for any particular application. If this method of functional
blocking is used, the analysis of a process does not require detailed
knowledge of all the aspects of each modification technology, but
rather a comparison of the appropriateness of how the process accom-
plishes the tasks of each of the functional blocks.

## REFERENCES

1.  Bunshah, R. F. et al. (1982). *Deposition Technologies for Films and Coatings*, Noyes Publications, N.J.
2.  Vossen, J. and Kern, W. (1978). *Thin Film Processes*, Academic Press, New York.
3.  Chapman, B. N. (1981). *Glow Discharge Processes*, Oxford University Press, London.
4.  Maissel, L. and Glang, R. (1970). *Handbook of Thin Film Technology*, McGraw-Hill, New York.
5.  Westwood, W. D. and Cuomo, J. J. (1985). *Sputter Deposition and Ion Beam Processes*, The Education Committee, American Vacuum Society, New York.
6.  Schwoekel, R. C. (1984). *Panel Report on Coatings and Surface Modification*, Council on Materials Science, Washington, D.C.
7.  Coburn, J. W. (1976). *J. Vac. Sci. Technol.*, *13*:1037.
8.  Coburn, J. W. (1971). *Thin Solid Films*, *64*:371.
9.  Thornton, J. (1982). *Deposition Technologies for Films and Coatings* (R. F. Bunshah, ed.), Noyes Publications, N.J.
10. Kramer, B. M. and Judd, P. K. (1985). *J. Vac. Sci. Technol.* *A3*:2439.
11. DuMond, J. and Youtz, J. P. (1935). *Phys. Rev. B*, *48*:703.
12. Kadin, A. M. and Keem, J. E. (1986). *Scripta Metallurgica*, *20*:443.

# 5

# Plasma Treatments

SIDNEY DRESSLER  *SECO/WARWICK Corporation,*
*Meadville, Pennsylvania*

## 1  DIFFUSION COATINGS

Diffusion coatings, which are quite distinct and different from overlay coatings, are becoming an increasingly important engineering
specification for the production of hard, corrosion-resistant, wear-
resistant, and fatigue-resistant surfaces on core materials that
can then be selected for their other properties. Diffusion coatings
are produced within the original boundaries of a workpiece and
are generally characterized by a concentration gradient of the chemical species added during the coating process with a maximum value
at the surface of the workpiece, decreasing to a minimum value
within the core of the original material.

Plasma surface treating is an extension of existing technologies
for the production of these diffusion coatings. Nitrogen, carbon,
and boron, among other chemical elements, can be added to the
surface of a workpiece by bringing an activated nitrogen-, carbon-,
or boron-bearing gas into direct contact with this surface while
it is being held at an elevated temperature. These chemical elements
can be added by using any one of several established surface-treating
technologies, each technology producing nearly identical metallurgical
results. Plasma surface treating is simply a new means to an already
familiar end.

It is an advantage that plasma surface treating is a normal
diffusion-limited process. After the introduction of the desired

chemical species, the metallurgy at the near surface, inside the workpiece, depends mainly on the concentration gradient of this species and only in a limited way on the surrounding plasma parameters. As an example, during nitriding, these surrounding plasma parameters would remain important only if variable compound zone characteristics were required. This chapter, then, can deal with just the production and maintenance of plasmas for surface treating, since there is no important need to review the already familiar mechanisms of surface treating or to describe a multitude of competing technologies. This more familiar information is readily available and can generally be obtained from a large body of organized literature.

Instead there will be an effort to describe just those features of plasma surface treating that make it different from older and more established technologies. This will require a limited review of the physics of the electrical glow discharge and then a detailed description of the equipment that is currently being used for these plasma processes. The special advantages of bell furnaces for independently heating the workpieces to the treatment temperature and of pulsed plasma power supplies for establishing the chemical activity of the species being added will be described, since these bear heavily on the current acceptance of this technology as a manageable one for surface treatment at the industrial level. Discussion will be limited to those aspects of the final metallurgy that are the direct result of this unique system. Some operating data will be presented to show what has already been accomplished and to serve as a guide for the preparation of new processing specifications for plasma surface treatments.

## 2   PLASMA SURFACE TREATING

Plasma surface treating differs from other surface treating technologies by using the phenomena of the electrical glow discharge to activate the gas species that is needed for each particular process. There are important consequences to activating the gas species in this manner.

1.   Plasma treating permits better control over the final work surface composition, its structure, and its properties. Plasma nitriding, as an example, can be accomplished without the formation of a mixed-phase, brittle, compound zone. Plasma nitriding can be at temperatures below those employed for conventional nitriding, retaining maximum core properties.

2. The electrical glow discharge permits faster deposition rates at lower surface temperatures. Plasma carburizing time, for example, can be shortened, without relying solely on a high treatment temperature, with its attendant potential for distortion, to increase the rate of carbon diffusion. There is evidence that a carbon-rich surface layer can be deposited much faster on a work surface that has first been sputter-cleaned in an electrical glow discharge. This high-carbon concentration gradient can be reliably produced, without formation of soot, and at every specific work surface temperature it promotes faster carbon diffusion.

3. Plasma treating can produce the required active species in a low-pressure gas mixture and, by this means, can eliminate the expensive postcleaning operations and the serious environmental problems that are always associated with treatment in a liquid salt. Plasma boriding, for example, can eliminate the requirement for a molten salt, the special work surface cleaning procedures that are required after this treatment, and then the eventual disposal of the cleaning solutions themselves in a manner spproved by the Environmental Protection Agency (EPA).

Nitriding, carburizing, and boriding with positive ions derived from the plasma of an electrical glow discharge is an effective surface treatment for cast iron and alloyed steel and, in some special cases, nonferrous metals to produce wear-resistant, fatigue-resistant, corrosion-resistant, and superficially hard surfaces. Plasma treating can improve metallurgical quality while reducing core material costs. Sputtering at low temperatures to remove superficial contamination from the workpiece permits short surface treatment times. And plasma treating can solve environmental problems, since, with few exceptions, the gas discharge from a plasma treating furnace is nontoxic and nonexplosive and can be vented directly to an outdoor location.

## 3 GLOW DISCHARGE FUNDAMENTALS

The glow discharge used for plasma treating occurs when an external voltage is applied between two electrodes, positioned within a gas mixture at some suitable partial pressure, as shown in Fig. 1. One electrode, called the anode, is the vacuum retort, electrically at ground potential. The other electrode, called the cathode, is the workpiece to be plasma treated. The workpiece is connected to operate at a negative potential with respect to the grounded vacuum retort. The voltage source in Fig. 1 supplies a controlled variable

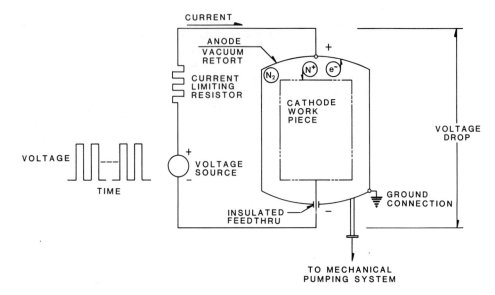

FIGURE 1    The electrical circuit for a glow discharge.

voltage. The current-limiting resistor permits varying the resistance in the external electrical circuit so that the current can be independently controlled at any voltage level. With this equipment arrangement, the current in the external circuit can be measured as a function of the voltage drop between the anode and the cathode.

In the operation, the space between the vacuum retort and the workpiece is filled to some partial pressure with a gas mixture selected for the process. The glow discharge occurs when molecular elements in this gas mixture are ionized by collisions with electrons traveling from the workpiece-cathode to the vacuum-retort-anode under the influence of the applied electrical voltage. Ionization of the partial pressure gas mixture permits a sustained electrical current—i.e., a negative electron flow from the workpieces to the vacuum retort and, more importantly, a positive ion flow from the ionized gas mixture to the workpieces being treated.

As shown in Fig. 1, after the gas has become ionized by electrons leaving the surface of the workpiece, the newly formed positive ions will be accelerated toward the workpiece where they can combine with the chemical elements of the work surface. If the partial pressure gas mixture is predominantly nitrogen, the work surface can be nitrided by these nitrogen ions. If the gas mixture is a hydrocarbon, the work surface can be carburized.

In summary, the work surface forms the cathode and the vacuum retort forms the anode for an electrical glow discharge. A regulated voltage pulse is applied between this cathode and the anode to produce a positive ion flow of some selected chemical species to the surface of the workpiece. Power supplies currently in use with this process permit the amplitude of the voltage pulse, its on-time duration, and its repetition rate to be independently selected and controlled to produce uniform coverage of the work surface with positive ions from the plasma without significant thermal heating. With these pulsed power supplies, the ion concentration gradient at the work surface and the work surface temperature can each be independently regulated by the process controller.

### 3.1 Partial Pressure Operation

The requirement that the gas mixture for the glow discharge be at a partial pressure is an important one that bears directly on the probability that molecular elements in the gas mixture will be ionized by a collision with an electron traveling from the work surface to the surface of the vacuum retort. Figure 2 shows how the gas mixture partial pressure effects the probability of ionization.

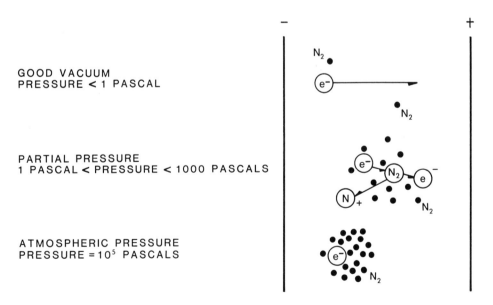

FIGURE 2   Number of gas molecules versus probability of ionization.

The distance that an electron will travel before colliding with a gas molecule is called its mean free path. This mean free path is proportional to the absolute temperature of the gas and inversely proportional to the gas pressure. In a plasma treating furnace, operating at a temperature well above ambient and at a pressure considerably below atmospheric, this mean free path becomes a process parameter with considerable significance.

In a vacuum retort pumped below 1 Pa, where 1 Pa = $1 \times 10^{-5}$ bar, the mean free path will be large, there will be few gas molecules in the electron path between electrodes, and there will be too few collisions between electrons and the gas molecules to sustain an electrical current. The number of electrons leaving the negative work surface is established, in part, by the applied voltage. A high voltage can be employed, if necessary, to produce a large number of electrons to increase the probability of ionization. However, as shown by Fig. 2, in a good vacuum, where there is considerable distance between gas molecules, the probability of an electron-gas molecule collision will still be low. If the gas mixture partial pressure is increased to between 1 and 1000 Pa, the mean free path will be reduced and there will be an adequate number of gas molecules to ensure gas ionization levels that can sustain a glow discharge with a current value suitable for plasma treatment. If the gas pressure is increased further, there will be more than enough gas molecules to ensure collisions with electrons. However, the electrons will now have such short mean free paths between collisions they will probably not acquire enough energy to cause ionization of any of the encountered gas molecules unless the voltage is made higher than values considered practical for plasma treating.

## 3.2  Paschen's Law

This dependence of the probability of ionization on the number of gas molecules between electrodes has been formally developed and is known as Paschen's law. Paschen's law, for planar parallel electrodes, is normally presented as a graph showing the minimum voltage that can be expected to produce a sustained glow discharge as a function of the gas pressure multiplied by the distance between electrodes. A typical graph is shown in Fig. 3.

In most recent technical publications, pressure measurements in millibars have been replaced by pressure measurements in pascals. In the International System of Units, 100 Pa = 1 mbar. In Fig. 3, the gas pressure, measured in millibars, multiplied by the electrode separation, measured in centimeters is proportional to the number of gas molecules in the space between electrodes at any fixed temperature. The minimum voltage for a glow discharge in any gas

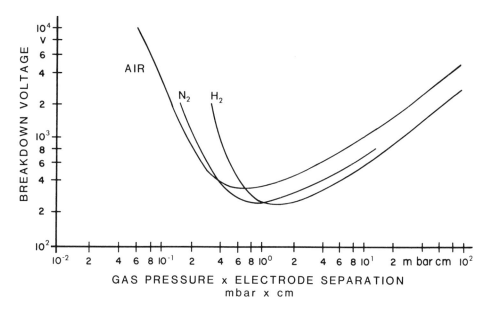

FIGURE 3  Paschen curve for some gases.

mixture is a function of the gas density and is not simply a function of the gas pressure. It is possible, for instance, for the gas pressure to be high but at a high gas temperature, and the system will not be at a gas density that can sustain a glow discharge at a specific voltage level.

### 3.3  Glow Discharge Electrical Characteristics

The voltage/current density characteristics of a glow discharge can be established with the scheme shown in Fig. 1. Actually the voltage/current measurements that can be directly made in the external circuit will be important only to the equipment manufacturer, who must establish a power supply rating for an anticipated application. The process engineer must be concerned with the phenomena within the discharge, at the surface of the work, and it will be necessary for his interests to divide the measured currents by the measured or calculated surface area of the workpiece to obtain the more meaningful voltage/current density relationship.

*The Nonmaintained Region*

Consider what happens when a voltage is applied between the vacuum retort and the workpieces in a gas mixture at a partial pressure of

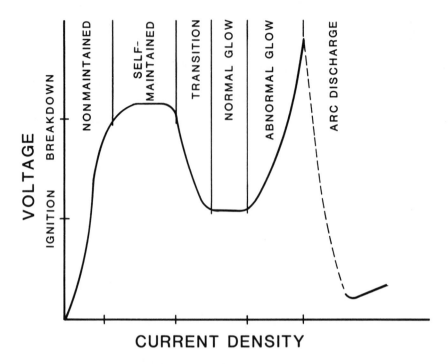

**FIGURE 4**   Glow discharge electrical characteristics.

several hundred pascals, as shown in Fig. 1. There will always be a few free electrons present in a gas mixture from random ionization produced by external agents like light and background radiation. With very low values of applied voltage these electrons will be accelerated toward the anode giving rise to a small but measurable current as shown in Fig. 4.

At some value of applied voltage, called the ignition voltage, these electrons can be accelerated toward the vacuum retort and will obtain sufficient energy to cause ionization when they collide with a gas molecule. With this additional gas ionization, the current can increase to a value higher than could be maintained by ionization from the external agents acting alone. In this electrical operating region, if the voltage between electrodes is reduced, the current will be reduced. This, then, is properly called the nonmaintained region with current densities in the $10^{-9}$ mA/cm$^2$ range.

*The Self-Maintained Region*

At some higher value of applied voltage, called the breakdown voltage, sufficient energy can be given to each of the electrons originating at

the cathode to cause an ionizing chain reaction. This breakdown
voltage is the voltage predicted by Paschen's law. More electrons
can be liberated by gas ionization than were present initially, and
many of these newly liberated electrons have sufficient energy to
cause additional ionization of the gas mixture. Since the distance
between the vacuum retort and the work surface is fixed by the
design of the equipment and the placement of the workpieces, at
any given temperature the number of gas molecules between the
two electrodes will be determined by the partial pressure of the
gas mixture. With applied voltages greater than the breakdown
voltage, new electrons can be formed by this ionization chain reaction
and the current in the external circuit can increase without any
appreciable increase in the voltage drop through the discharge
between the two electrodes. In fact, in this operating region the
current must be limited by the current limiting resistor in the
external circuit. This portion of the voltage/current density characteristic is called the self-maintained region, where current densities
in the $10^{-4}$ mA/cm$^2$ range can be measured.

### The Transition Region

If the current limiting resistance in the external circuit is now
decreased to lower values, the current density will increase while
the voltage drop between the vacuum retort and the work will
actually decrease. With this negative characteristic, where increasing
current density is accompanied by a decrease in discharge voltage
followed by a further increase in current density, it is difficult
to establish a stable operating point and this, therefore, is called
the transition region.

### The Normal Glow

When the external current limiting resistance is further reduced,
the voltage/current density characteristic enters the normal glow
region. In this operating region, a visible glow a few millimeters
in thickness, measured from the face of the workpiece will be seen
covering a portion of the work surface. The intensity of this visible
glow will appear uniform, evidence of a uniform current density
on those areas that are covered. As the current limiting resistance
is reduced further, the area covered by the visible glow will increase
until the entire work surface is finally covered with a visible glow
of higher, but still uniform, intensity. The current density at the
work surface, covered by the visible glow, will be uniform only
if all of the local parameters that can effect the discharge are themselves uniform.

The geometry of this visible glow is important to plasma surface
treatment. It can be seen that in those areas that are covered, the

visible glow accurately replicates the physical shape of the work
at a distance a few millimeters from its surface. Since the thickness
of this visible glow is very small compared to the overall distance
between the work surface and the vacuum retort, the operating
point on the voltage/current density curve will be determined mostly
by conditions that exist very near to the work surface. The fact
that the measured distance between the work surface and the vacuum
retort is different at different locations in the charge being plasma
treated is not critical to the electrical characteristics of the normal
glow discharge. If the work surface temperature and the composition
and partial pressure of the gas mixture are uniform near the work
surface, the conditions for a uniform current density over the entire
work surface will, in fact, exist.

This region, where the visible glow can spread from a small
local spot to cover the entire work surface, is called the normal
glow discharge region. The voltage drop from the vacuum retort
to the work surface remains nearly constant while the current density
is increased to about $10^{-1}$ mA/cm$^2$ in the normal glow discharge
region.

### The Abnormal Glow

A further decrease in the external current limiting resistance brings
the voltage/current density characteristic into the abnormal glow
discharge region. In spite of this name, which has historical implica-
tions, this is the region useful for plasma treating. It is only in
this abnormal glow discharge region that the work surface will be
completely and uniformly covered by the visible glow. The current
density, that is important for promoting the desired physical and
chemical reactions of plasma treating, will be uniform and at the
highest value that will still permit easy electrical control. Since
the visible glow already covers all of the work surface, an increase
in current density will now be accompanied by an increase in the
voltage drop through the resistance of the glow discharge. This
positive characteristic, where an increase in current density is
accompanied by an increase in voltage drop, is desirable because
it permits stable control of the operating point in the abnormal
glow discharge region. The current density in the abnormal glow
discharge region is generally between 0.1 and 5.0 mA/cm$^2$ for voltage
drops between 400 and 800 V.

### The Arc Discharge

If the power supply voltage is increased in the abnormal glow discharge region, the current density will increase. These increases
in both the voltage and the current density will produce an increase

in the power density and, therefore, the thermal energy delivered to the work surface through the plasma from the electrical power supply. If the delivered power is allowed to increase to values high enough to cause local overheating of the work surface, the resultant increase in electron emission will allow an additional increase in the current density. The glow discharge will concentrate itself in this overheated area, and a high thermal energy arc discharge will occur. This arc discharge, if it is allowed to persist, will cause noticeable pitting and even melting of the work surface. Proper equipment design and good plasma treating practice can limit the frequency and duration of arc formation.

These arc discharges can be turned off after a time interval that varies with the design quality of the power supply. In fact, arc discharge turn-off time is just one important indicator of power supply quality. Long switch-off times permit some damage to the work surface and switch-off times greater than 2 μs generally permit observable damage to polished work surfaces. Fortunately, state-of-the-art power supply software and hardware can interrupt an arc discharge, once formed, before a significant amount of power and thermal energy can be delivered to the work piece.

Marciniak and Karpinski have investigated the transition from the abnormal glow to the arc discharge region. They have measured the discharge power as a function of voltage as shown in Fig. 5. These measurements were made in a laboratory unit using a disk-shaped cathode 160 mm in diameter. Measurements were made in four operating regions.

1.  The normal glow discharge, at very low levels of discharge power, where the voltage remained constant with increasing discharge current and power
2.  The abnormal glow discharge, labeled glow discharge region, where stable operation was obtained
3.  A region where either the abnormal glow or arc discharges could occur
4.  The arc discharge, where arcs were consistently observed

The normal operating point for the experimental work, labeled a in Fig. 5, was located at the start of the abnormal glow region when the partial pressure of the nitrogen gas was between 100-500 Pa, where 1 hPa = 100 Pa. It was noted by these investigators that there is an optimum pressure for maximum discharge power without arcing, in this case 200 Pa, and that as the pressure is increased beyond this value, arc discharges will occur at lower discharge power levels. At higher partial pressures, a distinct transition from an abnormal glow to a stable arc discharge can be expected.

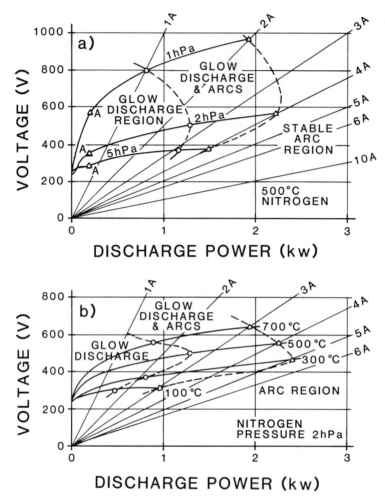

FIGURE 5   Characteristics of the glow discharge in a chamber for thermochemical treatment.

### 3.4 Additional Characteristics of the Glow Discharge

A qualitative description of the physical geometry of the glow discharge is helpful for understanding a special technique that can be employed for masking selected portions of the work surface from the effects of the plasma. In the abnormal glow discharge region the electrical and luminous characteristics of the plasma in the space between the work surface and the vacuum retort take the form shown in Fig. 6.

The plasma voltage is the primary power supply parameter that can be specified by the process programmer for a plasma surface treatment. Figure 6 shows that the major portion of this power supply voltage drops across the positive column and the Crookes dark space. When plasma nitriding, the losses in this positive column are normally small compared to the losses in the Crookes dark space and they can often be neglected. Field strengths in the positive column, for low pressures, are about 1 V/cm. At 200 Pa the field strength during plasma nitriding will be about 1-2 V/cm and at

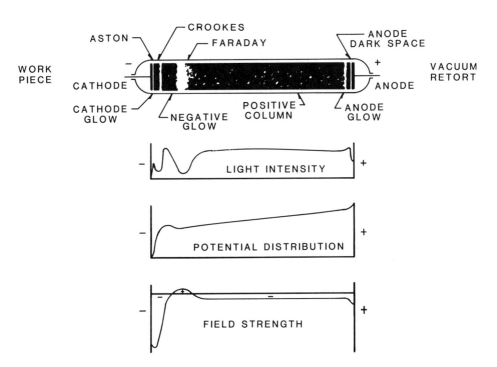

FIGURE 6    Electrical and visual characteristics of a glow discharge.

800 Pa about 5-10 V/cm. These values are typical only for gases with a low concentration of nitrogen. For example, with a 90% nitrogen gas mixture at 800 Pa, a field strength of 30 V/cm can be reached. This will not necessarily remain true for other plasma processes or at higher gas pressures where the voltage must be made higher to make certain that the discharge remains in the abnormal glow region. When nitriding at high pressures, the spatial arrangement of the work can become more critical and work cannot always be placed in the center of the plasma nitriding zone where it might be shielded from the anode potential of the vacuum retort. When nitriding at high operating pressures, it is sometimes necessary to use supplementary anodes to maintain a nearly uniform low voltage drop through the positive column.

The major portion of the total plasma voltage appears across the region labeled the Crookes dark space in Fig. 6. Electrons, originating at the work surface, are accelerated toward the vacuum retort by this voltage, called the cathode fall potential. The energy of these electrons as they enter the negative glow region will be nearly equal to this relatively high cathode fall potential. There is considerable evidence that a large fraction of the positive ions generated inside the plasma are produced by collisions within this negative glow region. Figure 6 shows that these positive ions, after moving across the interface between the negative glow and the Crookes dark space, will reach the high cathode fall potential and can then be accelerated toward the work surface. The impingement of these positive ions on the cathode work surface not only begins the chemical reaction for the plasma treating process but, in addition, cause the production of secondary electrons that are essential to the maintenance of the abnormal glow discharge. The important condition, that ensures the maintenance of the abnormal glow discharge, is that each electron initially released from the work surface must produce a positive ion in the gas mixture that can by itself yield at least one more new electron at the work surface by this secondary electron emission process.

Since the major portion of the plasma voltage drop occurs at the work surface the cathode fall voltage can generally be estimated as some large fraction of the total voltage drop between the retort and the work surface with reasonable accuracy. As a first approximation then, in most plasma nitriding applications, the current density can be considered to be simply a function of the total power supply voltage. It is important to recognize that this power supply voltage cannot be arbitrarily reduced to produce any desired value of current density. If the voltage is reduced too far, the discharge will not remain in the abnormal glow range. The workpiece will not be fully covered and plasma surface treating will not occur in those local areas not covered with the abnormal glow. The plasma voltage cannot be arbitrarily increased to accelerate the plasma process. If the voltage is made too high, an arc discharge will occur that can cause visible damage to the work surface.

### 3.5 Masking the Glow Discharge

Straemke has documented the relationships between the most important plasma parameters for a wide range of industrial operations (Straemke, 1988). This documentation, which will be used extensively throughout this chapter, includes dimensions for the thickness of the cathode fall region. This thickness is determined in part by the gas pressure in the system as shown in Fig. 7.

The interface between the Crookes dark space and the negative glow is located at this dimensional distance from the work surface. If the geometry at the work surface can be arranged so that the

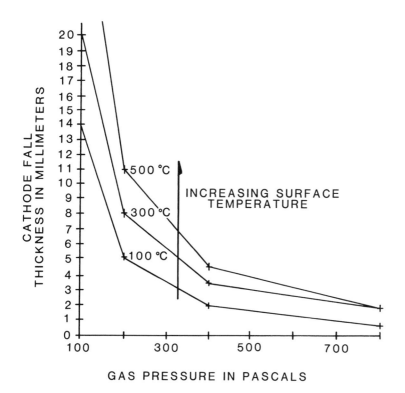

FIGURE 7  Cathode fall thickness = $f$ (gas pressure and surface temperature).

small space within this dimensional distance is obstructed, so that electrons leaving the work surface cannot reach their full kinetic energy before they encounter gas atoms for ionization, the conditions for sufficient secondary emission from the work surface will not be met and a sustained abnormal glow discharge will not occur on the work surface. The abnormal glow will be shifted to the external surface of the obstruction and the work surface will be effectively masked.

This feature of the glow discharge permits mechanical masking of portions of the work surface. A gas-tight mask is not essential and neither copper plating nor adherent painted masks are required in plasma processes to prevent work surface reactions in selected locations. A simple reusable sheet metal cover, a metal pin, or a threaded bolt that can be placed within the dimensional limits of the Crookes dark space, is the only requirement to prevent the formation of a covering abnormal glow, to prevent positive ion impingement, and to prevent an unwanted local surface reaction.

Low values of cathode fall thickness are generally desired to accurately replicate the shape of the workpiece. Figure 7 shows that these lower values can be obtained by increasing the gas pressure or by lowering the gas temperature to increase the gas density near the work surface.

## 4 PLASMA NITRIDING CHEMISTRY

Plasma treating in a partial pressure of nitrogen has been called glow discharge nitriding, ion nitriding, and plasma nitriding. The term *glow discharge* is now generally reserved for fundamental descriptions of the electrical process. For discussions of the metallurgical application, the term *plasma nitriding* is generally accepted as noted in Conybear (1985).

There are four chemical reactions at a work surface that are believed to be important when a plasma is used for the source for atomic nitrogen as shown in Fig. 8.

1.  Production of ionized and neutral nitrogen atoms by energetic electrons.

    $$e^- + N_2 = N^+ + N + 2e^-$$

2.  Sputtering of Fe and contaminants from the work surface by these ionized nitrogen atoms. The impact of the nitrogen ions on the work surface dislodges contamination that can then be removed by the pumping system. This effect, called sputter

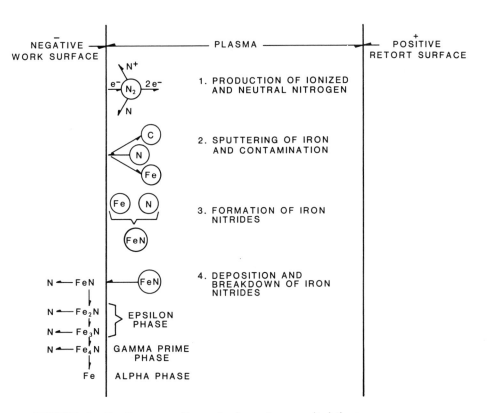

FIGURE 8   Surface reactions during plasma nitriding.

cleaning, removes a significant barrier to nitrogen diffusion from the work surface into the core.

$N^+ \rightarrow$ work surface = sputtered Fe and sputtered contamination

3. Formation of iron nitrides by the sputtered iron atoms and neutral nitrogen atoms. Although the predominant mechanism of plasma nitriding involves this reaction of iron atoms with nitrogen atoms in the gas phase near the work surface, and then their redeposition on the surface as a chemical compound, there is evidence that sputtering is not the only reaction mechanism. It can be shown that nitriding takes place even when the energy in the plasma is not high enough to cause sputtering.

Sputtered Fe + N = FeN

4. Deposition and breakdown of FeN on the work surface. The FeN produced from these chemical reactions is unstable and under the influence of the continuing ion bombardment from the plasma, it breaks down progressively into the epsilon phase $Fe_{2-3}N$, and into the gamma prime phase $Fe_4N$ forming an iron/nitrogen compound zone. At each stage of the breakdown, atomic nitrogen is released either to the plasma or to the work surface for diffusion inward and for the formation of an alloy nitride diffusion zone.

$$FeN \rightarrow Fe_2N + N$$
$$Fe_2N \rightarrow Fe_3N + N$$
$$Fe_3N \rightarrow Fe_4N + N$$
$$Fe_4N \rightarrow Fe + N$$

The use of nitrogen for surface treating metals is an established technology. Although the nonreactive nature of nitrogen has made it an important protective atmosphere, its dissociation makes it highly reactive and able to participate in surface treatments to produce high hardness, wear resistance, and corrosion resistance. Hochman (1986) has summarized the properties of some typical nitrides.

| Nitride | Properties and applications |
| --- | --- |
| AlN | Very refractory with good thermal shock resistance and a low coefficient of thermal expansion; very effective as a hardening agent in nitrided steels. |
| a-BN | Excellent refractory with good electrical resistance; a very good solid lubricant often called "white graphite." Used to contain some highly corrosive acid solutions. |
| b-BN | Very hard and often used as a diamond substitute; used in the composition of heat-resistant alloys; has excellent potential in tribological applications. |
| Cr and Fe Nitrides | Very good hardness and wear resistance; particularly effective as hardening agents in nitrided steels. |
| TiN and $Ti_2N$ | Good high-temperature materials with thermal shock resistance; good abrasive and tribological materials; very good corrosion resistance. |

## 4.1  Microstructures

Any core material that can be liquid or gas nitrided with conventional
techniques can be plasma nitrided. Badan et al. (1983) indicate
that the surface layers obtained by plasma nitriding are qualitatively
similar to those obtained using conventional methods. However,
the nitrogen distribution in the layers can be very different, permitting the formation of a variety of microstructures. One of the main
advantages claimed for plasma nitriding is the possibility of developing
an optimum combination of compound layer and diffusion zone properties to suit a specific application requirement. Generally, plasma
nitriding permits the production of surface layers in nitrided steels
with three configurations: epsilon compound and diffusion layers,
gamma prime compound and diffusion layers, diffusion layer only.

The epsilon $Fe_{2-3}N$ compound zone is commonly made 30 microns
thick to improve the corrosion resistance of the workpiece. The
gamma prime $Fe_4N$ compound zone thickness, however, is process
limited to less than 10 microns. It offers improved fatigue resistance.
Either compound zone can be produced without pores or cracks.
Generally a sharp transition can be seen between the compound
zone and the underlying diffusion zone. The structure and grain
size in the core material are not altered in the diffusion zone.
Hardening is caused by lattice distortion and a fine dispersion of
nitride precipitates. Hardness profiles show a continuous transition
from the work surface to the core. Brittle intergranular nitride
networks within the diffusion zone can be completely suppressed.

With each of these surface layers dimensional stability will be
excellent, there will be no tendency to spall brittle compounds,
and there will be no evidence of surface roughening. These features
permit work pieces to be used directly from the plasma nitriding
furnace without grinding or polishing. In fact, it is recommended
that workpieces be used directly to obtain maximum benefits from
this process. The production of these different layers and the relative thickness of the compound and diffusion zones, can be controlled
by varying the pretreatment sputtering time, the nitriding temperature and the time at temperature, the gas composition and total
gas pressure, and the plasma voltage and the current density.

## 4.2  Control of the Compound Zone

Conventional gas nitriders obtain active nitrogen from the dissociation
of anhydrous ammonia-$NH_3$. The nitrogen/hydrogen ratio, in this
case, is fixed at the 1/3 ratio of the ammonia supply. The *Metals
Handbook* describes the limitations of these systems.

Ammonia gas nitriding produces a compound zone that is a mixture of both epsilon and gamma prime structures. High internal

stresses result from differences in volume growth associated
with the formation of each phase. The interfaces between the
two crystal structures are weak. Thicker compound zones,
formed by ammonia gas nitriding, limit accommodation of the
internal stresses resulting from the mixed structure. Thickness,
internal stresses and weak crystal boundaries allow the white
layer to be fractured by small applied loads.

Under cyclic loading, cracks in the compound zone can
serve as initiation points for the propagation of fatigue cracks.
The single-phase gamma-prime compound zone, which is thin
and more ductile, exhibits superior fatigue properties. Reducing
the thickness of the ion nitrided compound zone further improves
fatigue performance. Maximization occurs at the limiting condition,
where compound zone depth equals zero.

Nitrogen and hydrogen for plasma nitriders are supplied from independent, separate, storage tanks and any nitrogen/hydrogen ratio
can be selected to meet the requirements for a single phased compound zone. Methane gas can be used to supply a controlled amount
of carbon to the plasma and the plasma carbon ion concentration,
which influences compound zone chemistry, can be easily independently controlled by regulating the partial pressure of this gas.
Plasma nitriders are equipped with mass flow controllers and total
pressure gages for the precise regulation of process gases. Total
system pressure, a result of the nitrogen, hydrogen, and methane
partial pressures, can be automatically controlled even when the
composition of the gas is being altered. Plasma nitriders then permit
accurate control over compound zone chemistry because the

1. Process gas mixtures can be accurately controlled over a wide
   range of partial pressures
2. Total gas pressure can be separately controlled
3. Concentration of nitrogen ions and, therefore, the nitrogen
   activity can be separately controlled

When cutting tools like drills, thread cutters, milling tools, and
punches are plasma nitrided in a gas mixture with low nitrogen
activity, only the diffusion layer will be formed and the service
life of the tool can be improved by a factor of 2 to 4. If the nitrogen
activity in the gas mixture is increased, a gamma prime compound
zone can be deposited over the diffusion zone for improved fatigue
resistance. If carbon is added to the gas mixture, the epsilon layer
can be formed over the diffusion zone. The epsilon layer will be
slightly harder and less ductile than the gamma prime layer but
will provide improved wear and corrosion resistance.

4.3 Advantages of Plasma Nitriding

Plasma nitriding permits better metallurgical control over the compound zone than competing processes.

1. The compound zone can be single phased with good wear, abrasive, and tribological properties.
2. Work can be reliably plasma nitrided with no measurable compound zone.
3. Plasma nitriding properties obtained by treatment below the tempering temperature and without a change in phase permits surface treatment without distortion.

Houvion (1980) has reported that these advantages have led to the following applications in Europe.

1. As a substitute for chrome plating for the corrosion protection of shock absorber rods. The plasma nitriding operation is pollution free and can be completed in one-third the time required for plating. Shock absorber rods manufactured from 0.38% carbon steel are given an epsilon layer about 10 microns thick.
2. As a substitute for carburizing and carbonitriding the synchonizer ring in automotive gearboxes. These gears manufactured from chrome-moly steels were gas carbonitrided to eliminate burring in operation, but growth during heat treatment caused dimensional tolerance problems that caused a high number of rejects. This problem was solved by plasma nitriding. A 10-micron epsilon layer with a superficial hardness of 550 HV5 was produced with a reduction in the cost of heat treating.
3. As a substitute for salt bath treatment for rocker arms and cam followers. Air pollution and salt disposal problems were eliminated with the use of plasma nitriding.

5  PLASMA CARBURIZING CHEMISTRY

Similar models have been proposed as the chemical mechanism for other plasma surface treatments, as for example, plasma carburizing,

$$e \rightarrow CH_4 = C^+ + 2H_2 + 2e^-$$

It has been noted that the surface cleanliness of plasma carburized workpieces can best be accounted for by the sputter-cleaning phenomena. Since the rate of carburizing is slow at low partial pressures of the carburizing atmosphere, as expected, but becomes very fast after the glow discharge has been established, it is believed

that the bombardment of the work surface by ionized carbon atoms
from the plasma is an accurate model of events at the work surface.
The carbon ions arriving at the sputter-cleaned work surface can
quickly create a high-carbon concentration gradient that is then
responsible for very fast diffusion of carbon into the work surface.

## 5.1  Plasma Carburizing Advantages

The commercial application of plasma carburizing continues to grow
to take advantage of the benefits already found in conventional
vacuum carburizing and to exploit the additional advantages peculiar
to plasma heat treating.

1.  Better case uniformity. The abnormal glow discharge mechanism
    provides a uniform concentration of active carbon at complex,
    nonplanar, work surfaces.
2.  Clean work surfaces. A high concentration of active carbon
    can be established and maintained at a work surface operating
    at cathode potential, without resorting to a high hydrocarbon
    throughput that can cause the formation of soot.
3.  Elimination of grain boundary oxides. Plasma carburizing furnaces
    operate at better vacuum levels than the levels employed with
    conventional vacuum carburizers that must utilize atmosphere
    recirculation to control case uniformity. These better vacuum
    levels permit plasma carburizing in an environment that has
    lower residual water vapor contamination.
4.  Less distortion. Work can be heated to the plasma carburizing
    temperature with gas convection for maximum temperature uni-
    formity. Temperature and carbon activity uniformity at the
    surface of the work produces a uniform case that can be quenched
    with less distortion.

Booth et al. (1983) attribute carburizing uniformity to the uniform
electrical field acting effectively normal in its direction to the work
surface. They have compared this effect during plasma carburizing
to the situation existing in a conventional vacuum carburizing furnace
as shown in Fig. 9. The arrows represent the random motion of
active gas molecules, which when they come into contact with the
hot work surface, can react to produce the required surface chemistry. Although this random motion can be improved for vacuum
carburizing by increasing the gas pressure and then directing the
gas flow by forced convection, it remains in sharp contrast to the
directional movement of an active gas species in an abnormal glow
discharge. In addition, since carbon ion generation is directly related
to current density, and since this current density is uniform in the

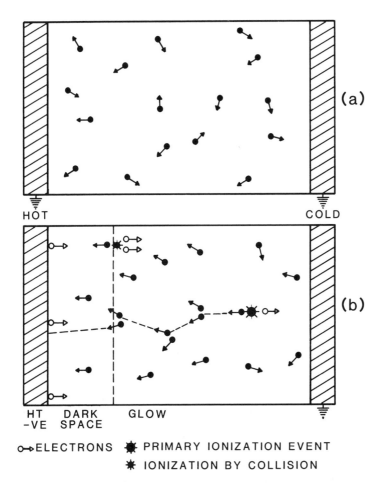

ELECTRONS ✹ PRIMARY IONIZATION EVENT
✸ IONIZATION BY COLLISION

FIGURE 9   Diagram comparing the random molecular motion in vacuum carburizing (a) with the flux to the cathode in plasma carburizing (b).

abnormal glow region, plasma carburizing need not rely on just the uniform recirculation of the carburizing atmosphere.

Grube (1978) has found that plasma carburizing rates can exceed the rates achieved by vacuum carburizing by a factor of 2 as shown in Fig. 10. The high-energy, high-temperature electrons moving at high speeds in the plasma can generate a carbon potential at the work surface in excess of the potential allowed by conventional thermodynamics. The hydrogen generated with the dissociation of this methane is available at the immediate surface for oxide reduction.

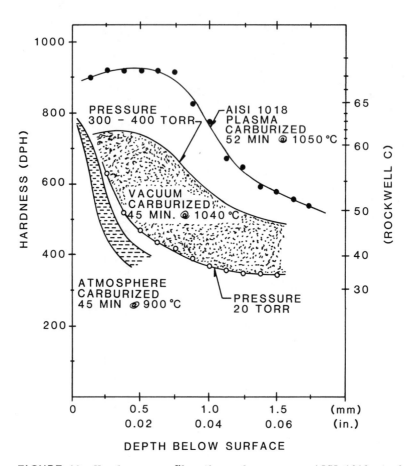

FIGURE 10   Hardness profiles through cases on AISI 1018 steel
comparing atmosphere carburizing at 900°C, vacuum carburizing
at 1040°C, and plasma carburizing at 1050°C.

Since the dissociation of methane occurs only in the cathode fall
region, at the hot work surface, the carbon potential in other areas
of the furnace remains quite low, eliminating the problems of sooting
that usually limit operation at high carbon potential in conventional
atmosphere and vacuum systems.
    Grube has found that enough carbon to produce an effective
case depth of 1 mm on standard carburizing steels can be produced
within 10 min when followed by a 30-min diffusion cycle. These
high carburizing rates result from the high rate of carbon infusion
into the work surface rather than from anomalously high rates of

carbon diffusion. A diffusion time/carburizing time ratio of 3/1
was found to produce the most favorable results as shown in Fig. 11.
    It can be seen that the abnormal glow discharge, with uniform
current density over the work surface, produced exceptionally
uniform case depths. Conventional vacuum carburizing requires
the uniform recirculation or the periodic removal and replacement
of the carburizing atmosphere within the furnace to achieve case
uniformity. Plasma carburizing allows better control of case uniformity
by permitting control of carbon potential and carbon activity with
the set points of the programmable glow discharge process microprocessor.
    There was a significant savings in the amount of natural gas
consumed due to the shorter process times, from the elimination
of a carrier gas for the carburizing atmosphere, and because of
the small amount of methane needed to produce the desired carbon
potential. Plasma carburizing at 2500 Pa, if generally adopted as
the commercial means for carburizing, would have a significant
impact on the amount of natural gas consumed for heat treating.
    Although it is sometimes possible to obtain sufficient thermal
energy from the plasma for a low temperature process like nitriding,
carburizing in the temperature range between 850-950°C invariably
requires the use of auxiliary furnace heaters. These furnace heaters
permit the selection of an operating point in the abnormal glow
discharge region well below the arc transition point.
    Grube (1979) has described the use of diluted propane for
the plasma carburizing atmosphere. Tests were made with
nitrogen/propane ratios between 4/1 and 50/1. It was found that
nitrogen/propane ratios less than 8/1 caused the formation of soot.
Ratios between 30/1 and 40/1 produced favorable carbon profiles
without the formation of soot. The carbon profile was not materially
different from the profile obtained at a high propane content until
a 50/1 ratio was employed as shown in Fig. 12. Grube suggests
that since propane dilution with relatively large amounts of nitrogen
permits carburization with less than the maximum carbon concentration, this method could be applied for single-stage carburization
without the use of a diffusion step.
    Tonchev et al. (1987) have compared gas and liquid feed systems
for the production of the plasma carburizing atmospheres. A mixture
of propane-butane was found to be an active medium for plasma
carburizing, however, even when very low quantities of this mixture
were fed into the system sooting was observed and carbide networks
were formed. A mixture of methane and argon was employed both
to decrease the carbon input by inert gas dilution and to take advantage of the known beneficial effects of argon sputtering. The thicker
layers obtained with this mixture were attributed to the argon

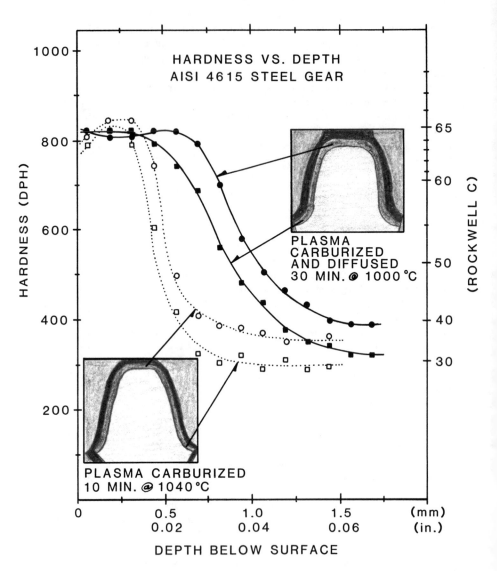

FIGURE 11   Hardness profiles through plasma carburized and diffused cases on AISI 4615 steel gear comparing depth of case at tip of tooth and that in the fillet area. Insets: photomicrographs of teeth.

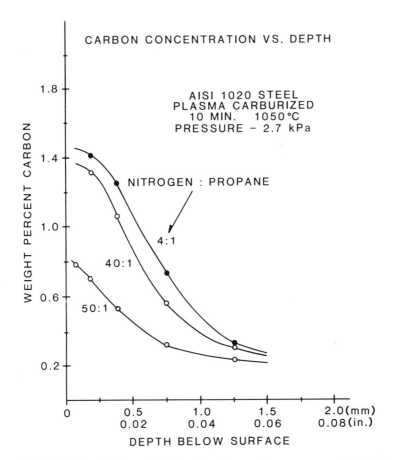

FIGURE 12  Carbon distribution through cases on AISI 1020 steel plasma carburized for 10 min at 1050°C at various nitrogen to propane ratios as indicated.

sputtering and the generation of a large number of surface vacancies conducive to the carbon diffusion. Carbide networks were observed after 120-min cycles at temperatures between 920° and 1000°C.

Tonchev et al. believe that the use of a liquid media provides a number of advantages: the feeding system for saturated gases is simple and the danger from handling explosive mixtures can be eliminated. Ethyl alcohol was employed for plasma carburizing, however, regardless of the treatment cycles employed, soot was observed on the work surfaces. Sooting caused surface melting at temperatures over 1000°C. When methyl alcohol was substituted directly, some

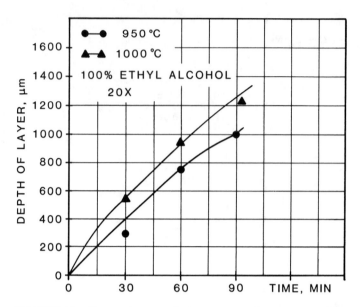

FIGURE 13  Dependence of the depth of the carburized layer on time at various temperatures.

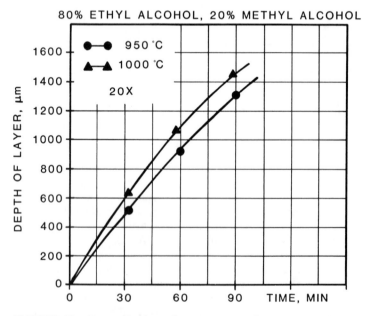

FIGURE 14  Dependence of the depth of the carburized layer on time at various temperatures.

surface decarburization was observed. A mixture of ethyl and methyl alcohols produced desirable results with no loss in performance as evidenced by the data in Fig. 13 for a 100% ethyl alcohol feedstock and Fig. 14 for an 80% ethyl alcohol-20% methyl alcohol feedstock.

## 6 DESCRIPTION OF THE EQUIPMENT AND SUPPORT SYSTEMS

Equipment supplied for plasma treating in the abnormal glow discharge region generally resembles the conventional hot wall vacuum furnace shown in Fig. 15. The equipment normally will include a bell furnace heated with resistance elements and lined with light-weight thermal insulation to limit heat loss. A mechanism will be

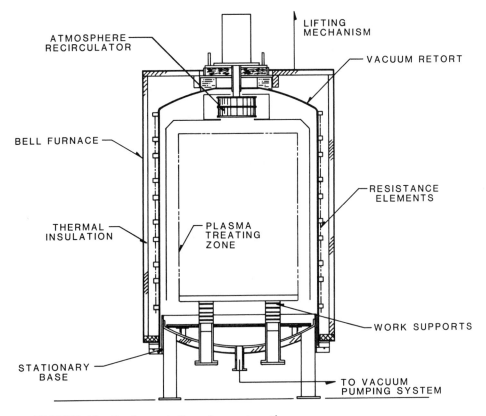

FIGURE 15  Equipment for plasma treating.

provided for lifting the bell furnace and the vacuum retort from the stationary base to provide complete access to the work supports for top or front loading. The base will contain electrically insulated hearth rails and piers to support the work in the center of the plasma treating zone. The stationary base design allows permanent utility connections, including connections to the process gas inlet and pumping system outlet lines. The fixtured work can remain stationary throughout the plasma treating process and can be instrumented with direct contact temperature sensors to monitor all heating and cooling cycles.

A vacuum pumping system will be supplied to initially purge the retort and then to maintain the partial pressures needed for the plasma operation in the abnormal glow discharge region. The retort can be backfilled with an inert or protective atmosphere after vacuum purging. With this insulated, hot wall arrangement, the retort wall can be efficiently operated at all temperatures at full atmospheric pressure. A motor driven, internal atmosphere recirculator can then be employed for heating the work by gas convection inside the vacuum retort. The work can be quickly heated by convection using the inert or protective atmosphere at nearly full atmospheric pressure.

The selection of lightweight fibrous insulation makes it possible to lift and move the bell furnace at all operating temperatures, since the independent vacuum retort will be able to contain the workpieces within the protective atmosphere after the bell furnace has been removed. After the bell furnace has been removed, the work can be efficiently cooled by gas convection, again driven by the internal atmosphere recirculator.

## 6.1 Power Supply Duty Cycle

Separate power supplies are generally provided for heating the bell furnace and for maintaining the plasma voltage and current parameters. Both continuous dc and pulsed power supplies have been supplied for maintaining the plasma parameters. With a dc power supply, a continuous voltage is applied to the system and, therefore, the plasma current density will also be continuous. The power density on the work surface, that determines the amount of heating energy transferred through the plasma and the resulting workpiece temperature, will be proportional to the area under the voltage and current time curve as shown in Fig. 16.

In this case, an increase or a decrease in the plasma voltage, to produce some desired variation in the positive ion density, must be accompanied by an increase or decrease in the heating energy transferred from the plasma, whether this additional variation was

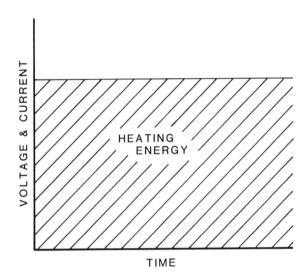

FIGURE 16   Voltage and current time with a DC power supply.

desired or not. The *Metals Handbook* notes this limitation. It is
first recognized that a hot wall furnace is desirable because of
its energy efficiency, requiring only about 20% of the power that
would be used by a similar cold wall furnace at 1060°F. But it is
also noted that a dc power supply, operating without control of
the power duty cycle, can overheat the workpiece in a hot wall
furnace at high current densities and that, in this case, adequate
surface activation of high alloy and stainless steels may not be
possible.

The heating effect of the plasma can be independently controlled
with a system that permits regulation of the time the plasma is
on and the time the plasma is off. Pulsed plasma power supplies,
with a variable timed separation between pulses, permit the furnace
operator to select a variable power duty cycle that can be defined
as

$$\text{power duty cycle} = \frac{\text{time on}}{\text{time on + time off}}$$

With a pulsed power supply, the plasma voltage can be applied
continuously for a 100% power duty cycle, or with equal on/off
timed pulses, a 50% duty cycle. Figure 17 shows the voltage and
current relationship with time with a 10-kH$_2$ pulse frequency producing a 50% power duty cycle. This duty cycle can be reduced

to 5% by extending the off-time interval without any undesirable effect on the positive ion density at the work surface as shown in Fig. 18.

Since the power density remains the area under the voltage and current curve, the energy input through the plasma will now be so small the work piece will not be significantly heated by plasma energy alone. Independently controlled resistance heaters can now be employed to bring the work to the temperature needed for surface treatment. In normal operation, with good programming practice, the current density and therefore the power density will not be a function of this power duty cycle. The current density can remain essentially independent of the power duty cycle as long as the power from the plasma is not allowed to heat the gas and the work remains completely covered by the glow discharge. Figure 19 shows the continuing direct relationship of the current density and voltage that remains nearly independent of the power duty cycle for changes in the duty cycle between 5% and 50% with an abnormal glow discharge. It will be shown later that in contrast to this the duty cycle can affect the current density in a hollow cathode discharge.

## 6.2 Microprocessor Control

A dedicated microprocessor will control the work temperature, the gas pressure and mixture composition, and the plasma operating parameters. In most cases, the microprocessor can be set by selecting process parameters in response to operator interface prompts, by selecting a previously installed menu, or by downloading process parameters from an external device. A recorder will plot the operating parameters, the system status, and alarm messages in real time.

The equipment will shut down in the event of a utility failure. After loss of electrical power the plasma will be interrupted, all motors will stop, and all valves will close to protect the hot work surfaces. When power is restored the equipment can be restarted by the operator with a pushbutton, with software providing the optimum restart plasma parameters. After loss of cooling water pressure, air pressure for valve operation, or gas backfill pressure, the microprocessor will indicate the nature of the utility failure, and an alarm will sound. The alarm can be silenced with a pushbutton while the microprocessor continues to indicate until the utility failure is corrected. If the utility failure becomes critical to the safety of the operator or the process, the equipment will shut down.

## 6.3 Workpiece Arrangement

Plasma treating begins with cleaning the workpiece to remove loose dirt, machining oil films, and nonadherent surface contamination.

PULSED POWER SUPPLY

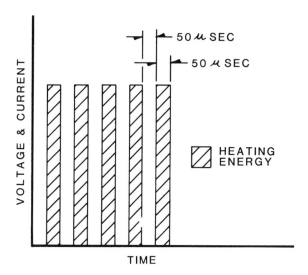

FIGURE 17   Voltage and current/time with a 50% duty cycle.

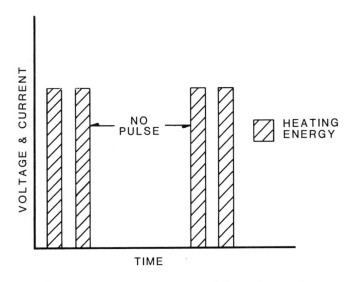

FIGURE 18   Voltage and current/time with a 5% duty cycle.

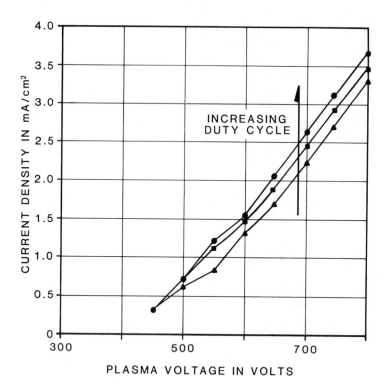

FIGURE 19   Current density = *f* (plasma voltage and duty cycle)
for fixed gas pressure, surface temperature, and gas composition.

Since plasma treating is a technology involving chemical reactions
at a work surface, proper surface cleaning is fundamental to consistent metallurgical results. The clean work is then positioned
within the plasma treating zone in a manner that will permit the
plasma to reach all of the surfaces to be treated. Generally a distance of 0.5 in and larger is allowed between adjacent parts.

Two important aspects of a work piece are its area/volume ratio
and the thermal emissivity of its surface. At every level of power
density, the surface area of the work determines the thermal energy
transfer from the plasma while the volume establishes the total thermal
input that will be required to reach a specified operating temperature.
The emissivity of the work surface determines the energy that will
be lost by reradiation to colder surfaces. These two parameters,
the area/volume ratio and the emissivity, together determine an
equilibrium workpiece temperature for every level of plasma power
input. This workpiece temperature will be different from the average

gas temperature in almost every operating situation. At every location on the surface of the work, the temperature will be the result
of a balance between the incoming and the outgoing thermal energies.
The incoming energy will be the sum of the thermal radiation from
the vacuum retort and the plasma energy from the cathode fall
region. The outgoing energy will be mainly from thermal reradiation
from the work surface to colder locations on the vacuum retort.

## 7 TEMPERATURE UNIFORMITY

The temperature uniformity throughout a charge will depend upon
the manner in which the work is arranged in the plasma heating
zone. If heat is provided only by the plasma power supply, the
temperatures of the hottest and coldest pieces will be a function
of the ratio of surface area/volume of the workpiece receiving this
plasma energy. Unlike other heating systems, heat input from the
plasma remains essentially independent of the work surface temperature. In contrast to this, in an insulated, hot wall, plasma treating
furnace, designed for convection heating with a recirculating atmosphere, the rate of heat transfer to the work will be a function
of both the surface area of the work and the temperature difference
between the recirculating atmosphere and the work surface:

$$H_c - K_1 A(T_{atm} - T_{work})$$

where $K_1$ is a factor that depends upon the geometry of the system
and the surface coefficient of heat transfer by convection, and A
is the work surface area.

When work is heated by radiation in a hot wall retort in a vacuum
the heat transfer will be a function of the surface area and the
temperature difference between the radiating hot wall and the work
surfaces:

$$H_r = K_2 A(T_{wall}^4 - T_{work}^4)$$

where $K_2$ is a factor that depends on the emissivity and the geometry
of the system and where A is again the work surface area. In both
cases, workpieces that momentarily become hotter than the average
workpiece because of the dependence of thermal input on surface
area, will immediately begin to receive less heat because of this
dependence on the temperature difference. The work will eventually
reach the temperature of the recirculating atmosphere or of the
radiating hot wall if the heat loss from the system is made small.
In a hot wall plasma treating furnace, if the bell furnace zone heaters

are themselves in equilibrium and these zones are at nearly equal temperatures, there will be no significant temperature nonuniformity in the workpiece. At typical nitriding temperatures the influence of wall temperature on work temperature is normally found to be small. Wall temperature differences of 50°F cause work temperature differences of less than 10°F.

However, if the work is heated by plasma energy alone, the rate of heat transfer to the work does not depend in any direct way on the temperature difference between the work surface and any measurable furnace temperature. With plasma heating

$$H_p = AVI_d$$

where A is the surface area, V is the cathode fall voltage and $I_d$ is the work surface current density. Workpieces with larger surface areas generally will become hotter than workpieces with smaller areas. The only mechanism leading to temperature uniformity within the charge will be a thermal energy exchange between workpieces initially at different temperatures. Since the location of workpieces within the charge cannot be altered after the cycle has been started, if the work is heated by plasma energy alone, the operator must in each case prearrange the workpieces in a manner that will provide a thermal radiation balance that will ensure the required temperature uniformity. If the only source of heating energy is the plasma, this will require mixing the workpieces, by locating pieces with relatively large surface areas close to pieces with small surface areas. In some cases it will be necessary to introduce dummy workpieces into the charge to act as receptors for plasma energy with the expectation that these pieces, after they have been heated by the plasma, will reradiate their thermal energy to the lower temperature pieces. Although it is possible that an experienced operator may be able to obtain temperature uniformity in this manner, it will generally be necessary for him to first make a large number of experimental runs to develop essentially empirical data.

The local work surface temperature will have an important influence on the final metallurgy at the surface of the work. In most cases the plasma parameters themselves can have a significant effect on the temperature of the work surface and it is, therefore, generally desirable to limit the plasma energy input to just that level needed to cause the desired physical or chemical reaction without raising the work temperature unnecessarily. This technique avoids temperature nonuniformity.

Plasma energy delivered to the workpiece is directly proportional to its actual local surface area and is nearly independent of the spatial relationship between the work and the retort wall. However,

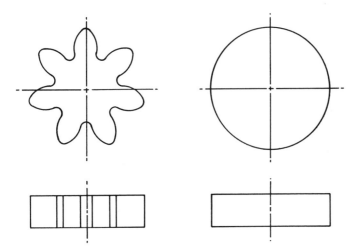

FIGURE 20 Plasma energy input is a function of the actual surface area.

heat loss from the work by thermal reradiation is directly proportional to the projected external local surface area of the work and
can be an important function of this spatial relationship. This means
that a gear as shown in Fig. 20, which may have twice the surface
area of a smooth cylinder of the same external dimensions, will
receive twice the plasma energy input of a cylinder. With approximately the same projected external surface area as the cylinder,
and, therefore, nearly the same heat loss as a function of temperature, the gear must have a significantly higher temperature at
thermal equilibrium before the higher thermal input can come into
balance with a higher thermal loss.

If a workpiece is heated by plasma energy alone it will be necessary, in many cases, for the operator to develop a charging scheme
for each different furnace size since temperature uniformity will
be influenced by the distance of a workpiece from the furnace wall
as well as the distance to an adjacent piece. Since documentation
of the position of workpieces within the charge is dependent upon
the furnace operator's measurements this loading technique has
proved to be a difficult one to document for quality assurance purposes. This need for documentation, with systems that provide
only plasma energy for heating is recognized by Conybear (1985),
which, in part, indicates the following:

4.2.2.4 Establishment of processing procedure. For each part
number, a loading, operating, and testing practice shall be

established and approved by the cognizant quality assurance
organization. This will include, for each furnace in which the
parts to be processed, the following:

Loading diagram showing the location and spacing of each
    identical part number.
Minimum and maximum number of parts which can be run with
    the same practice.
Other part numbers or geometries which can be run in the
    same load.
Specification of fixturing used including auxiliary anodes or
    cathodes.

4.5 Furnace log entries and recorder chart entries shall be
in accordance with AMS 2759, and in addition, will include the
following, for each furnace load:

Sketch, diagram or photograph of load arrangement and work
    thermocouple locations, indicating location of control point.

The auxiliary cathodes, noted in this proposed draft, are the dummy
workpieces needed with systems of this type. If a separately heated
bell furnace is not employed, this documentation and considerable
experience with the anticipated thermal balance may be required
to ensure temperature uniformity during the course of the treatment.
The separately heated bell furnace will, however, permit bringing
the work to the treatment temperature with a uniformity entirely
consistent with current protective atmosphere and vacuum furnace
processing procedures already approved by this and many other
organizations.


## 8   DESCRIPTION OF THE PROCESS

After the workpieces have been arranged, the vacuum retort and
the bell furnace can be positioned to begin the process. Plasma
treating cycles are relatively long and require close supervision
of the process parameters if consistently good results are to be
obtained. Plasma treating requires the following processing steps
that must be monitored in real time to make certain that a satisfactory
end point has been reached before the next programmed step is
started. Although every specific plasma treating requirement must
be individually characterized, the following description is generally
valid and can serve as a guide to the understanding of most plasma
processes.

1.  Pump the system to remove air, water vapor, and residual gases.
2.  Heat the work, while pumping, to accelerate the removal of volatile liquid contaminants.
3.  Sputter-clean the work surface to remove the more adherent solid contaminants.
4.  Begin adding the desired ion species to the work surface.
5.  Regulate the ion concentration to control the work surface chemistry.
6.  Control the work surface temperature to establish a rate for diffusion consistent with other process requirements.
7.  Start the inert atmosphere convection cooling cycle.
8.  Signal the operator to unload the furnace after the work has cooled to a satisfactory unloading temperature.

### 8.1  Vacuum Purging to Remove Surrounding Air and Water Vapor

The first step before plasma nitriding or carburizing is vacuum purging to remove air and water vapor from the vacuum retort to preclude surface oxidation while the work is being heated to the operating temperature. The diffusion rate during plasma nitriding will then not be limited by the passivating effect of surface oxidation. Work surfaces can be plasma carburized without intergranular oxidation. This intergranular oxidation can be caused by residual air and water vapor that commonly cannot be completely removed from a protective atmosphere carburizing furnace with a carrier and enrichment atmosphere flowing through the furnace at atmospheric pressure.

The final chemistry of a plasma nitrided surface will be sensitive to the amount of hydrocarbon in the residual atmosphere after vacuum purging. This hydrocarbon content can be destructively high for two reasons.

1.  If the system, including work surfaces that may not have been adequately precleaned, are not initially vacuum pumped to a sufficiently low hydrocarbon partial pressure.
2.  If vacuum pump oils are allowed to backstream, or if a power interruption permits pump oil to be forced into the vacuum retort by ambient pressure when pump rotation stops.

Therefore, a well-trapped mechanical pump or a mechanical booster pump is employed to pump the system to approximately 2 Pa for the vacuum purging step. When the system reaches 2 Pa the total residual contamination in the system will be less than 100 parts per million. The residual gas remaining in the vacuum retort will

be predominately water vapor and plasma treating at this vapor
pressure will be equivalent to heat treating at a -50°C dew point.


## 8.2  Convection Heating to Remove Adsorbed Water

When work is first placed into a plasma treating furnace, it has
several monolayers of water adsorbed on its surface. The gas and
water vapor surrounding the work can be removed very quickly
by the vacuum pump. After the initial removal of this relatively
accessible gas and water vapor, the residual contamination level
in the furnace will become dominated by the rate at which the
adsorbed water can be removed. Since the rate of surface desorption
is strongly dependent on surface temperature, it is a significant
benefit if the equipment used for plasma treating has a hot wall
design.

A hot wall furnace design permits backfilling the vacuum-purged
retort to nearly atmospheric pressure with an inert atmosphere
for convection heating. At ambient temperature, convection is the
only important means for heat transfer. In the temperature range
between ambient and 200°C, with normal convection coefficients
and emissivities, 80% of the heat transfer will be by convection
and only 20% by radiation. In the range between 200° and 300°C
the situation changes only slightly to 70% by convection and 30%
by radiation. Since it is considered unwise and unnecessary to
apply plasma power to a work surface before it has been properly
cleaned by desorption, convection heating remains the most efficient
means for heating work from ambient temperature.

A hot wall design has another important advantage. All surfaces,
including workpiece and vacuum retort surfaces that will be ultimately
exposed to the plasma, will be brought to the same elevated temperature for uniform degassing during the convection heating cycle.
In cold wall systems water, from cooler furnace locations behind
the radiation shields, will become continuously available as a vapor
as the work temperature increases. This water vapor will generally
cause work surface and intergranular oxidation as it travels from
its desorption location to the pumping port, even when the vacuum
gauges indicate a satisfactory vacuum level is being maintained.

A water-cooled cold wall will increase in temperature as the
work temperature increases. This increase in cold wall temperature
will lead to a release of other adsorbed gases that can contaminate
the furnace atmosphere during the critical high temperature portion
of the surface treatment cycle. Although this contamination can
sometimes be overcome by increasing the gas flow, it is difficult
to ensure that the gas flow distribution will be sufficiently uniform
to prevent contamination of some parts in heavily loaded furnaces.

## 8.3 Sputter Cleaning to Remove Solid Barriers to Diffusion

Plasma treating cycles can be made shorter if superficial barriers to diffusion are first removed by sputtering. After convection heating has brought the work to an elevated temperature and removed volatiles, the furnace is again vacuum purged to 2 Pa and then backfilled with an inert or reducing atmosphere to 10 Pa for sputter cleaning. At this time, when the work surfaces are relatively clean, the plasma power can be safely turned on, placing the work at cathode potential, and the vacuum retort at anode potential. The last traces of solid contamination can then be removed by sputtering the work surfaces with energetic, positively charged, hydrogen and nitrogen ions. This controlled sputter cleaning must be at a low current density in the abnormal glow discharge range. Properly managed, it will not produce any observable surface modification or damage. Only angstroms of contaminating materials will be dislodged from the surface to be carried away by the pumping system. Sputter cleaning should not be confused with the flashing on and off of the visible glow discharge that is sometimes observed through the vacuum retort viewport when a deficient power supply cannot prevent a shift from the abnormal glow discharge to an arc discharge.

The binding energies of chemically adsorbed atoms are typically less than 10 eV. Physically adsorbed atoms have binding energies less than 1 eV. The energies of the bombarding ions from the abnormal glow discharge can be several hundred electron volts and can efficiently sputter-clean the work surface. Since plasma treating is a diffusion limited process, it is sputter cleaning, and the efficient removal of surface contamination, that is responsible for short plasma treating cycles.

In a cold wall system recirculating atmospheres cannot be confined to just the heated section of the furnace. The hot atmosphere needed for convection heating will contact the cold wall before re-entering the heated section causing unacceptable work piece temperature nonuniformity. The *Metals Handbook* notes that

> The vacuum vessel could be constructed with a single wall as shown in the schematic, with cooling water provided at the O-Ring seal. Usually, the vacuum vessel itself is water jacketed. . . .
> Water-jacketed vessels with few or no reflective baffles usually are not provided with a circulating fan.

If the equipment has a cold wall design it is generally not possible to employ convection heating for the removal of adsorbed contamination

at an elevated temperature. The furnace must then be vacuum
pumped to the lowest possible pressure at ambient temperature
and the plasma power supply switched on to begin the sputter-
cleaning cycle. Sputter cleaning at ambient temperature, although
less efficient, will also remove adsorbed gas layers and contamination
from the work surface. This ambient temperature sputtering process
has been called off-sparking in the literature because visible arcs
can, in most cases, be observed. If the off-sparking discharge
is made too intense, the surface can be visibly etched, the geometry
can be altered, and the work surface can be damaged. Intense
bombardment of a work surface by ions from the plasma can often
be made strong enough during off-sparking to remove significant
amounts of alloying elements and iron from the surface. Sputter
cleaning is a uniform process because the plasma is uniformly distributed in the abnormal glow discharge. However, sputtering can
still alter the composition of the work surface because of the different
sputtering yield of different materials. Sputtering yield is measured
as atoms of sputtered work material divided by ions of the bombarding species. Carbon generally has a low sputtering yield while nickel
and cobalt have high sputtering yields.

In a furnace where the only means of heating the work is plasma
energy this sputtering procedure must, by necessity, begin at
ambient temperature and may take several hours to complete. The
well-documented benefits from outgassing surfaces at higher temperatures cannot be realized until most of the surface contamination
has first been removed by off-sparking. If the work has not been
thoroughly precleaned, off-sparking must limit the initial rate of
plasma power input to very low values that will extend the plasma
treatment cycle to avoid unacceptable surface damage.

## 8.4 Adding the Desired Ion Species

The programming step after sputter cleaning is the addition of
the desired positive ion species to the hot work surface. The
parameters that must be controlled during this plasma treating step
include

1. Temperature of each furnace zone
2. Temperature of the work surface
3. Total system gas pressure
4. Mass flow of several reactive gases
5. Plasma voltage, current, and power levels

All of the plasma treating parameters are interrelated and can react
with each other. However their individual adjustment must still
permit control of the two important work-related parameters.

1.  The ion potential that controls the activity of the species being
    added during the plasma treatment by establishing the concentra-
    tion gradient at the work surface
2.  The thermal energy transferred to the work surface that deter-
    mines the surface temperature, the rate of diffusion of the
    added species, and core tempering effects

It is important to control the ion potential and the thermal input
independently. For example, increasing the ion potential simply
by increasing the current density, allowing the input power and
the work surface temperature to increase, could produce a reduction
in core hardness by tempering during plasma nitriding, or grain
coarsening during plasma carburizing. Plasma treating power supplies,
therefore, employ a pulsing technique with variable duty-cycle
control to separate the effects from these two control parameters.
Pulsed plasma power supplies must have a control mechanism for
increasing the amplitude of the current pulse to increase the plasma
ion density at the work surface. Separate automatic controls ensure
a compensating reduction in the pulse duration time and the pulse
repetition rate to hold the plasma power and thermal input constant,
at its original value.

## 8.5  Furnace and Work Temperature Control

It is important to recognize that although furnace temperatures
and work surface temperatures are related, they are usually quite
different. To meet process requirements for accuracy, the temperature measurement system for the work must include thermocouples
in direct metal-to-metal contact with the work surface. Since the
work surface is at cathode potential, the measurement system must
prevent this relatively large cathode potential from obscuring the
much smaller thermocouple signal. If the temperature measurement
system supplied for control of the plasma treating furnace cannot
reject this large cathode potential, the thermocouples used for work
surface measurement must be electrically insulated and, as a result
of this special need, will be improperly thermally insulated from
the workpieces. This thermal insulation will cause some loss in
measurement accuracy.

If the thermocouple is isolated, it may become necessary to
rely on just visual estimates and optical pyrometer readings, with
empirical corrections for viewport interference, to establish workpiece temperature as noted in the *Metals Handbook*:

> Workpiece temperature is usually measured directly by an inserted
> sheathed thermocouple that is electrically isolated. Observing the
> work through the viewport, with the glow extinguished, provides

confirmation of the thermocouple measurement. At temperatures close to the limit of visibility, 500 to 510°C (930 to 950°F) visual estimates are quite accurate.

Conybear (1985) recognizes this requirement for some plasma treating systems as follows:

> 3.2.1.1 Thermocouples: Shall be protected in tubes to prevent deterioration due to furnace atmospheres. Thermocouple tubes must be isolated electrically from both the anode and cathode to prevent heating of the thermocouple by the glow discharge.

## 8.6  Gas Mixture and Plasma Parameter Control

The bell furnace, with its separately controlled resistance heaters, generally provides the major portion of the thermal energy needed to reach the desired work surface treating temperature. Ideally, the bell furnace zone temperature setpoints control the workpiece temperature. If the equipment does not include a separately heated bell furnace, and only plasma parameters can be changed to control the workpiece temperature, the operating procedures can become a great deal more complex.

The *Metals Handbook* indicates the following situations, which can lead to significant operating problems:

> Work loads of practical size require large amounts of power to reach nitriding temperature; power used to maintain the glow discharge usually is the only source of process heat. Resistance heating elements, if available, are not used at the nitriding temperature.
>
> Resistance heating elements may be used to bring the work load to nitriding temperature; at that temperature the elements are then shut off. Plasma discharge alone provides the necessary energy to maintain the work at temperature.

If the system utilizes only plasma power for heating, a major problem can be encountered when the furnace charge reaches the surface treatment temperature and it becomes necessary to change the power input to stabilize the system at this final temperature. A change in plasma power input can generally be accomplished by changing the gas mixture, by changing the gas pressure, or by changing the plasma voltage.

As an example, during a plasma nitriding cycle, the current density, and, therefore, the plasma power density, can be reduced by reducing the nitrogen content in the gas mixture. However, reducing the nitrogen content will also reduce the nitrogen activity. Since this will extend the time required to meet the metallurgical specifications, this is not a method that can be regularly used to stabilize the temperature.

The gas pressure can be reduced to reduce the current and power density, however, a pressure reduction can lead to the formation of a damaging hollow cathode. When the charge reaches the surface treatment temperature, a reduction in gas pressure may permit the cathode fall thickness to make a transition from a value that initially provided hole penetration to a value that does not. This transition is described in more detail during the discussion of hollow cathodes.

Generally, if a bell furnace with resistance heaters has not been provided for the control of work temperature, the plasma power must be reduced by reducing the plasma voltage. However, if the charge has a large surface area or if the geometry is complex, reducing the voltage can transfer operation from the abnormal glow discharge to the normal glow discharge region. Work surfaces that were completely covered with a uniform abnormal glow discharge during the heating cycle will now be only partially covered when the surface treatment temperature is finally reached. If this is allowed to happen, uniform treatment cannot be expected.

For a furnace charge with a limited work surface area and with simple geometry there is perhaps less danger that this reduction in plasma voltage will cause a transition to the undesirable normal glow discharge. A reduction in plasma voltage will reduce the plasma power input and will permit the operator to stabilize the system at the desired temperature. In some cases an experienced operator can recognize that a charge will have limited surface area and that he should operate with low plasma power to avoid a transition to an arc discharge. When loading the cold wall system, the operator can install some auxiliary radiation shields to limit the heat loss from the system so the specified treatment temperature can be reached at a low plasma power density. However, the installation of these radiation shields can lead to temperature nonuniformity if the charge itself has some sections with a high surface-to-volume ratio and other sections with a low surface-to-volume ratio. Generally a conservative operator will avoid the use of auxiliary radiation shields and will instead choose to operate with a low power density at a low treatment temperature with an extended cycle time.

## 8.7 Convection Cooling and Liquid Quenching

After the completion of the plasma treating cycle, the work can be cooled as fast as possible within the limits of acceptable dimensional

distortion. The removable bell furnace design permits fast cooling
without the thermal inertia normally associated with other hot wall
furnace designs. A cold inert atmosphere can be recirculated within
the vacuum retort to transfer heat from the work surface to the
retort inside wall. Cold air can be circulated over the outside wall
of the retort to carry this heat to a ducting system where it can
be discharged away from the heat treating operating. Carburized
work can either be quenched with liquid directly from the plasma
treating temperature or first cooled in a recirculating atmosphere,
reheated and stabilized at an intermediate temperature and then
quenched from this intermediate temperature. Workpiece transfer
to the liquid quench can be through air or within a transfer hood
under a protective or inert atmosphere depending upon process
requirements.

### 8.8   Process Parameters for Plasma Nitriding

There are a number of power supply, gas mixture, and workpiece-
related parameters that can be independently selected by a process
programmer for a plasma nitriding cycle to be run with microprocessor
control:

1.   Pulse voltage amplitude
2.   Gas mixture partial pressure
3.   Nitrogen, hydrogen, and methane mass flow rates
4.   Pulse time duration and repetition rate
5.   Bell furnace zone temperatures
6.   Nitriding time to the end point

The setpoints selected for the pulse voltage amplitude, the gas
mixture partial pressure, and the nitrogen, hydrogen, and methane
mass flow rates will interact through the mechanism of the abnormal
glow discharge to establish the activity of nitrogen at the surface
of the workpiece. Considerable operating experience indicates that
at least the nitrogen activity and the work surface temperature
must be independently controllable for a plasma nitriding process
to be truly manageable at an industrial level by a furnace operator.
Interdependence of these parameters can make the specification
of a plasma nitriding program extremely complex and often a matter
requiring empirical determinations. As an example, it is often necessary to vary the gas composition during the course of a single run
to produce the desired surface layers. However, the voltage/current
relationship, and therefore the thermal input from the glow discharge,
depends in large measure upon this same gas composition. It is
therefore necessary to separate the heating effect from the require-

ments for surface chemistry so that a desired change in gas composition is not inadvertently accompanied by an undesirable change in the temperature of the work surface. This operating requirement can be met with pulsed power supplies that are now state-of-the-art.

The pulse time duration and repetition rates initially selected by the furnace operator can limit the thermal energy transfer from the plasma to values that will not appreciably influence work surface temperature. The work surface temperature can then be continuously monitored with direct contact thermocouples to ensure this limited thermal energy transfer. If the plasma parameters selected for nitrogen activity conspire to increase the work surface temperature, the microprocessor software can adjust the pulse duration and the pulse repetition rate to lower the thermal energy transfer to an acceptable value. Under these conditions the bell furnace zone temperatures will control the workpiece temperature and, therefore, the rate of nitrogen diffusion from the surface of the work to the core. The Bell furnace temperature and the nitriding time to end point will determine the final nitride metallurgy.

*Gas Mixture Partial Pressure and Mass Flow Rates*

Mass flow controllers and downstream pressure control are most commonly provided for the control of the plasma atmosphere. The use of bottled gas permits operation with any ratio of nitrogen to hydrogen as opposed to operation with dissociated ammonia where this ratio is fixed by the extent of dissociation. Mass flow controllers are generally supplied in place of volume flow controllers to avoid process variation from changes in gas supply pressure and temperature. A partial pressure monitor in the vacuum retort provides a feedback signal for comparison with the gas mixture partial pressure set point. The microprocessor, then, regulates a variable conductance exhaust valve to maintain this partial pressure set point even when the mass flow rates of the process gases are changed to produce a desired work surface chemistry.

There are several different gas temperatures that can be measured in a plasma. In a vacuum of several millibars free electrons can be accelerated to very high velocities because of their light weight and their long mean free path between collisions. When a collision does occur, these electrons can transfer some of this high kinetic energy to a heavier particle for example to an atom of nitrogen. The plasma will then contain a mixture of fast electrons at 20,000°F and slow heavy ions and atoms at an average temperature of 1000°F. It is this thermal nonequilibrium that permits surface modifications at relatively low average furnace temperatures by reactions that are thermodynamically possible only at much higher temperatures. The final operating temperature for any furnace depends only on

the balance between power input and heat loss. If this thermal
balance can be maintained, the lowest work surface plasma treatment
temperature can be ambient if the process itself is not diffusion
limited.

Before the development of the glow discharge, the gas pressure,
the gas temperature, and the molecular density were related by
the equation of state

$$PV = nRT$$

where P is the pressure, V is the volume, n is the number of moles,
T is the temperature, and R is a constant. This relationship will
remain generally useful during the abnormal glow discharge if the
temperature is interpreted as the average thermal temperature of
the gas mixture. However, controlling just the average furnace
temperature during a plasma nitriding cycle is not sufficient. A
workpiece can be plasma nitrided at a normally acceptable average
temperature and still be locally overheated or even melted by high
current and power densities. High local temperatures can produce
an unacceptable work surface even when the furnace overtemperature
setting and the nominal power supply parameters have not been
exceeded.

### Nitrogen Activity

The density of each molecular species in the gas mixture becomes
important during an abnormal glow discharge because of its individual
contribution to electrical conductance and to the final metallurgy
of the surface. As an example the activity of nitrogen is a measure
of the nitrogen that is actually available for the nitriding reaction.
Although nitrogen activity is a function of many parameters that
cannot always be accurately calculated, at the present time a good
working assumption is that this activity is proportional to the partial
pressure of nitrogen and to the current density at the work surface.
This dependence of nitrogen activity on current density has an
important implication and a system for plasma surface treating must
have special features not found in a conventional vacuum furnace.
Since the temperature of a work surface is governed by the laws
of plasma physics as well as the laws of thermal radiation, the geometrical relationship of the workpiece and the vacuum retort will
be as important as the relationship between their emissivities. As
an example, the best process parameters for plasma nitriding a
2-ft-diameter cylindrical workpiece in a 3-ft-diameter furnace will
be different from the parameters for plasma nitriding a 0.5-ft-
diameter workpiece in the same furnace. Figure 21 shows the influence
of the workpiece/vacuum retort wall diameters and their emissivities

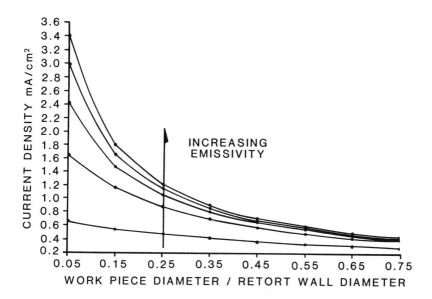

FIGURE 21  The current density necessary to reach a temperature of 1000°F as a function of the workpiece diameter/retort wall diameter for different emissivities.

on the plasma current density and, therefore, the power density that is required to bring a workpiece to 1000°F when a separate bell furnace is not employed to establish work surface temperature.

This same dependence on geometry will remain true even after the small diameter and large diameter workpieces have reached the 1000°F operating temperature. The total heat loss from the furnace at 1000°F, since it is only a function of the insulating qualities of the furnace and not a function of workpiece geometry, will be the same. The measured current and power input to both of these systems must be the same to maintain an identical operating temperature. The current and power densities, then, must be different because their work surface areas are different. The current density difference will cause a difference in the nitrogen activity at the work surface that will produce different process end points. Again, a separately heated bell furnace completely avoids this operating problem.

*Current Density: A Function of Gas Pressure*

The large number of interrelated plasma parameters indicates that the current density can be analyzed best with a graphical presentation

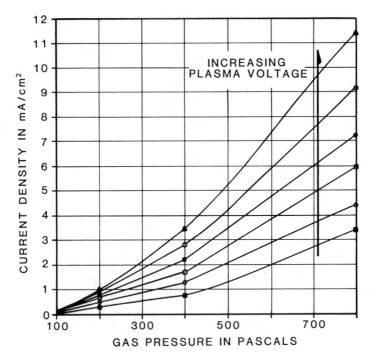

FIGURE 22   Current density = $f$ (gas pressure and plasma voltage) for a fixed gas composition, duty cycle, and surface temperature.

of collected empirical data. Current density is responsible for most of the metallurgical and thermal effects at a work surface and is the parameter most advantageously plotted as the dependent variable. The power supply current is an easily measured parameter, however, for the understanding of plasma surface treatment phenomena it is apparent that the more important parameter will be the current density at the surface of the work. Since the work geometry is known its surface area can always be calculated and a relationship between power supply current and work surface current density can be easily established for analytical work. The current density in a plasma process is a function of the gas density or the number of molecules per unit of volume. In Fig. 22, however, the gas pressure has been used as the independent parameter because of its good measurability. Figure 22 shows that increasing the gas pressure increases the current density by increasing the number of available charge carriers if other plasma parameters are fixed.

*Current Density: A Function of Gas Composition*

The gas composition has an important influence on current density. Hydrogen and nitrogen gas mixtures are generally used for most plasma nitriding processes. Gases which are needed to supply carbon to the system are used only in low concentrations. Those gases that are more easily ionized will produce higher current densities if the other parameters remain fixed. Figure 23 shows that increasing the concentration of nitrogen at any fixed gas pressure lowers the resistivity of the plasma and allows an increase in the current density. Although percent $N_2$ is an easily measured quantity, it should be remembered that the more important physical parameter is the partial pressure of the gas, not the percentage of the gas.

*Current Density: A Function of Plasma Voltage*

The current density is an important function of plasma voltage as shown in Figs. 24, 25, and 26 where current density has been measured for different gas compositions.

*Calculating Current Density*

If the gas composition remains constant, the equation of state can be used in the following form to calculate similar plasma conditions where current densities can be expected to be nearly the same:

$$\text{current density}_1 \times \frac{P_1}{T_1} = \text{current density}_2 \times \frac{P_2}{T_2}$$

where P = gas pressure in pascals and T = surface temperature in degrees Kelvin. As an example the current density in a plasma discharge at 100 Pa and at 100°C will be nearly equal to the current density in a discharge at 200 Pa at 473°C as shown in Fig. 27. Since the gas density decreases with decreasing gas pressure and increasing temperature, the current density will also decrease with increasing temperature as shown in Fig. 27.

## 8.9 The Hollow Cathode

The dimensions of holes and slots in a workpiece are of special concern during plasma surface treatments because of a glow discharge phenomena called the hollow cathode effect. In a stable abnormal glow discharge, the conductance of the plasma depends upon the balance between charge carrier generation and their annihilation. Charge carriers are generated by the electrical field within the plasma and annihilated by several different mechanisms, but principally by recombination at the work surface and the surface of the

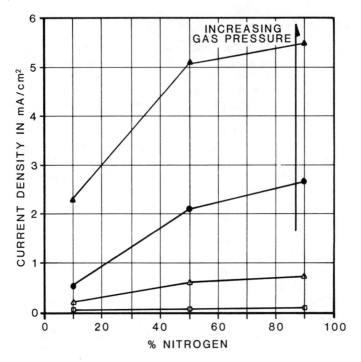

FIGURE 23   Current density = $f$ (% $N_2$ and gas pressure) for a
fixed plasma voltage, duty cycle, and surface temperature.


vacuum retort. The balance between charge carrier generation and
annihilation can be upset by deep holes in a work surface. In a
deep hole, a larger number of charge carriers can be generated
at the large cylindrical surface defining the hole than can escape
to the vacuum retort through the comparatively small circular opening
leading from the hole. The resulting increase in the charge carrier
population will lead to an increase in electrical conductance and
consequently to an increase in the gas temperature from the resulting
higher plasma current. The increase in gas temperature itself can
lead to an additional increase in conductance which can quickly
lead to a concentration of the entire discharge in the hollow space
defined by the deep hole. The concentration of the discharge can
produce temperatures sufficiently high to melt the work surface
within the hole. This destructive hollow cathode effect must, of
course, be prevented by the design of the equipment and the regulation of the plasma operating parameters by the microprocessor.

The hollow cathode effect on the local current density is always important. This effect is not limited to just the enclosed space within deep holes as its name might imply. Any variation in work geometry that effects the balance between charge carrier generation and annihilation can change the local current density. This influence of work geometry is best analyzed by first normalizing the workpiece dimensions with respect to the cathode fall thickness. For example, although an intense hollow cathode can be developed in a 2-cm-diameter hole at 300 Pa, the same size hole at 600 Pa will not cause problems. The following is generally true, as shown in Fig. 28:

A.  Hole diameter < 2 × cathode fall thickness: There will be only a slight penetration of the hole by the glow discharge. Generally

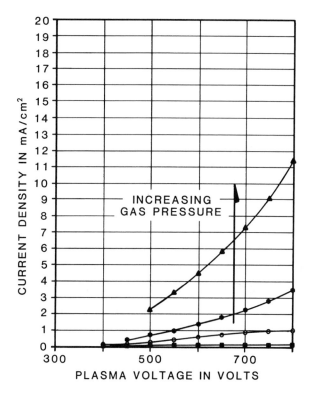

FIGURE 24  Current density = $f$ (plasma voltage and gas pressure) with 10% nitrogen for a fixed duty cycle and surface temperature.

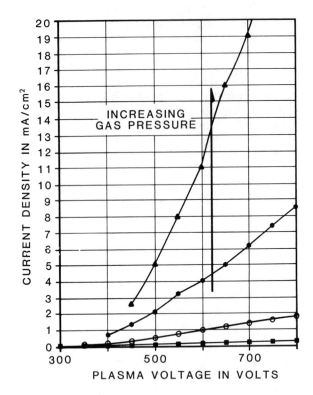

FIGURE 25  Current density = $f$ (plasma voltage and gas pressure)
with 50% nitrogen for a fixed duty cycle and surface temperature.

the depth of this penetration will be less than two hole diameters
and plasma treatment will not be found below this depth.
B.  2 × cathode fall thickness < hole diameter < 4 × cathode fall
    thickness: There will generally be an intense and destructive
    hollow cathode effect.
C.  Hole diameter < 4 × cathode fall thickness: The glow discharge
    will penetrate the hole and plasma treating in the abnormal
    glow range can be obtained.

Although the work geometry is of primary importance, all of the
process parameters must be considered to determine the potential
for a hollow cathode. For example the composition of the gas has
an important influence on the cathode fall thickness and, therefore,
the potential for a hollow cathode as shown in Fig. 29.

At a low temperature of 100°C a destructive hollow cathode might occur that would not occur if the gas were heated to 300°C as shown in Fig. 7. Figure 7 also shows that deep holes in the work surface that were not penetrated by the abnormal glow discharge at low pressure can present a problem when the gas pressure is increased. Every hole can meet the conditions for a hollow cathode if there is a large step change, either an increase or a decrease, in gas pressure. If the hollow cathode from this change in pressure is allowed to persist without current interruption or plasma parameter changes by the microprocessor, the surface of the work can be damaged.

Now it can be seen that the power duty cycle can have an important effect on the potential for the development of a hollow cathode. With a pulsed power supply operating with a low power duty cycle, gas heating can be kept low to prevent the formation

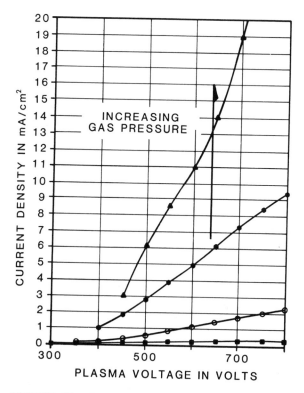

FIGURE 26  Current density = $f$ (plasma voltage and gas pressure) with 90% nitrogen for a fixed duty cycle and surface temperature.

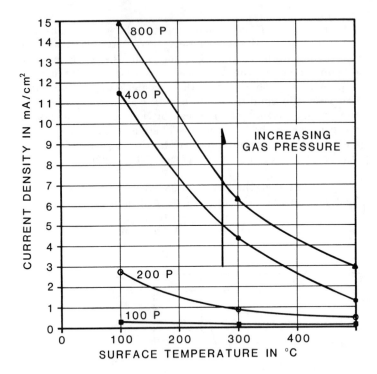

FIGURE 27  Current density = $f$ (surface temperature and gas pressure) for a fixed gas composition, plasma voltage, and duty cycle.

of a hollow cathode. With a low power duty cycle gas temperature will not increase and the plasma will maintain its electrical resistance. In contrast to this, with a continuous direct current power supply, if the gas temperature increase cannot be prevented, the electrical resistance will decrease and a destructive hollow cathode will develop. This effect is so important that almost all plasma nitriding systems now include a power supply that permits the control of the power duty cycle. If the power duty cycle cannot be automatically controlled by the microprocessor, the furnace operator must be held responsible for preventing workpiece damage as noted in the *Metals Handbook*:

> Periodically, the glow is extinguished and the load inspected for possible visibly hot areas. As the work reaches the visible temperature range, the glow is extinguished to permit visual confirmation of the measured temperature and satisfactory temperature distribution within the load.

## Dense Charging and the Hollow Cathode

Overly dense charging of complex shapes can always produce a
hollow cathode effect. When charging is too dense, the current
density cannot remain independent of the power duty cycle. For
example, the current density can increase by a factor of 4 for
two parallel mounted plates as the distance between them is allowed
to decrease. Plasma voltage and the resulting current versus time
for a pulsed discharge are shown in Fig. 30 for a parallel plate
system.

Figure 30 shows an increasing hollow cathode influence. The
current versus time clearly shows the rise in conductivity from
the hollow cathode effect. Superimposed on this increase in conduc-
stivity there can be, of course, a decrease in conductivity because
of a decrease in gas density if the gas temperature is allowed to
increase. This increase in the current density can be happening

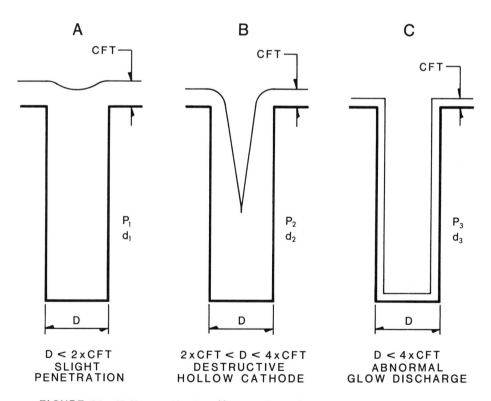

FIGURE 28   Hollow cathode effect and work geometry.

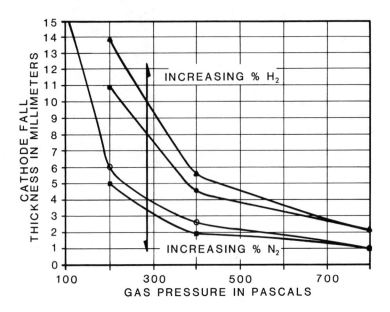

FIGURE 29  Cathode fall thickness = $f$ (gas pressure and gas composition) for a fixed plasma voltage, duty cycle, and surface temperature.

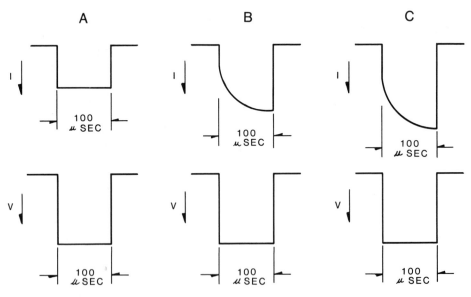

FIGURE 30  (A) Voltage and current versus time without a hollow cathode effect. (B) Voltage and current versus time with a hollow cathode effect with parallel plates separated by 1.5 cm. (C) Voltage and current versus time with the hollow cathode effect more fully developed with parallel plates separated by 1.0 cm.

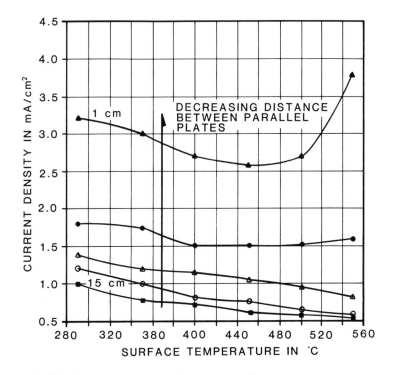

FIGURE 31 Current density = $f$ (surface temperature and work spacing) for a fixed gas pressure, gas composition, plasma voltage, and duty cycle.

even when the typical visual appearance of the hollow cathode is not present. Figure 31 permits estimating this effect.

Some basic trends of the abnormal glow discharge can now be defined.

If the voltage, pressure, or percent $N_2$ are increased, the current and power density will increase.

If the temperature is increased, the current and power density will decrease.

These trends occur in a stable abnormal glow discharge. If these trends are not observed, it is a good indication of the presence of a hollow cathode or that the discharge is not in the abnormal glow range where the work will be fully covered and where uniform plasma treating can be expected.

## 8.9   Plasma Nitriding Deep Holes

Kwon et al. (1986) have compared the efficiencies of the pulsed
and dc power supplies for nitriding deep blind holes in a work
surface. Nitralloy specimens were prepared with drilled holes as
shown in Fig. 32. Specimens were plasma nitrided for 6 h using
the parameters shown in Fig. 33. The specimens were longitudinally
sectioned after plasma nitriding, metallographically prepared, and
the penetration depth measured. Penetration depth was defined as
the distance from the hole entrance to the point with a 0.5-micron-

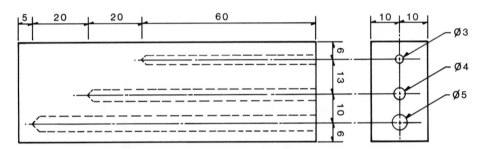

FIGURE 32   Specimen geometry.

| POWER MODE | GAS MIXING RATIO (H : N ) | PRESSURE (PASCAL) | VOLTAGE (VOLT) | TEMPERATURE (°C) |
|---|---|---|---|---|
| PULSE | 1:2 | 660 | 410 | 500 |
| D.C. | 1:2 | 660 | 395 | 500 |

FIGURE 33   Experimental parameters for ion nitriding.

| POWER MODE | HOLE DIAMETER | | |
|---|---|---|---|
| | 3 | 4 | 5 |
| D C | 13 | 29 | 50 |
| PULSE | 31 | 47 | 66 |

FIGURE 34   Penetration depth in hole specimen.

thick nitride layer, determined optically. Figure 34 shows the penetration depth in millimeters as a function of hole diameter in millimeters for the pulsed and dc power supplies. This penetration depth was normalized by dividing by the hole diameter and the following relationships established for dimensions in mm.

$$\frac{\text{Pulsed power supply penetration depth}}{\text{diameter}} = 1.7 \text{ (diameter} + 4.9$$

$$\frac{\text{DC power supply penetration depth}}{\text{diameter}} = 2.8 \text{ (diameter)} - 4.25$$

Both pulsed and dc power supplies are available for plasma treating. Studies continue to show the advantages that can be expected from the use of the pulsed power supply.

## 9  PLASMA NITRIDING: CASE HISTORIES

Plasma nitriding can make an important contribution to applications requiring superficial hardness or improved wear, fatigue, or corrosion resistance. The metallurgical changes made by plasma nitriding are most often characterized by an identification of the compound zone and with a hardness profile measured from the work surface to the core.

### 9.1  Carbon Steels

Carbon steels and cast iron that do not contain alloys for the formation of nitrides can be hardened by the formation of an epsilon compound zone and a diffusion zone. This epsilon compound zone can be very dense and uniform and will contribute improved wear resistance to applications with light loads and broad areas of load contact. A thick epsilon compound zone can provide good corrosion resistance. Figure 35 shows a hardness profile obtained with a 1045 carbon steel plasma nitrided for 20 h at 450°C. A nearly identical profile can be obtained for this same steel by plasma nitriding for 4 h at 550°C. Typically, carbon steels can be plasma nitrided to produce a Vickers hardness HV.2 250-400 with an epsilon compound zone 10-15 microns thick and a diffusion zone 0.375-0.750 mm thick.

### 9.2  Alloy Steels

Alloy steels can be hardened by the formation of either a gamma prime or epsilon compound zone and a diffusion zone. Almost 80% of all application requirements can be met with a gamma prime compound

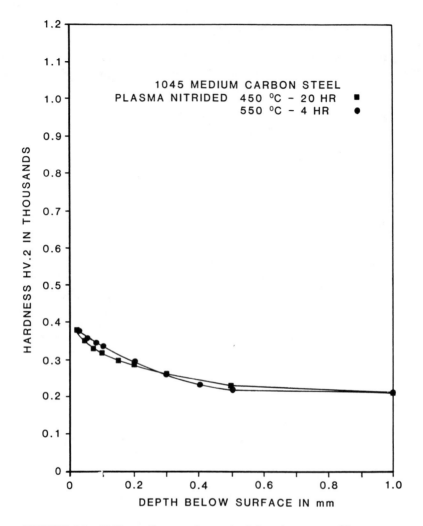

FIGURE 35   1045 medium carbon steel hardness profiles.

zone contributing principally fatigue resistance and an improvement
in wear resistance to the workpiece. The remaining requirements
can be met with an epsilon compound zone contributing principally
corrosion resistance and some wear resistance. A mixed compound
zone, with both the epsilon and gamma prime phases, is seldom
recommended because of its brittle nature and its tendency to spall
from the work surface under light loads. The hardness profiles

for a 4140 alloy steel, plasma nitrided at two different temperatures for the same time, are shown in Fig. 36.

Since plasma nitriding is a diffusion limited process, higher plasma nitriding temperatures produce a greater depth of diffusion and a lower surface hardness. Figure 37 shows the hardness profiles that can be obtained at higher temperatures and with longer plasma nitriding times.

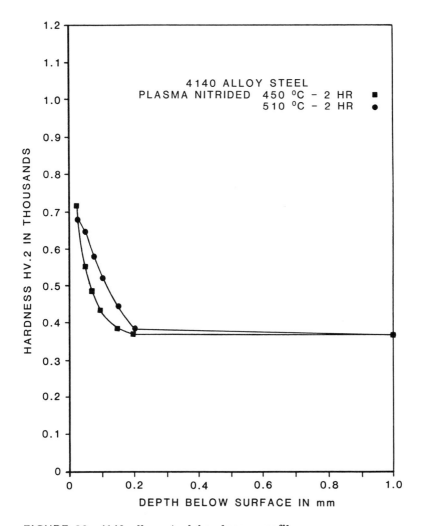

FIGURE 36   4140 alloy steel hardness profile.

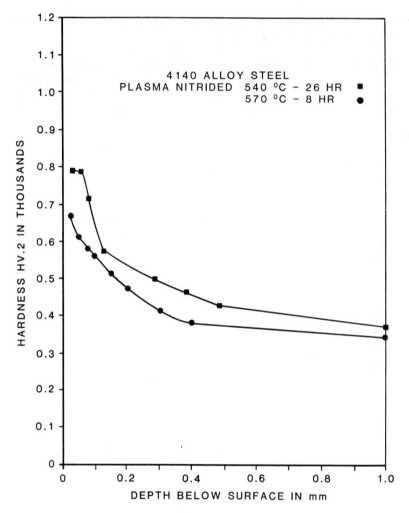

FIGURE 37   4140 alloy steel hardness profiles.

Figure 38 shows the hardness profiles for a 14CrMoV69 alloy steel plasma nitrided at different temperatures for different times to permit estimating the effects of these two process parameters.

Figure 39 shows the nearly identical hardness profiles that can be obtained with a 16MnCr5 alloy steel by increasing the plasma nitriding temperature while decreasing the time. Alloy steels can be plasma nitrided to produce a Vickers hardness HV.2 500-900 with either an epsilon compound zone 10-15 microns thick or a gamma prime compound zone 5-8 microns thick, and a diffusion zone 0.375-0.850 mm thick.

## 9.3 Hot Work Steels

Forging and extrusion dies manufactured from hot work steels can be hardened by the formation of a gamma prime compound zone and a diffusion zone to reduce erosive wear and improve fatigue

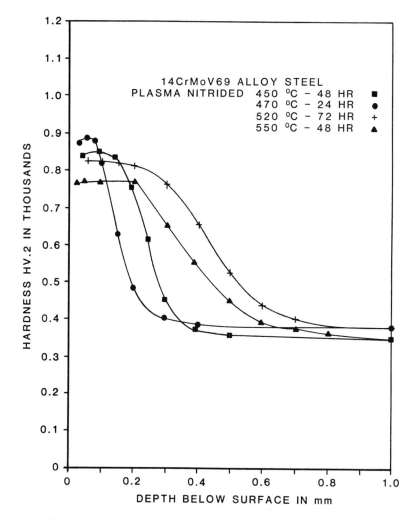

FIGURE 38   14CrMoV69 alloy steel hardness profiles.

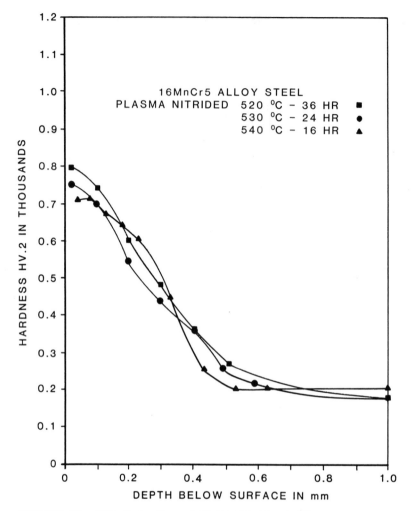

FIGURE 39    16MnCr5 alloy steel hardness profiles.

resistance. Figure 40 shows the effect of increasing the plasma
nitriding temperature with fixed time on the hardness profiles for
an H-13 hot work steel. Figure 41 shows the effects of increasing
the plasma nitriding time with fixed temperature on the same H-13
hot work steel. The hardness profiles that can be obtained after
tempering and plasma nitriding at the highest temperatures recommended for this material are shown in Fig. 42. Figures 43 and 44
show hardness profiles that can be obtained with other hot work
steels H-10 and H-12. Hot work steels can generally be plasma

nitrided to produce a Vickers hardness HV.2 900-1200 with a gamma prime compound zone 2.5-5 microns thick with a diffusion zone 0.100-0.500 mm thick.

## 9.4   Cold Work Die Steels

Cold work die steels are readily plasma nitrided to increase their superficial hardness either with a thin gamma prime compound zone or with no compound zone and only a diffusion zone. Figure 45

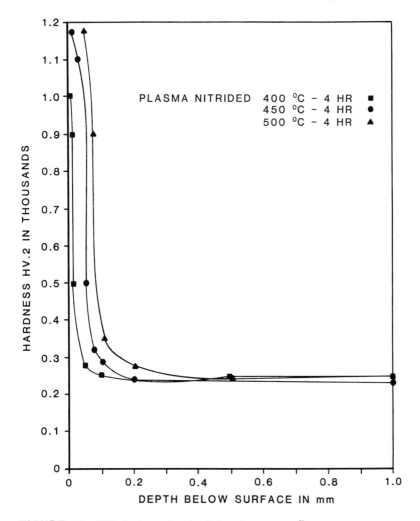

FIGURE 40   H13 hot work steel hardness profile.

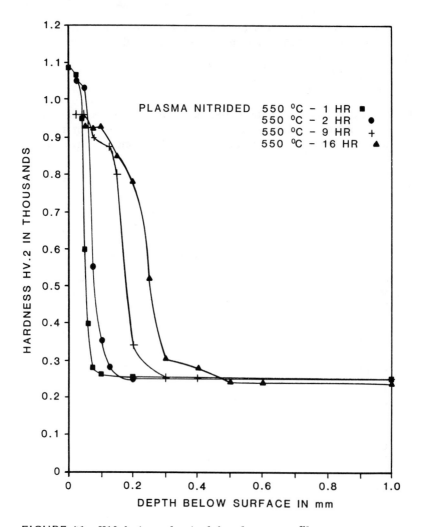

FIGURE 41   H13 hot work steel hardness profile.

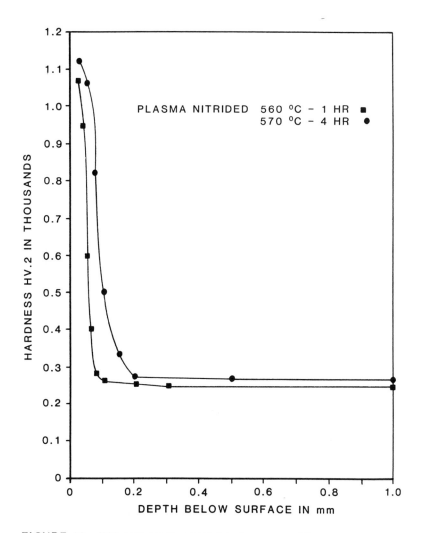

FIGURE 42   H13 hot work steel hardness profile.

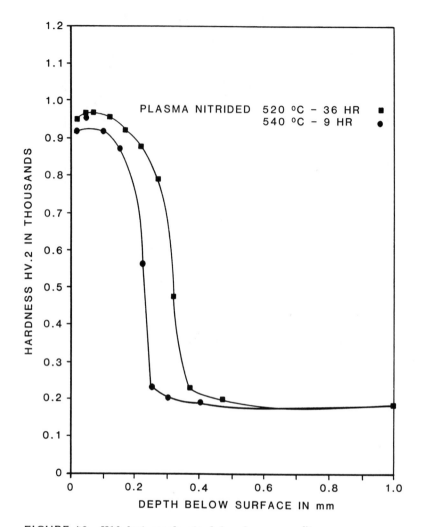

FIGURE 43   H10 hot work steel hardness profile.

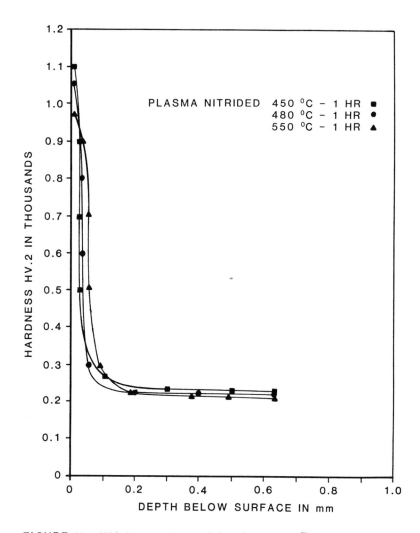

FIGURE 44   H12 hot work steel hardness profile.

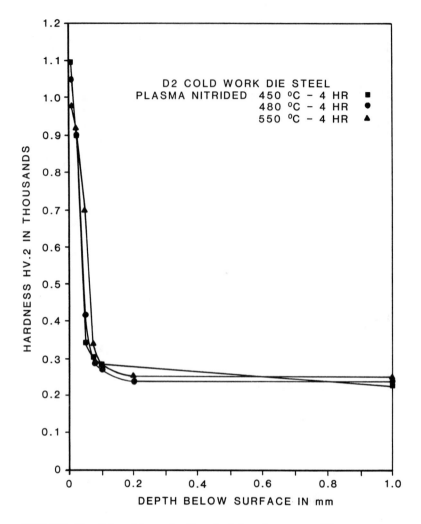

FIGURE 45    D2 cold work die steel hardness profile.

shows the effect of increasing the temperature through the normal
plasma nitriding temperature range at constant time. A hardness
profile obtained at the high end of the plasma nitriding temperature
range that still retains maximum core properties is shown in Fig. 46.
Figure 47 shows the nearly identical hardness profiles that can be
obtained with an X210CrW12 steel plasma nitrided at different temperatures for different times. Cold work die steels can be plasma nitrided

to produce a Vickers hardness HV.2 950–1200, generally with no compound zone and a diffusion zone 0.100–0.200 mm thick.

## 9.5 Nitriding Steels

Gears manufactured from nitriding steels with a high proportion of nitride formers as alloying elements can be plasma nitrided with no compound zone and no nitride networks for critical load bearing

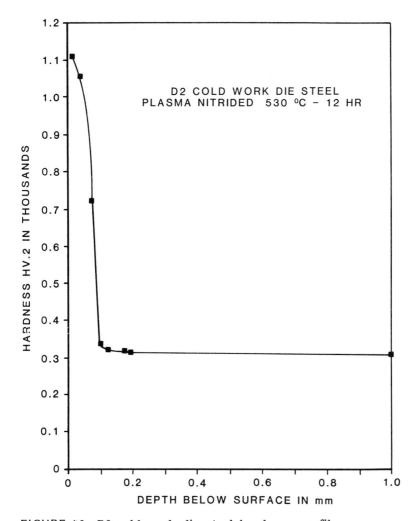

FIGURE 46  D2 cold work die steel hardness profile.

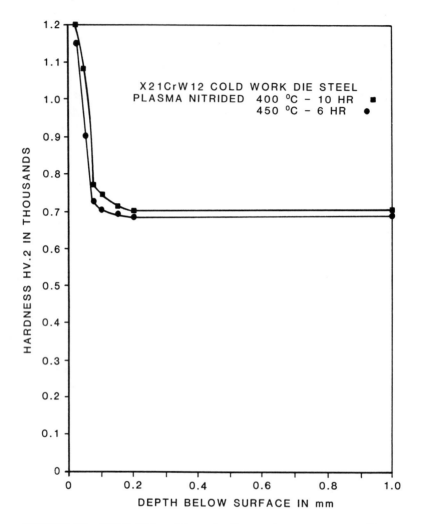

FIGURE 47   X21CrW12 cold work die steel hardness profile.

applications. Figure 48 shows the hardness profile obtained on a
Nitralloy 135M ring gear. This hardness profile was obtained with
no evidence of a white layer or nitride network in any portion of
the material. Hardness measurements made at the tip and the root
of a Nitralloy N gear are shown in Fig. 49. The hardness profiles
for a 31CrMoV9 nitriding steel as a function of plasma nitriding
temperature and time are shown in Fig. 50. Figure 51 shows the

hardness profiles for a 34CrAlNi7 alloy nitrided at two different temperatures and for different times to permit estimating the effects of these two parameters. Nitriding steels are generally plasma nitrided to produce a Vickers hardness HV.2 850-1200, either with no compound zone or a gamma prime compound zone 5-10 microns thick. A diffusion zone 0.375-0.750 mm thick is normally produced.

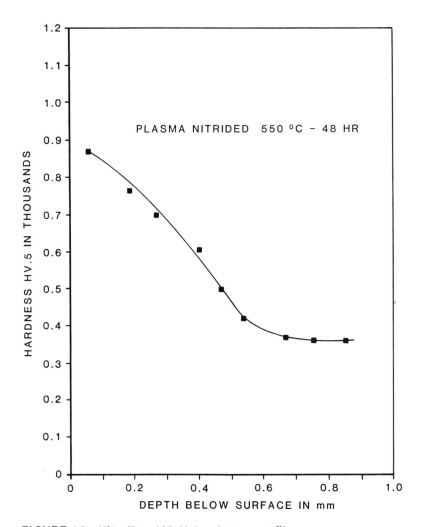

FIGURE 48   Nitralloy 135 M hardness profile.

## 9.6    Stainless Steels

Stainless steel workpieces can be plasma nitrided to improve their
superficial hardness and wear resistance with only limited and often
no loss of corrosion resistance. Figure 52 shows the superficial
hardness that can be obtained on 304 stainless steel. Figure 53
shows the hardness profiles that can be obtained by increasing
the time at constant plasma nitriding temperature, and Fig. 54 the

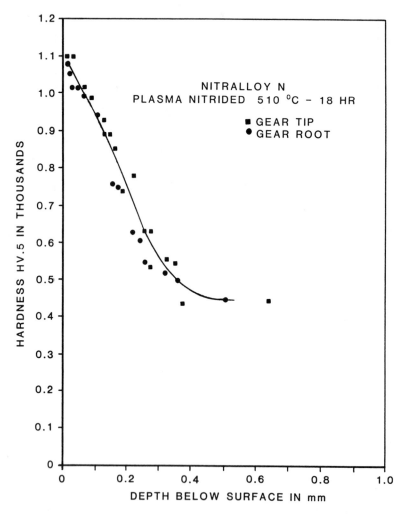

FIGURE 49    Nitralloy N hardness profile.

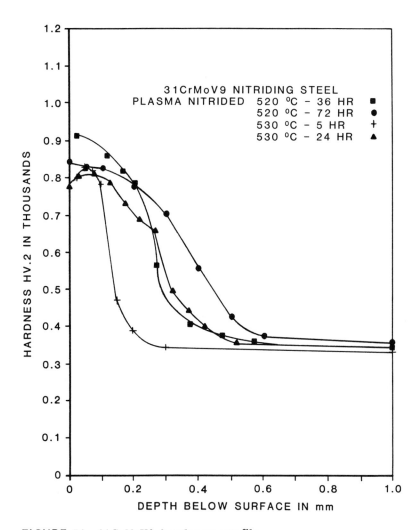

FIGURE 50   31CrMoV9 hardness profile.

hardness profile by increasing the temperature with constant plasma nitriding time.

### 9.7   Diesel Engine Timing Gears

Plasma nitriding of diesel engine timing gears, to reduce operating noise levels, has been investigated by Stosic and Ziatanovic (1986). Plasma nitrided nodular cast iron spur gears were tested as a

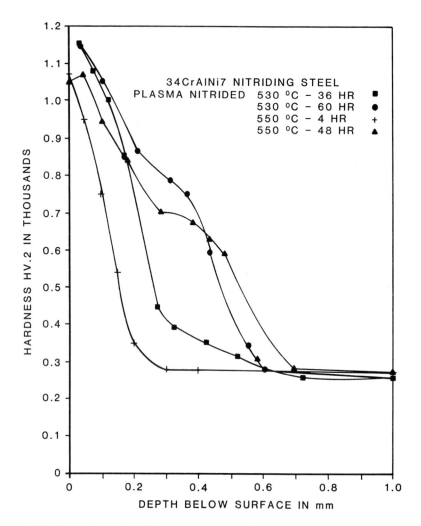

FIGURE 51    34CrAlNi7 hardness profile.

replacement for carburized and through-hardened alloy steel gears. Gears were treated for 18 h at 520°C in a 30% nitrogen-70% hydrogen gas mixture at 600 Pa to produce a 5-10-micron gamma prime compound zone and a 0.3-mm diffusion zone with the microhardness profile shown in Fig. 55. In this application plasma nitriding reduced material and heat treating costs and permitted the production of gears without distortion that could run with less noise than the

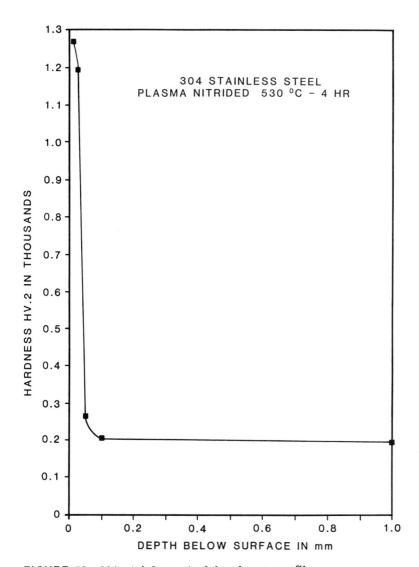

FIGURE 52   304 stainless steel hardness profile.

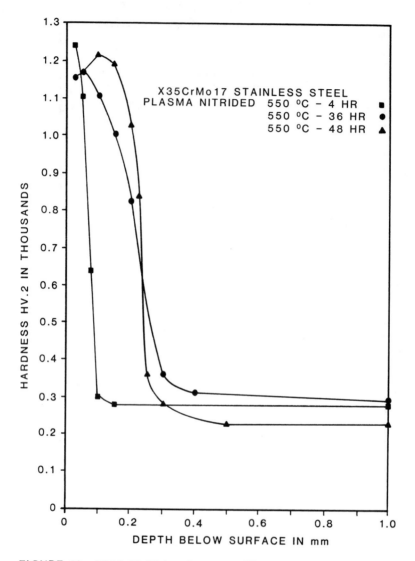

FIGURE 53   X35CrMo17 hardness profile.

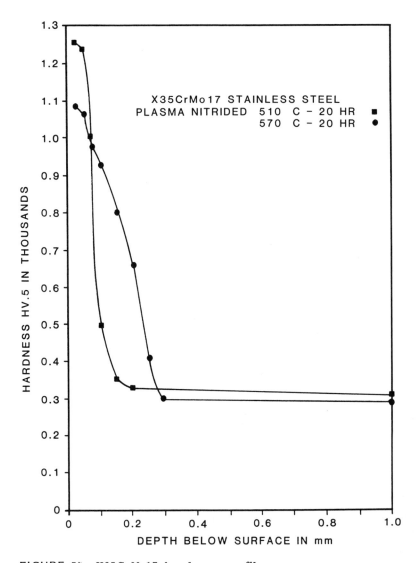

FIGURE 54   X35CrMo17 hardness profile.

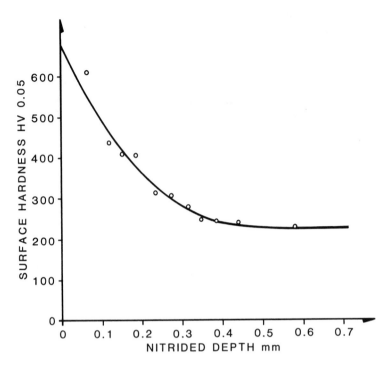

FIGURE 55   Microhardness profile of nitrided layer.

original carburized or hardened gears. It was determined by field
testing that the plasma nitrided timing gears had the same service
life as the more conventionally heat treated gears.

### 9.8   Oilwell Downhole Tooling

Dickens (1986) has tested the effectiveness of plasma nitriding
to improve the corrosion resistance of oilwell downhole drilling tools.
Dickens indicates that plasma nitriding significantly improves the
corrosion resistance of 4145 alloy steels and that this corrosion
resistance remains effective in oilwell downhole tooling despite considerable loss of surface material by abrasive wear.
        It was found that tooling, corroded from previous service,
could be restored to a working condition by additional plasma nitriding
if the dimensions of the surface pits and scratches were large compared to the thickness of the cathode fall and favored the formation
of a covering glow. Only thick oxides in the bottom of corrosion
pits formed a diffusion barrier that completely masked the plasma

nitriding effect. Both the sulfide and chloride stress corrosion
cracking resistance of 4145 alloy steel was improved by plasma
nitriding as shown in Fig. 56. However the corrosion resistance
of plasma nitrided 17-4 PH stainless steels tooling was considerably
decreased. This loss in corrosion resistance was attributed to
chromium alloy depletion from the formation of chromium nitrides.

## 9.9 Maraging Steels

High nickel maraging steels achieve full hardness, 50-54 Rc, by
an aging treatment, usually 3 h at 480°C. Since hardness does
not depend on the quenching rate, full hardness can be achieved
in very large cross sections making these steels suitable for aluminum
die casting dies and cores, aluminum hot forging dies, dies for
molding plastics and for various support tooling used in the extrusion
of aluminum. Maraging steels are less prone to heat checking than
H-13 tool steel and can be used at a higher temperature.

Since this aging temperature is within the operating limits of
plasma nitriding equipment, Ozbaysal and Inal (1986) were able
to study the possibility of combining the aging treatment with a

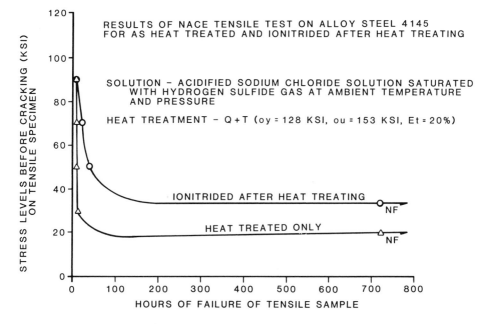

FIGURE 56   Results of NACE tensile test on alloy steel 4145.

plasma nitriding treatment to improve surface hardness while increasing core hardness to normal values. Solution treated, water quenched
samples were used in all tests. The nitrogen content during plasma
nitriding was 25%, and the total gas pressure was 2000 Pa for all
tests. A thin compound layer was produced in all cases that was
mechanically removed prior to surface hardness measurements. The
surface and core properties of grades 250, 300, and 350 maraging
steels were measured after plasma nitriding for 2-10 h at temperatures
between 440 and 520°C as shown in Figs. 57, 58, and 59.

It can be seen that when plasma nitriding maraging steel in
this temperature range, where there is a core aging process and
a surface hardening process occurring simultaneously, the higher
core hardness and the higher surface hardness occurred at the
lower plasma nitriding temperatures and with the shorter plasma
nitriding times. This is generally in contrast to experience with

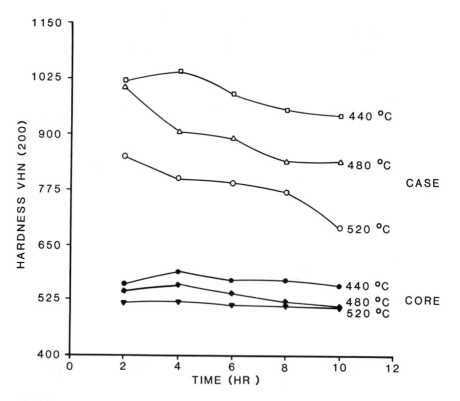

FIGURE 57   Surface and core hardness as functions of ion nitriding
temperature and time for 250 grade.

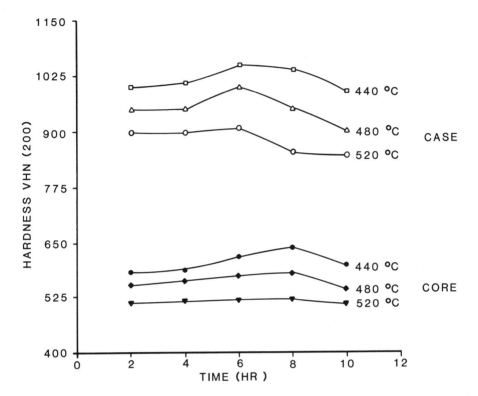

FIGURE 58  Surface and core hardness as functions of ion nitriding time and temperature for 300 grade.

plasma nitriding fully hardened and tempered steels. They conclude that it is possible to simultaneously age the core and harden the surface of maraging steels, while avoiding a reversion to austenite, by imposing reasonable limits on the temperatures and times used for plasma nitriding.

## 9.10  Aluminum and Its Alloys

Arai et al. (1986) have shown the possibility of nitriding most commercial aluminum alloys. Aluminum alloys, important because of their low density and good machinability, are limited in their application if they cannot be surface hardened. The adherent aluminum-oxide layer on finished aluminum parts had previously acted as a barrier to nitriding. Their research has shown that this barrier to nitriding can be successfully removed by sputtering in argon.

Tests showed that sputtering in hydrogen or nitrogen gas, by
comparison, did not remove the aluminum-oxide film or permit subsequent successful plasma nitriding.

After surface activation by argon sputtering aluminum and its
alloys were plasma nitrided at temperatures lower than 500°C. Layers
were grown at a rate of several microns/hour with an indicated
hardness greater than 1000 HV as shown in Figs. 60 and 61. They
note that Al-Cu, Al-Mg-Si, and Al-Zn-Mg alloys can be subjected
to $T_4$ or $T_6$ heat treatment after plasma nitriding without spalling
of the layer or loss of the developed wear resistance from the
finished part.

Their process begins with pumping the plasma nitriding system
to $10^{-3}$ Pa to remove air and water vapor. Argon sputtering for
1 hr at 100 Pa at a part temperature between 400 and 500°c eliminated
the superficial aluminum-oxide barrier to nitrogen diffusion. In fact,

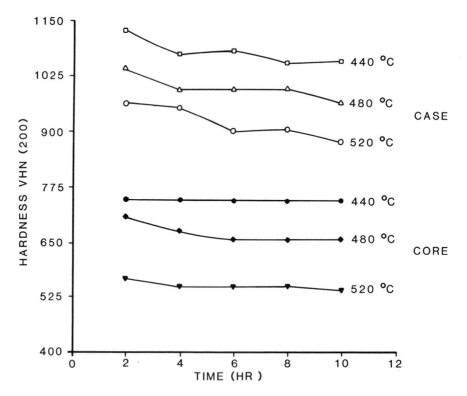

FIGURE 59  Surface and core hardness as functions of ion nitriding
temperature and time for 350 grade.

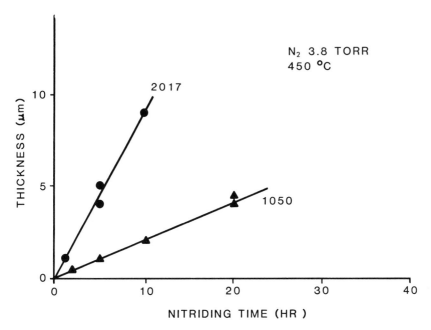

FIGURE 60   Relationship between thickness of nitride layers formed on 1050 and 2017 and nitriding time.

| ALLOYS | HARDNESS. HV * |
|--------|----------------|
| 1050   | 1000           |
| 2017   | 1250           |
| 5052   | 1600           |

∗ WITH LOAD 10 GR.

FIGURE 61   Microvickers hardness values obtained on cross section of nitrided layers.

tool marks on the surface prior to sputtering were removed and
a surface with conical projections was produced. Plasma nitriding
was then accomplished at 185-500 Pa in a temperature range 400-600°C.

X-ray analysis revealed that a polycrystalline phase of AlN had
been formed. There was strong bonding to the surface and wear
properties were judged to be excellent. The conical projection produced with argon sputtering were no longer in evidence and the
part surface was smooth.

## 9.11  Titanium and Its Alloys

Titanium and its alloys have a tendency to seize and gall under
conditions of wear. Therefore, any application that requires good
wear properties necessitates a surface treatment. Metin and Inal
(1986) have investigated plasma nitriding to increase the surface
hardness of pure titanium and the Ti 6242 alloy as shown in Figs. 62
and 63.

Plasma nitriding and gas nitriding can both be employed to
improve the wear characteristics of titanium and its alloys, however,
gas nitriding, due to slow nitrogen diffusion requires high nitriding
temperatures which eventually lead to the degradation of the mechanical properties of the bulk product. The plasma nitriding technique,
on the other hand, with faster nitriding kinetics, avoids this problem.
Plasma nitriding forms nitride layers at lower temperatures and in
shorter times and can be successfully employed when thin nitride

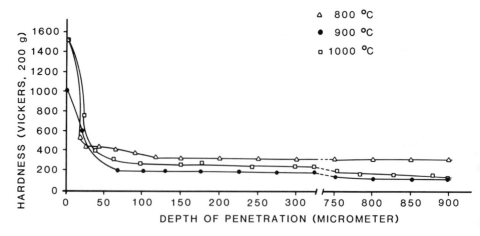

FIGURE 62  Hardness penetration curves for pure titanium ion
nitride for 8 hours at specified temperatures.

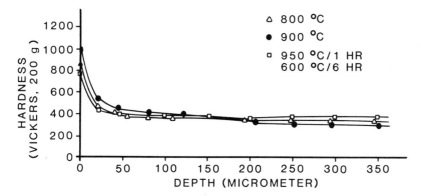

FIGURE 63   Hardness versus depth of Ti 6242 alloy for 8 hours.

layers at the surface are desired. To avoid grain coarsening and
the degradation of the bulk properties due to high nitriding temperatures, the complete heat treating cycle can be performed in the
plasma nitriding furnace if the furnace is designed to achieve the
necessary cooling rate.

Pure titanium and the Ti6242 alloy were plasma nitrided with
pure nitrogen at 665 Pa for 4-20 h at temperatures between 800
and 1000°C. Layers of TiN and $Ti_2N$ were formed at the surface
followed by a nitrogen-stabilized a-Ti. The growth of these layers
increased parabolically with time indicating a diffusion-controlled
process as shown in Figs. 64(a) and 64(b). Figure 65 shows the
hardness profiles obtained with a TiAl6V4 alloy plasma nitrided
at different temperatures for different times.

## 9.12   Powder Metal Parts

Chen has shown that plasma nitriding can replace gas and vacuum
carbonitriding and improve the wear and fatigue resistance of powder
metal parts. Carbonitride properties are derived by quenching to
produce a transformation to martensite. This transformation can
cause work distortion and dimensional changes that are difficult
to predict in powder metal parts—especially parts with irregular
cross sections, or parts at less than full density. Plasma nitride
properties are obtained at lower temperatures with slow cooling
and, therefore, with less distortion. They are obtained with no
change in phase and, therefore, with essentially no change in dimensions.

Most importantly, during gas and vacuum carbonitriding, interconnected porosity in parts at less than full density can lead to

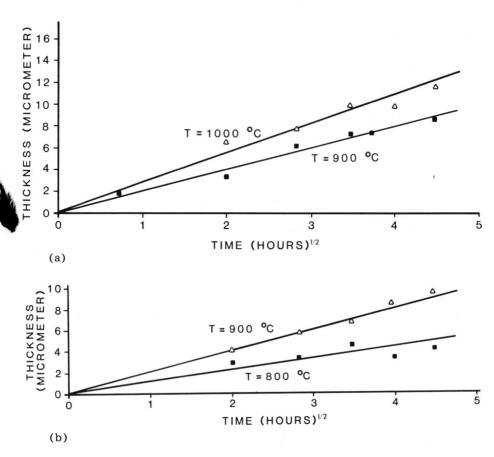

FIGURE 64 (a) Thickness of the nitride layer versus square root of time, pure titanium specimen. (b) Thickness of the nitride layer versus square root of time, Ti 6242 alloy.

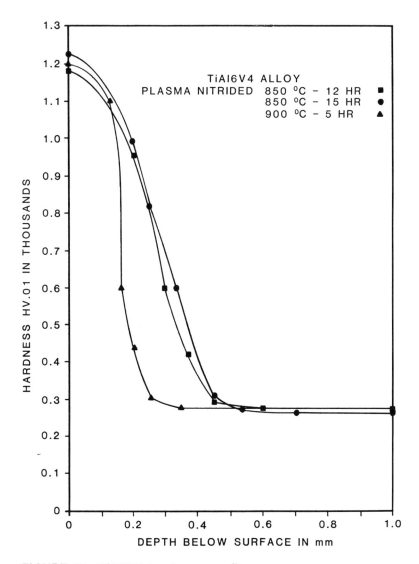

FIGURE 65   TiAl6V4 hardness profiles.

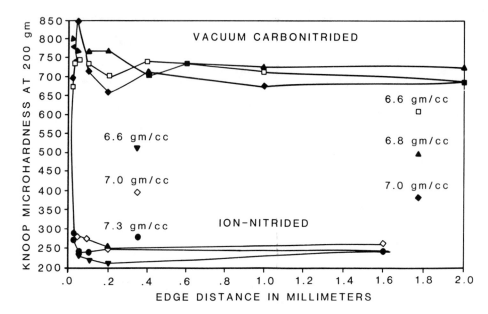

FIGURE 66  Microhardness profile of low alloy P/M steel by vacuum carbonitriding and ion nitriding.

carbon penetration through the entire cross section, producing irregular and unpredictable hardness profiles. Special materials must often be added to the powder mix to reduce interconnected porosity, or the sintered parts must be copper infiltrated or steam oxidized to block this carbon penetration. The fact that the abnormal glow will not penetrate an opening smaller than the cathode fall thickness has an important consequence. Since this thickness is larger than the dimensions of the interconnected pores in powder metals, plasma nitriding can be reliably confined to the outer, exposed surface of the parts. Chen has found that sharp transitions between core and case properties are possible as shown in Fig. 66. This permits the production of tough powder metal parts with good wear and fatigue properties.

## 10   PLASMA CARBURIZING: CASE HISTORIES

Although the fundamental requirements for plasma carburizing are now well understood, its commercial development has lagged behind the development of plasma nitriding. Since carburizing accounts

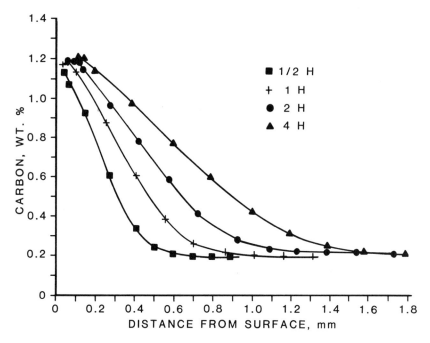

FIGURE 67  Plot of carbon profiles obtained after plasma carburizing.

for well over 80% of the total requirement for case hardened products, the commercial development of plasma carburizing will be of significant value.

Booth et al. (1983) have shown the time dependence of the plasma carburizing process by measuring the carbon profiles of mild steel obtained at 950°C with a methane/hydrogen gas mixture in the pressure range 200 to 2000 Pa. Figure 67 shows that high levels of carbon can be introduced by this process before the start of the diffusion cycle. Figure 68 shows the carbon profile obtained with an AISI 8620 steel both before and after the diffusion cycle. Figure 69 shows the hardness profile of the AISI 8620 sample after the diffusion cycle. They conclude that the plasma carburizing process permits the rapid introduction of carbon in a controlled manner that leads to the formation of a uniform case and that this process will compare favorably with conventional atmosphere and vacuum carburizing.

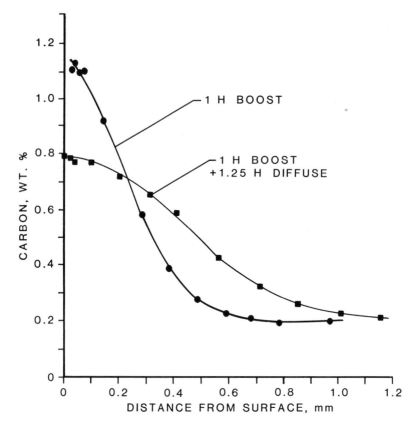

FIGURE 68   Comparison of typical carbon profiles obtained after
plasma carburizing a sample of AISI 8620 both with and without
a subsequent diffusion treatment.

## 10.1   Tungsten Penetrators

Tungsten alloys are being used as an armor penetrator because of
their high density. Pehlivanturk and Inal (1986) have tested plasma
carburizing as a mechanism for increasing the surface hardness
of 96.1% W, Fe, Ni, Co alloys used in this application.

   Plasma carburizing of ferrous alloys is carried out in the austenitic phase region so that a martensitic structure can be formed by
quenching. The solubility of carbon in tungsten at conventional
carburizing temperatures is practically nil, however, a compound
zone containing tungsten carbides, one of the hardest compounds,
can be formed with plasma carburizing.

Tungsten penetrators were heated to temperatures between 900 and 1100°C in a partial pressure hydrogen atmosphere with the abnormal glow. At the carburizing temperature, a 5% methane-95% hydrogen mixture was introduced at a total gas pressure of 1333 Pa for plasma carburizing.

A compound zone of tungsten carbides, predominantly WC, was formed at the surface. The thickness and hardness of this compound zone, as expected, increased with increasing time and temperature. At 1100°C a compound zone 30 microns thick at full theoretical hardness, 2400 VPN, was produced in 14 hours. A sharp drop in hardness was observed below the compound zone since the core was not subjected to any additional hardening mechanism as shown in Figs. 70, 71, and 72.

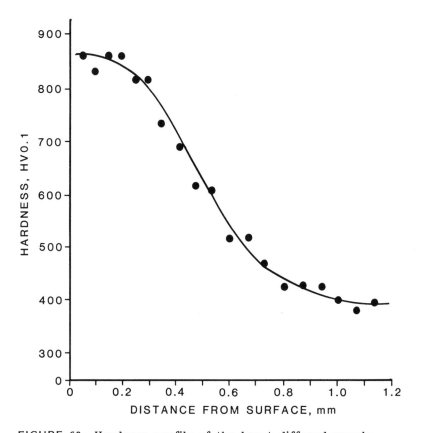

FIGURE 69  Hardness profile of the boost-diffused sample.

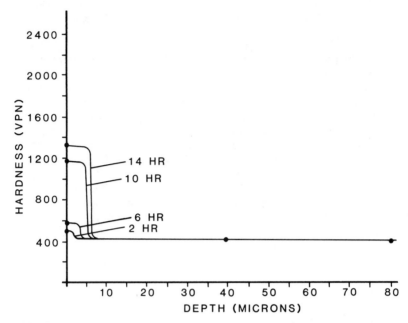

FIGURE 70 Hardness versus depth characteristics of tungsten ion carburized at 900°C.

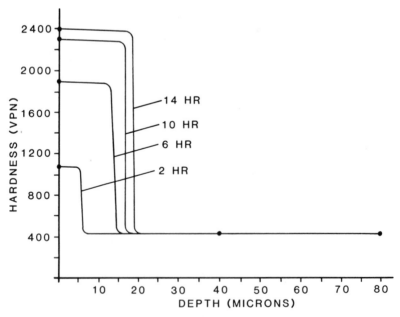

FIGURE 71 Hardness versus depth characteristics of tungsten ion carburized at 1000°C.

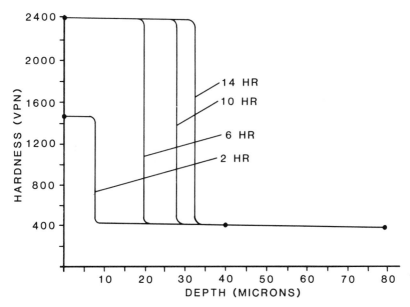

FIGURE 72 Hardness versus depth characteristics of tungsten ion carburized at 1100°C.

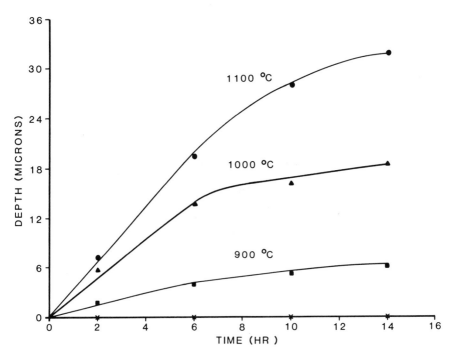

FIGURE 73 Case depth versus time characteristics of ion carburized tungsten.

Their investigation revealed that a high concentration of carbon could be introduced initially without the formation of soot. An initial linear increase in thickness with time indicated that, in the early stages of plasma carburizing, the process is interface controlled. It was found that after an incubation period the process became diffusion limited, with the thickness increasing as a function of the square root of the plasma carburizing time as shown in Fig. 73. These ivnestigators believe that tungsten penetrators can be surface hardened with highly uniform crack-free cases with the plasma carburizing technique.

## 11  PLASMA BORIDING

The boriding of steel produces very hard surfaces that are resistant to abrasive wear, and show excellent resistance to corrosion. Many investigators, including Rashkov (1980), have already reported on the boronization of steel in activated powders. Plasma boriding has been investigated as a replacement for the current practice of boriding in powder or salt by Wierzchon et al. (1980). The plasma assisted chemical vapor deposition process they describe can be cost effective when post cleaning operations to remove adherent powder and salt can be eliminated.

They indicate that plasma boriding, when compared to the powder and salt processes, can more reliably reach all surfaces on workpieces with complex geometries. Their laboratory tests with $BCl_3$ in an $H_2$ carrier showed the possibility of controlling the phase composition of the deposited layer, and indicated that treatment time could be shortened and treatment temperatures lowered with plasma assistance. Plasma kinetics permitted the development of a high concentration of boron at the surface, increasing the concentration gradient and the diffusion speed. Sputter cleaning removed the surface barrier to diffusion, activated the surface, and permitted the shortened cycle time.

Carbon steels were borided in equipment similar to that shown schematically in Fig. 74. Steels with carbon contents of 0.20 and 0.45% were borided in the 700-850°C temperature range, at 300-800 Pa, with a 10-50% volume percent of the process vapor in a hydrogen carrier gas. Process times ranged up to 4 hours. FeB and $Fe_2B$ layers were produced. Figure 75 shows a typical microhardness profile as a function of depth. An investigation of the influence of process parameters upon the thickness and phase composition of these layers showed the following dependence.

An increase in the process time and temperature increased the thickness of the layer and the amount of the $Fe_2B$ in the layer.
An increase in the total gas pressure also increased the thickness of the layer, but at the same time it increased the relative

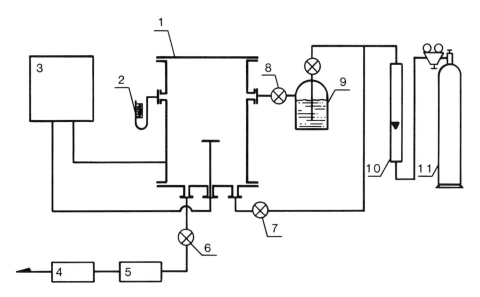

FIGURE 74 Plasma assisted CVD equipment for boriding. 1, Vacuum chamber; 2, pressure gauge; 3, power supply; 4, vacuum pump; 5, exhaust scrubber; 6, vacuum valve; 7, carrier gas valve; 8, process gas valve; 9, vapor generator; 10, carrier flowmeter; 11, carrier supply.

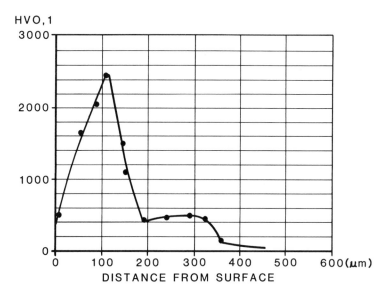

FIGURE 75 An exemplary profile of changes in microhardness of a boron treated layer in carbon steel with a carbon content of 0.2% C.

amount of FeB in the layer making the layer more brittle with
a tendency to spall.
An increase in the $BCl_3$ content in the atmosphere also increased
the relative amount of FeB in the layer.

These investigators conclude that in order to obtain thick layers
with a minimum content of the FeB phase, a phase of higher hardness but a phase that is more brittle, the process temperature
and time should be increased and a low ratio of $BCl_3$ to $H_2$ should
be used. Boride layers up to 100 microns thick were obtained after
3 hours of treatment at 800°C with 20-25% $BCl_3$ in a hydrogen carrier.

## 12   PLASMA TITANIZING

Plasma-assisted chemical vapor deposition (CVD) for titanizing has
been investigated by Wierzchon et al. (1983). Hard, corrosion-
resistant layers of titanium carbide were produced on steels using
equipment similar to that shown in Fig. 76. They report the use
of the abnormal glow discharge in a $TiCl_4$ vapor with a $H_2$ carrier
gas at gas pressures between 300 and 1000 Pa can be a viable process
and that both the temperature and the time for plasma-assisted

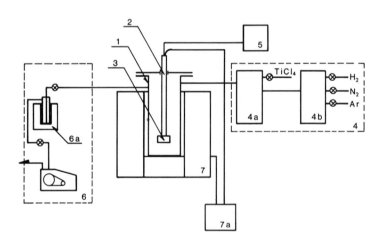

FIGURE 76   Schematic diagram of diffusion titanizing installation.
1, Reaction chamber (anode); 2, cathode; 3, object under treatment;
4, proportioner of gaseous medium; 4a, mixer for reactive gas;
4b, preparatory gas mixer; 5, electric feeder for glow discharge;
6, vacuum system; 6a, cold traps; 7, resistance furnace; 7a, feeder
to furnace.

CVD can be lower than values required for conventional CVD processes. The electrical ionization of the gas mixture and the activation of the work surface by sputter cleaning are cited as the reasons for this observed difference.

Hydrogen was used as the carrier gas for the vapor to promote the reduction of the titanium chloride to produce titanium at the work surface. The presence of ionized hydrogen increases the availability of $TiCl_2$ for the production of titanium. This titanium, then, can react with carbon diffusing from the work for the production of titanium carbide. It was found that minor impurities from the gas supply could adversely affect the hardness and ductility of the resulting layers and gas purification may be a requirement for this process. At the temperatures employed for plasma-assisted CVD the presence of water vapor can result in the destructive oxidation of the titanium carbide layer. They conclude, that with the abnormal glow discharge, TiC layers up to 20 microns thick can be obtained after 4 hours at 900°C.

## 13 CONCLUSION

Diffusion coatings are being produced using the electrical glow discharge phenomena for the production of the active positive ions in a plasma surrounding the workpiece. Plasma processes have demonstrated their ability to produce metallurgically superior coatings with high deposition rates using environmentally neutral operating procedures. Proper plasma surface treatment includes both vacuum degassing and sputter cleaning steps before the addition of the desired ion species. A study of the abnormal glow discharge leads to the conclusion that the plasma processes can be most reliably managed at the industrial level when two plasma parameters can be independently set and controlled by the system microprocessor, the plasma current density that establishes the activity of the positive ions at the work surface, and the plasma power duty cycle that determines the final temperature of the work surface. Plasma power supplies with variable pulse time duration and pulse repetition rates permit this independent control. With this supporting equipment and microprocessor control a furnace operator must assume no unusual responsibility for predicting temperature uniformity within the charge or for preventing a damaging hollow cathode discharge.

Plasma nitriding is already a commercially viable process for the production of superficially hard, corrosion-resistant, wear-resistant, and fatigue-resistant surfaces. Plasma carburizing, while less developed at the present time, promises even greater commercial benefit because of its greater field of application. Both plasma boriding and plasma titanizing processes are being investigated on a experimental scale as commercial interest in plasma surface treatments for diffusion coatings continues to expand.


## BIBLIOGRAPHY

Arai, T., Fujita, H., and Tachikawa, H. (1986). "Ion Nitriding of Aluminum and Aluminum Alloys," Proceedings of an International Conference on Ion Nitriding, Cleveland, Ohio, pp. 37-41.

Badan, B., Magrini, M., Ramous, E., and Bavoro, M. (1983). "Microstructural Examination of Industrial Ion-Nitriding," Proceedings of the Third International Congress on Heat Treatment of Materials '83, Shanghai, China, pp. 1.35-1.40.

Booth, M., Farrell, T., and Johnson, R. H. (1983). *Heat Treatment of Metals, 2*:45-52.

Chen, Y. T. "Surface Treatment of Powder Metal Steels by Ion Nitriding,"

Conybear, J. G. (1985). *Proposed Draft, Aerospace Material Specification*, Society of Automotive Engineers, Inc.

Dickens, D. R. (1986). "Corrosion Properties of Ionitriding by an Ionitriding User/Customer," Proceedings of an International Conference on Ion Nitriding, Cleveland, Ohio, pp. 149-159.

Grube, W. L. (1978). *High Rate Carburizing in a Glow-Discharge Methane Plasma, Metallurgical Trans. A, AIME, 9A*:1421-1429.

Grube, W. L. (1979). *J. Heat Treating, 1*:95-97.

Hochman, R. F. (1986). *"Effects of Nitrogen in Metal Surfaces"*, Proceedings of an International Conference on Ion Nitriding, Cleveland, Ohio, pp. 23-30.

Houvion, J. (1980). "The Industrial Development of Ionic Nitriding Units," Proceedings of the 18th International Conference on Heat Treating of Materials, Detroit, Michigan, pp. 289-295.

Kwon, S. C., Lee, G. H., and Yoo, M. C. (1986). "A Comparative Study between Pulsed and D. C. Ion Nitriding Behavior in Specimens with Blind Holes," Proceedings of an International Conference on Ion Nitriding, Cleveland, Ohio, pp. 77-81.

Marciniak, A. and Karpinski, T. *Some Aspects of the Glow Discharge Heating*, Institute of Materials Science and Engineering, Warsaw Technical University, Poland.

*Metals Handbook*, 9th ed., Vol. 4, *Heat Treating*, American Society for Metals, pp. 213-216.

Metin, E. and Inal, O. T. (1986). "Characterization of Ion Nitrided Structures of Ti and Ti6242," Proceedings of an International Conference on Ion Nitriding, Cleveland, Ohio, pp. 61-65.

Ozbaysal, K. and Inal, O. T. (1986). "Surface hardening of Marage Steels by Ion Nitriding without Reduction in Core Hardness," Proceedings of an International Conference on Ion Nitriding, Cleveland, Ohio, pp. 97-115.

Stosic, P. and Ziatanovic, M. (1986). "Plasma Nitriding of Timing Spur Gears," Proceedings of an International Conference on Ion Nitriding, Cleveland, Ohio, pp. 139-142.

Straemke, S. (1988). *Introduction to Plasma Heat Treatment*, prepublication notes.

Tonchev, T., Toshkov, V., and Madzarov, T. (1987). *Carburization in Glow Discharge: Considerations in Controlling Results*, Paper translated from Russian and edited. Industrial Heating.

Pehlivanturk, N. Y. and Inal, O. T. (1986). "Carburizing of Tungsten in a Glow Discharge Methane Plasma," Proceedings of an International Conference on Ion Nitriding, Cleveland, Ohio, pp. 179-188.

Rashkov, N. (1980). "Boronization of Steel Details in an Electrothermal Rotary Layer," Proceedings of the 18th International Conference on Heat Treatment of Materials, Detroit, Michigan, pp. 25-28.

Wierzchon, T., Bogacki, J., and Karpinski, T. (1980). "Ion Boriding from the Viewpoint of the Applied Gaseous Medium," Proceedings of the 18th International Conference on Heat Treatment of Materials, Detroit, Michigan, pp. 13-24.

Wierzchon, T., Michalski, J., and Karpinski, T. (1983). "Ion Titanizing," Proceedings of the Third International Congress on Heat Treatment of Materials '83, Shanghai, China, pp. 3.38-3.43.

# 6
# Surface Alloying Using Lasers

P. A. MOLIAN  *Iowa State University of Science and Technology,*
*Ames, Iowa*

## 1  INTRODUCTION

Most engineering failures occur at surfaces, due to fatigue, corrosion,
friction, and wear, because (i) stresses are often highest at surfaces,
and (ii) surfaces are exposed to the environment. The engineering
solution is to provide the material with surface properties that are
different from those of the bulk. Currently there are two methods
available to achieve this. The first method involves a modification
of surface composition by carburizing, nitriding, cyaniding, applying
coatings by physical and chemical deposition, cladding, surface
alloying, etc. The second method is surface-hardening through
phase transformation, which includes conventional flame and induction
heating. Recently, energy sources such as electron beams and lasers
have been used to surface-harden steels and cast irons to enhance
fatigue, wear, and corrosion properties.

Directed energy sources, such as lasers, are used more than
ever for a variety of materials processing applications that include
cutting, welding, and surface treatment. Laser surface modification
processes (Fig. 1) are designed to alter the compositions and microstructures of surface layers and thereby improve and tailor the
surface properties to the bulk. Alloying surface layers with a laser
is a novel approach because laser parameters can be controlled to
obtain the desired surface alloy compositions with unusual microstructures. Laser surface modification processes for changing the
surface composition can be divided into (1) laser surface alloying
(LSA) and (2) laser cladding (LC).

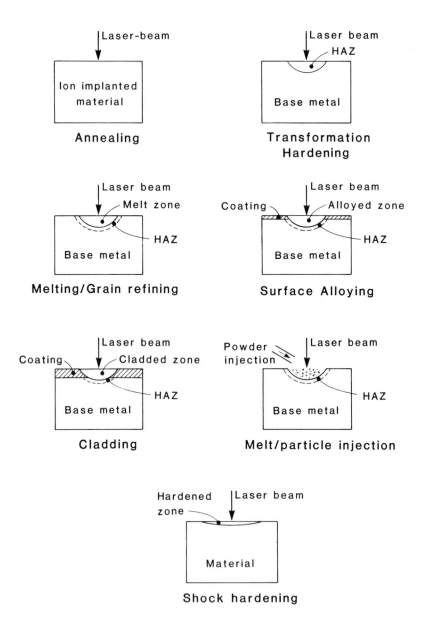

FIGURE 1   A schematic of laser surface treatment processes.

The beneficial effects of employing a high-power laser for surface treatment are conservation of strategic and expensive alloying elements, formation of nonequilibrium crystalline and amorphous phases, refinement of grains, homogenization of microstructures, increased solid solubilities of alloying elements, and modification of segregation patterns. Laser surface alloying has some advantages over its competitors, conventional arc hard-facing and plasma spraying. In these latter processes, the radiant energy is not coherent and, hence, cannot be transmitted through air over large distances. The heat distribution is much broader than the sharply defined pattern of the laser. The power density is generally low for both processes, but can approach that of the laser in plasma spraying. These two processes introduce uneven heating, nonuniform cooling, and thermal shock during quenching, which result in distortion and cracking; subsequent straightening and/or grinding operations are often needed. The major advantages of laser surface alloying are the ability to control the power density precisely, ensuring heating to a controlled depth and thereby reducing distortion; the ability of the laser to alloy inaccessible and localized areas; efficient energy usage through rapid processing; and the production of uniform alloyed case depths on irregular parts as a result of the laser's depth of focus. Thus, for many applications, laser surface alloying is an excellent alternative to conventional thermal spraying techniques for generating the surface properties.

Lasers are sometimes advantageous over other directed energy sources such as ion beams and electron beams. Although ion implantation is the standard practice in electronics industry for doping of P, Sb, Bi, B, etc., into semiconductors, it has limited acceptance in metal industries for extending fatigue and wear. Electron beam processing of materials is identical to that of laser beam processing but requires a high degree of vacuum.

Many effects of laser surface alloying may be explained on the basis of rapid heating and subsequent quenching rates. During laser heating (or cooling), large temperature gradients exist across the surface layer and the underlying substrate, which provide heating and cooling rates of about $10^6$-$10^8 °C/s$. The substrate by itself is an excellent heat sink, so additional cooling mediums, such as water, are not needed. As a result of rapid heating and cooling, considerable changes in chemical and microstructural features occur and tailor the surface for improved properties.

## 2  LASER SOURCES

Lasers used for surface alloying, in decreasing order of importance, are carbon dioxide ($CO_2$), neodymium-doped yttrium aluminum garnet

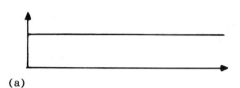

(a)

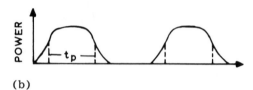

(b)

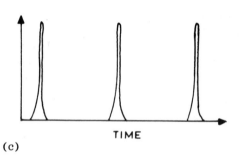

(c)

FIGURE 2   A schematic of time-dependence of output power for
three laser types. (a) Continuous, or CW, (b) pulsed, (c) cavity
spoiled or Q-switched. (After Draper and Poate, 1985.)

(Nd:YAG), neodymium-doped glass (Nd:glass), and chromium-doped
aluminum oxide (ruby). Both pulsed and continuous wave (CW)
modes of lasing can be used. Q-switching of lasers is sometimes
done to produce high peak power pulses of very short duration.
Figure 2 schematically describes the time dependence of continuous,
pulsed, and Q-switched lasers. The high-power leading edges in
pulses and Q-switched lasers tend to minimize reflectance of the
laser beam and to reduce thermal conductivity of the substrate
during laser processing (Draper and Poate, 1985).
    Important factors for selecting a laser for surface alloying are

1.   Output power
2.   Beam diameter

TABLE 1 Typical Laser Variables Used in Surface Alloying

| Laser | Wave-length, μm | Average power, W* | Focused optical spot mm | Focused power density MW cm$^{-2}$ | Dwell/pulse length, ns | Repetition rate, Hz | Spatial distribution |
|---|---|---|---|---|---|---|---|
| CW-CO$_2$ | 10.6 | 200-15,000 | 0.1-10 | 1-10 | $10^7$-$10^{10}$ | — | Multimode, top hat, or gaussian |
| Q-switched Nd:YAG | 1.06 | 0.5-3 | 20-40 | 80-200 | 130 | 11,000 | Gaussian |
| Frequency-doubled Q-switched Nd:YAG | 0.53 | 0.05-0.3 | 10-30 | 30-100 | 120 | 5,000 | Gaussian |
| Homogenized Q-switched ruby | 0.69 | 1-10* | 6 | 10-200 | 2-80 | 0.03 | Top hat |
| Pulsed ruby, Nd:YAG | 0.69, 1.06 | 2-50* | 0.5 | 0.5-10 | $2 \times 10^6$-$10^7$ | 0.1-10 | Multimode |

*Output capabilities of low repetition rate lasers are normally measured and quoted in energy content (J) per pulse.

*Source:* After Draper and Poate, 1985.

3.  Beam configuration
4.  Wavelength
5.  Pulse length
6.  Pulse repetition rate

$CO_2$ gas lasers are preferred because they are electrically more efficient and produce higher powers than other lasers in continuous mode. Table 1 compares the characteristics of various lasers used in surface alloying. The spatial distribution of laser beams will be discussed in the next section.

## 3  PROCESS VARIABLES

In laser surface alloying processes, a high-power laser beam interacts with a precoated specimen (or workpiece) while a high relative speed is maintained between the two. The workpiece is usually mounted on an x-y table, the speed of which is controlled by a computer numerical controller. For convenience, the laser beam is kept stationary, and the specimen is moved at high speed. The independent variables are

1.  Laser power
2.  Beam size of Fresnel number
3.  Beam configuration from the laser cavity
4.  Optical integrator for producing the desired beam profile
5.  Optical scanner to generate an oscillating beam
6.  Multiple-overlapping laser passes
7.  Travel speed of the workpiece
8.  Precoating method
9.  Precoating powder size and flow rate
10. Precoating thickness
11. Precoating composition
12. Preheating the substrate
13. Surface condition of the substrate

The dependent variables are

1.  Absorption coefficient
2.  Alloyed case depth and coverage rate[*]
3.  Composition gradients in the alloyed casing
4.  Microstructures of surface and near surface layers
5.  Hardness

---

[*]Coverage rate is the product of alloyed case width and travel speed, $in^2/s$.

6. Residual stresses
7. Heat-affected zone (HAZ) size
8. Corrosion
9. Wear
10. Fatigue
11. Impact
12. Alloyed zone defects, including cracks and porosity
13. Surface roughness and integrity

Laser power necessary for surface melting of metallic materials is generally high due to reflectance of the laser beam and high thermal conductivity. Reflectivity of the metal surface is actually related to electrical conductivity. For instance, it is difficult to surface-melt aluminum or copper with a $CO_2$ laser beam because these materials absorb little laser power. For metallic materials, the laser power for surface alloying should exceed 0.5 kW. For nonmetallic materials, such as ceramics and polymers, low powers are sufficient. Increasing laser power increases absorption and melt penetration, ensures uniform mixing, introduces surface rippling, increases HAZ, alters the shape of the fusion zone, and increases the tendency to crack.

The beam size determines the power density on the specimen surface (power density is defined as the power divided by the cross-sectional area of the laser beam). Typical values of power density for surface alloying are $10^4$-$10^8$ W/cm$^2$. The beam size is a difficult dimension to measure, since most measurement techniques are affected by the exposure time and processing variables. A simple method is to measure the distance of the focal point from the workpiece surface and relate it to the beam size. For the purposes of calculating power density and interaction time (beam size/travel speed) of the beam, the beam size must be known. Optical diffraction theory equations (Fig. 3) may be used to convert the distance d to the beam size. However, we should bear in mind that these equations are applicable only to the laser beam that has a gaussian energy distribution. Increasing the beam size by defocusing or other means decreases the heat input per unit volume, reduces melt penetration, and increases melt width.

The energy distribution and shape of the beam is described by the beam configuration or beam profile. Four beam profiles are made available by the lasers in conjunction with beam delivery systems. These include gaussian, multimode, square (or rectangular), top hat (Fig. 4). It is seldom possible to produce gaussian beam configuration ($TEM_{00}$) at high powers, although many high-powers, although many high-power laser manufacturers claim it can be done. A gaussian beam is most suitable for cutting and welding applications rather than for surface treatment because, being a "sharp tool," it

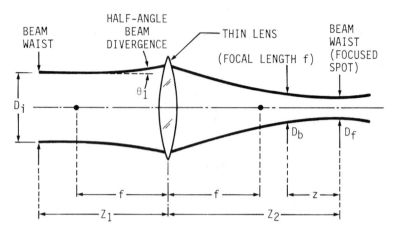

FIGURE 3   Beam size calculations for a gaussian beam. Spot size
at focal point ($D_f$): $D_f = 4\lambda f/\pi D_i$. Spot size away from focal point
(equation relating $D_f$ and $D_b(Z)$) is: $D_b(Z) = D_f [1 + (4\lambda Z/\pi \, D_f^2)]^{1/2}$.

tends to vaporize and melt the substrate deeply. In contrast, multimode, top hat, and square profiles ("blunt tools") are preferred
for surface alloying. These beam profiles offer alloyed casings with
wider coverage rates and uniform case depths.

Square and rectangular beam profiles are generated by using
an optical integrator or scanner. The integrator device consists
of small, segmented square mirrors mounted on a spherical surface
and enclosed in a black box. The gaussian beam is sent through
this integrator, where each segmented mirror produces a square
beam of the desired size on the image plane. Oscillating a gaussian
beam at high frequency using a rotational mirror in the beam delivery
system may produce a square or rectangular profile because the
thermal lag response of the material will act to even out the laser
power. Figure 5 describes the experimental setup for the scanner.

The travel speed of the workpiece during laser surface alloying
(also called scan rate) is a convenient parameter for the operator
to vary. It affects the mechanism of surface alloying by controlling
the diffusion time of the alloying elements, changes the melt depth,
width, and profile, and determines the microstructures of surface
alloys to a greater extent than other variables.

Alloying elements in the laser melt pool may be introduced by
means of predeposition methods or codeposition methods. Predeposition methods include electroplating, thermal spraying, vacuum evaporation, boronizing, carburizing, nitriding, laying of loose powders,

thin sheets, or rods on the surface of substrate, etc., all of which are done prior to laser melting. Codeposition involves injecting powder, wire, or rod forms of alloys into the laser-generated melt pool of the substrate. Obviously, the one-step codeposition method is more attractive than the two-step predeposition technique. Additionally, the codeposition method offers the advantage of real-time control on the supply of alloying elements. Precoating variables include thickness, particle size, composition, and method of application. Preheating of the substrate is sometimes done to minimize residual stresses and to increase the laser beam absorption during surface alloying. Substrate surface condition is an important consideration in codeposition because it affects the absorption of laser power. Usually the substrate is sandblasted or shot-peened prior to laser alloying.

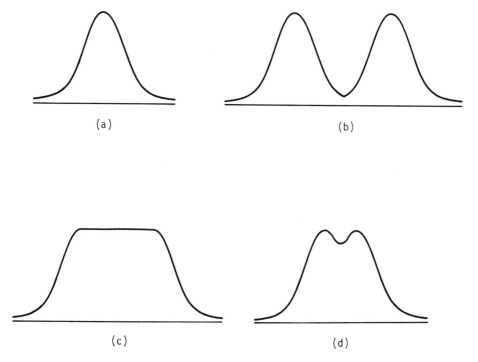

FIGURE 4  Beam profiles used in surface alloying. (a) Gaussian mode, (b) multimode, (c) square mode, (d) ring mode ($TEM_{01}*$).

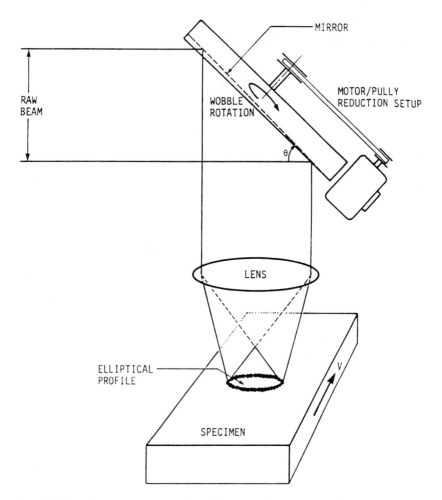

**FIGURE 5** Experimental setup for scanning a gaussian beam at high frequency.

## 4  LASER SURFACE ALLOYING

Figure 6 schematically describes the laser surface alloying process using the predeposition method. The laser induces melting of the alloying elements and a shallow portion of the substrate. Mixing of the substrate and alloying elements to form a suitable alloy requires the proper balance of heat and fluid flow in the melt pool, which are controlled by surface tension forces, temperature gradients, diffusion times, convective forces, and thermal properties of the melt pool. The following paragraphs describe the results obtained by the author and other investigators on the surface alloying of various materials. Particular attention is paid to the chromium-alloyed system for evaluating the influence of process variables because surface alloying rather than bulk alloying of Cr, Co, and Ni into lower-grade, inexpensive steels conserves these strategic metals.

### 4.1 Surface Alloying of Chromium and Nickel

Figure 7 depicts the transverse sections of electrodeposited Cr and the laser-alloyed fusion zone typically observed in single-laser-pass processed samples. Two parameters, penetration depth and melt width, which determine the composition, cooling rate, micro-structure, hardness, residual stresses, etc., are defined in Fig. 7(b).

*Laser Beam Power*

Penetration depth and melt width of the fusion zones of chromium-alloyed iron as a function of laser beam power at different scan

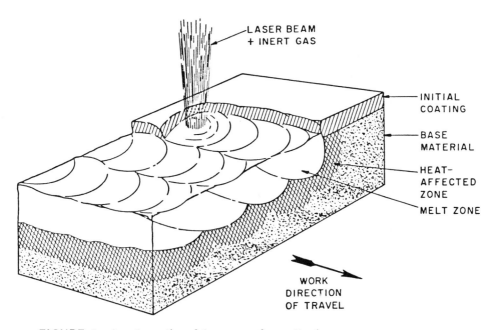

FIGURE 6   A schematic of laser surface alloying process.

## (A) As Deposited

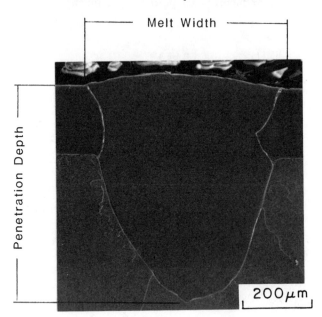

## (B) Laser  treated

FIGURE 7   Transverse sections of electrodeposited and laser-alloyed specimens.

rates are shown in Figs. 8 and 9. Penetration depth increases
linearly with power, but melt width remains almost independent
of power. The sensitivity of laser beam power to penetration depths
rather than melt width is attributed to a higher power-density peak
at the center of the focal spot. The penetrations are significantly
greater and such "deep penetration" melts are primarily caused
by small beam size and high laser power. Deep penetration melting
occurs because of "keyholing" in which the beam delivers energy
more rapidly than it can be removed by thermal conduction. A hole
is then drilled so that laser energy can penetrate deep into the
workpiece. A flow of molten material then recloses the hole after
the beam passes to a new area. Deep penetration melts, characterized
by an hourglass shape with a high aspect ratio (aspect ratio =
penetration depth/melt width), have been shown by many investigators in laser and electron beam welds. It has been suggested, and
substantiated experimentally, that under the conditions of deep
penetration melting the laser power is completely absorbed. This
was explained by the multiple reflections that occur in the cavity,
which increase the absorption of light. According to Breinan and
Banas (1975), when a laser beam is focused on a material surface
with sufficient power density, an in-depth vapor column is created
and maintained in equilibrium with a liquid pool which surrounds
it. Evidence for vapor formation during laser surface alloying of
chromium on steels in deep penetration melts had been verified
experimentally by collecting and analyzing the vapors on filter paper.

*Focal Position/Beam Size*

The shape of laser-alloyed fusion zones is determined by several
characteristics. One such controllable characteristic is the location
of the focal point relative to the surface of the specimen (Fig. 10).
The results of experiments performed on the effect of the focal
position on penetration depths is given in Fig. 11 for chromium-
alloyed iron. It can be seen that maximum penetration occurred
when the beam was focused slightly (0.025 in.) inside the surface.
Similar results were reported by Locke and Hella (1974) for 304
stainless steel (0.060 in. below the surface) and Engel (1976) for
1018 steel (0.050 in. below the surface). This plot also indicates
that higher penetrations were always obtained when the beam was
focused below the specimen surface rather than above for a given
beam size. The shape of this plot appears parabolic, but essentially
depends on focal length of the lens, laser power, scan rate, and
material system. The shape of the fusion zones indicated that the
aspect ratio decreases and the shape changes from hourglass to
semicircular through triangular when the specimen surface is moved
away from the focal point. The hourglass shape corresponds to deep

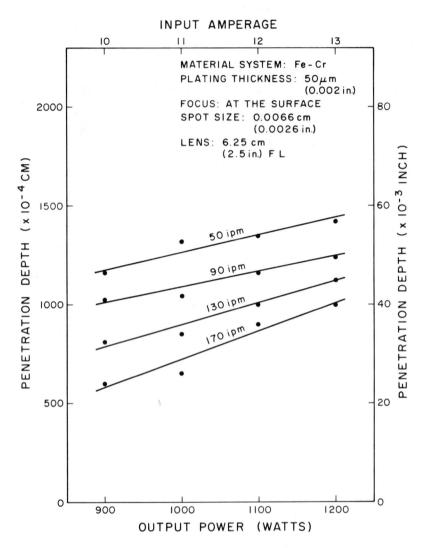

FIGURE 8   Effect of laser power on laser-alloyed depth.

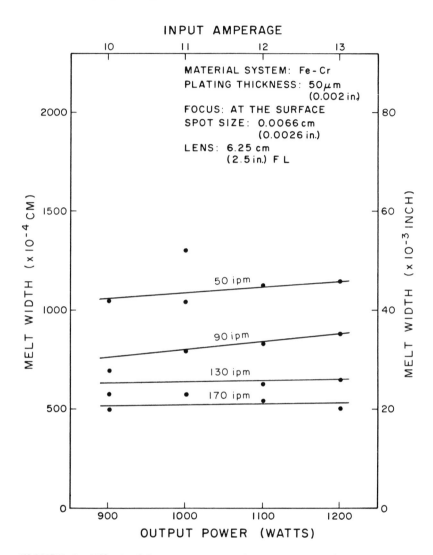

FIGURE 9  Effect of laser power on laser-alloyed width.

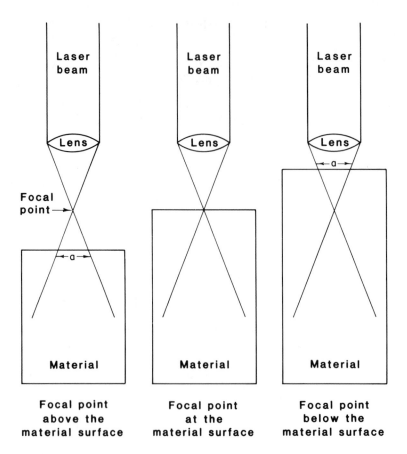

FIGURE 10  Illustration of focal point position with respect to work-piece surface.

penetration melting for conditions of focusing to a higher power density such that substantial vaporization occurs. The triangular and semicircular fusion zones correspond to conduction melting for conditions of focusing such that no substantial amounts of vaporization occurs (Fig. 12).

*Scan Rate*

Figure 13 summarizes data on the penetrations achieved in chromium-alloyed iron at various rates of beam traverse for values of laser power from 900 to 1200 W. Both penetration depth and melt width decrease with an increase in scan rate. Similar results were obtained

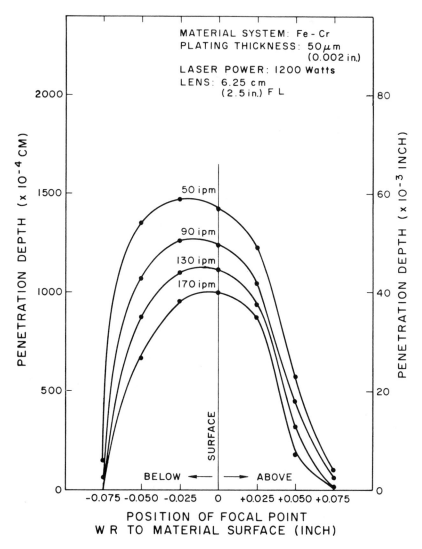

FIGURE 11   Effect of focal position on laser-alloyed depth.

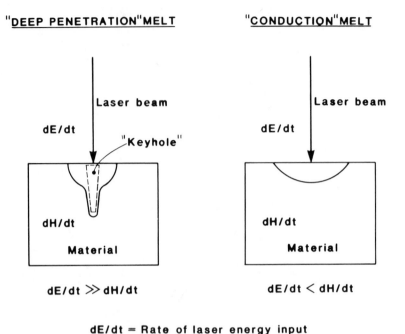

FIGURE 12   Deep penetration and conduction modes of melting.

by several investigators for different materials. Thus, laser beam parameters, such as laser power, focal position/beam size, and scan rate can be manipulated to obtain desired penetration depths (50–1500 µm), melt widths (200–1200 µm), and fusion zone shapes (from hourglass to semicircular). Consequently, the compositions, cooling rates, microstructures, and mechanical properties of the alloys produced were controlled.

*Coating Composition*

The final alloy composition of the fusion zone is primarily determined by the initial coating composition. Since penetration depth also controls the resulting alloy composition, it is imperative to study the effects of coating composition on penetration depth. Figure 14 shows the transverse sections of chromium, nickel, and chromium deposited over nickel coatings on pure iron. Figure 15 presents the data on the effects of these coatings on penetration depths as a function of laser beam parameters. It is evident that laser alloying of the

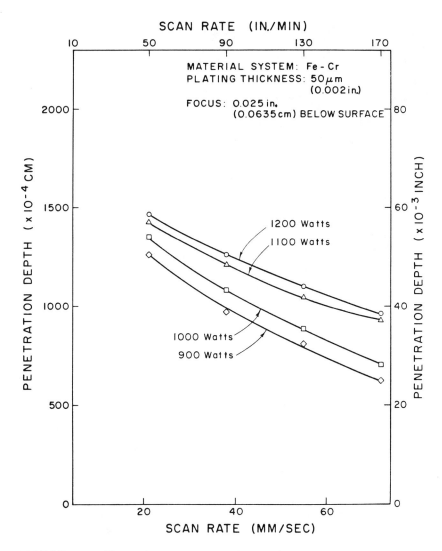

FIGURE 13  Effect of scan rate on laser-alloyed depth.

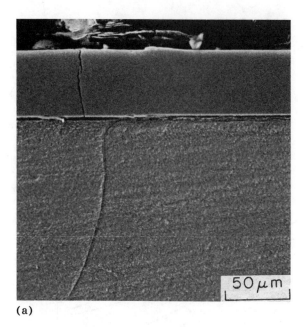

(a)

FIGURE 14  Transverse sections of electrodeposited chromium and nickel. (a) Chromium, (b) nickel, (c) chromium-nickel.

chromium/nickel coating exhibited a greater penetration compared to laser alloying of either chromium or nickel. This increase was obtained under all laser beam conditions. This also implies that higher penetration in Fe-Cr-Ni systems over Fe-Cr and Fe-Ni was achieved independently of melting modes such as deep penetration or conduction. However, the amount of increase in penetration appears to be influenced by laser beam parameters. For given laser parameters, there is no appreciable difference in penetration depths between Fe-Cr and Fe-Ni systems. Another interesting aspect of Fig. 15 is that the Fe-Cr-Ni system is more sensitive to the position of the focal point relative to the material surface than are the Fe-Cr and Fe-Ni systems, thereby indicating the differences in laser response to individual material systems. These observations suggest that the probable mechanism in controlling penetrations is a combined effect of the initial surface condition (which controls the initial absorption) and the thermal properties of alloys formed (which controls the heat flow) during solidification.

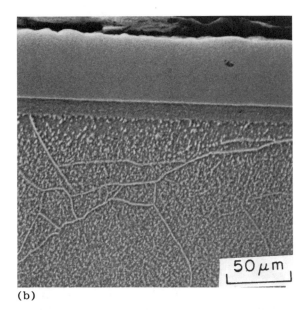

(b)

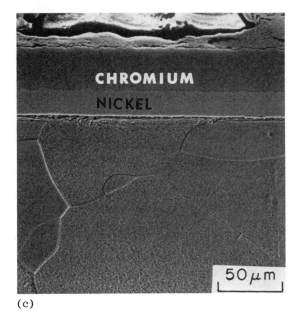

(c)

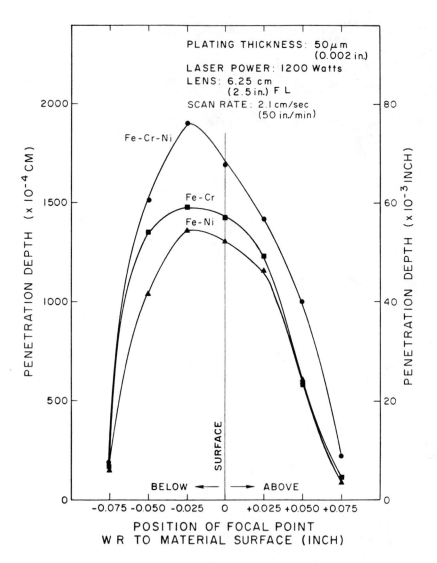

FIGURE 15   Effects of initial composition of coating on laser-alloyed depth.

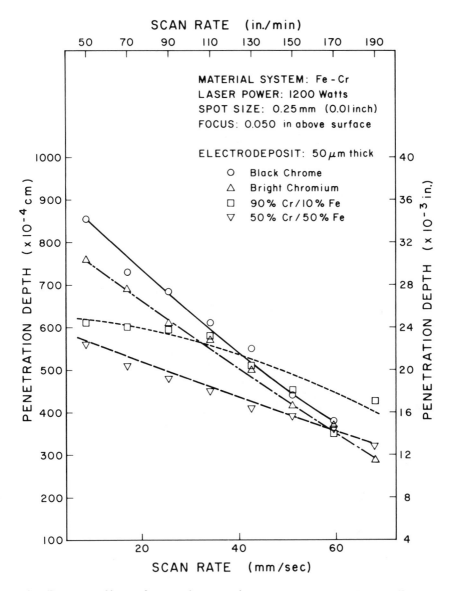

**FIGURE 16** Effect of chromium coating morphology on laser-alloyed depth.

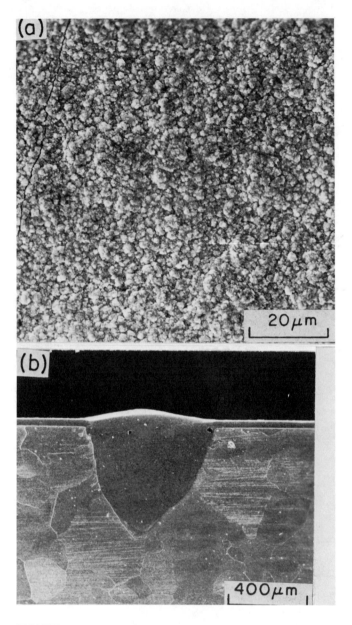

FIGURE 17  Comparison of bright and black chromiums on the laser-
alloyed fusion zone geometry.

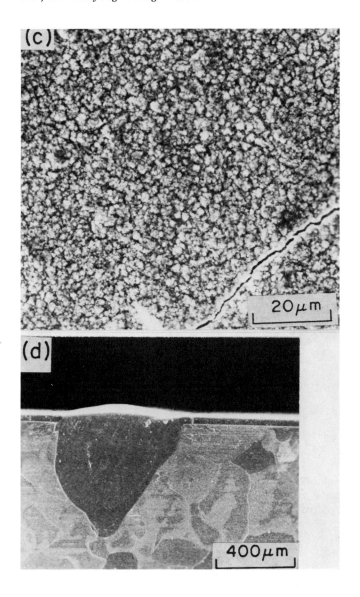

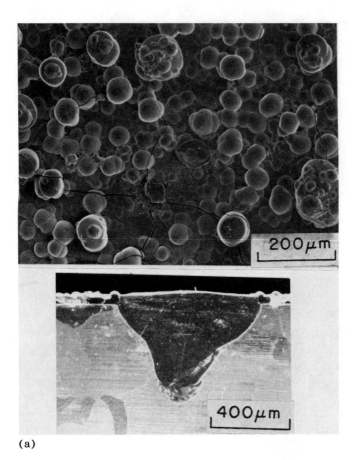

(a)

FIGURE 18    Effect of chromium-iron alloy plating on the laser-alloyed
fusion zone geometry. (a) 90% Cr/10% Fe alloy plating. (b) 50%
Cr/50% Fe alloy plating.

*Coating Morphology*

To date there has been no reported data on the response of coating
morphology to high-power laser beam absorptivity in the laser surface
alloying process, though it has been frequently said that the initial
surface appearance alters the penetration through its effects on
absorption. The results of test conducted on the influence of chromium
and alloy coating morphologies on penetration depth and fusion zone
shape are given in Figs. 16 to 18, which also show the coating
morphologies and the laser-alloyed surfaces of black and bright
chromium and codeposited iron-chromium coatings for the given

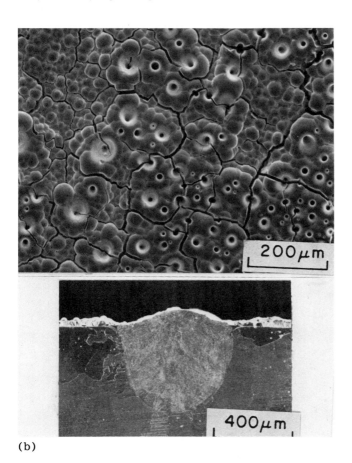

(b)

laser parameters. The coating morphologies, as seen in these micrographs, largely differ from each other in terms of particle size
and surface defects, such as porosity, cracking, pitting, oxidation,
and roughness. The results obtained indicate that there is no significant effect of coating morphology on penetration depth and melt
width except for a small change in fusion zone geometry. This suggests further that the heat transfer properties of the coating are
more important in deciding the laser beam absorptivity rather than
the coating morphology.

*Coating Thickness*

Initial coating thickness controls the extent of dilution of the alloying
element and is a versatile parameter for obtaining the desired alloy
composition. The results of studies performed on the influence of

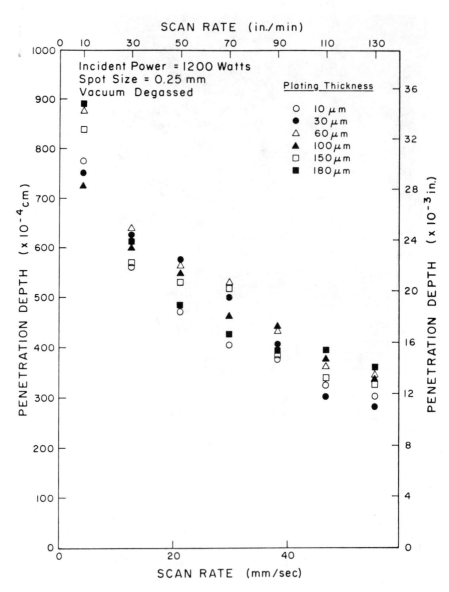

FIGURE 19   Effect of chromium coating thickness on laser-alloyed depth.

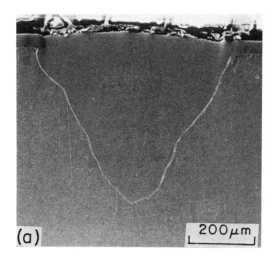

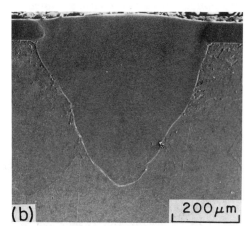

Figure  19  (continued)

chromium coating thickness on penetration depths is given in Fig. 19. These experiments were repeated on several steel substrates to assess and confirm that the effects are caused only by chromium coating thickness variations. As seen from these plots, there is no substantial influence of chromium coating thickness on penetration depth and melt width. However, the amount of variation in penetration suggests that chromium coating thickness has some role in controlling the fusion zone size, the reasons for which are not fully understood. It is believed that the thermal properties of the alloys formed and their variation with temperature are responsible for the observed variation in penetration depth with respect to coating thickness.

*Coating Defects*

Hydrogen entrapped during the application of the coating appears to play a significant role in determining the porosity formation. Figure 20 shows the top views of laser-alloyed surfaces with and without trapped hydrogen. It is apparent that hydrogen had been expelled from the melt toward the free surface and remained as a surface pore. Contrary to vacuum-degassed samples [Fig. 20(b)], considerable amounts of porosity were observed in samples that were not degassed. The pore shapes are elongated and irregular. A significant amount of hydrogen is introduced into the coating during the chromium electrodeposition process as a result of high current density ($0.3$ A/cm$^2$) and high temperatures ($65°C$). When the laser beam rapidly heats the samples to the melting temperature, and above, of the coating, the hydrogen trapped within the pores of the coating experiences an increase in temperature and pressure, particularly if the pores are the closed type. Consequently, the expanding and pressurized gases tend to escape the fusion zone prior to the onset of solidification. If the rate of solidification is less than the rate at which these gas bubbles rise to the surface, then they either remain as enlarged residual pores or they force out portions of the resulting fusion zone surface.

　　Figures 21(a)-(d) show the longitudinal and transverse sections of laser-alloyed fusion zones of nondegassed samples. Porosity in the fusion zones is aggravated by either thick chromium coatings or rapid scanning of the laser beam on the material surface. The former is due to the increased concentration of hydrogen gas, and the latter is due to lack of time for the gases to escape. These results suggest that hydrogen gas porosity in laser-alloyed surfaces may be avoided by employing either a thin coating or a low scan rate. If these conditions are not provided, then vacuum degassing prior to laser irradiation is necessary to eliminate the gas porosity. In addition to hydrogen porosity, another form of porosity was also observed, but only in deep penetration melts (melt depth > 700

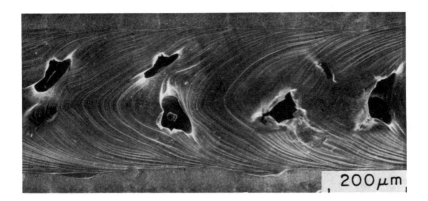

(a)

(b)

FIGURE 20   Laser-alloyed surfaces of (a) nondegassed and (b) degassed specimens.

μm), as a consequence of a key hole effect. Figure 22 shows this type of porosity at the bottom of fusion zones.

High-energy density ($10^6$-$10^8$ W/cm$^2$) and short interaction times ($10^{-6}$-$10^{-2}$ s) involved in this process raise the question of whether uniform composition can be achieved in laser-alloyed fusion zones. Weinman, DeVault, and Moore (1979) had shown by electron microprobe analysis of laser-alloyed chromium on AISI 1018 steel that uniform mixing of chromium had occurred in the melt pool within 50 μs. The composition uniformity had been studied as a function of fusion zone shape. Figure 23 shows the transverse sections of laser-alloyed zones produced under different laser beam conditions

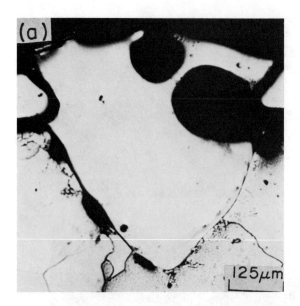

FIGURE 21  Effects of chromium coating thickness and scan rate on porosity: (a), (b) scan rate = 50 in/min; (c), (d) scan rate = 100 in/min.

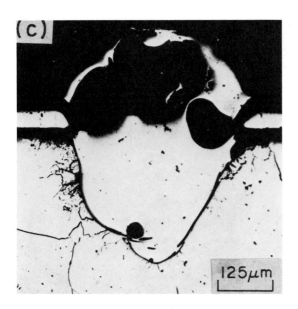

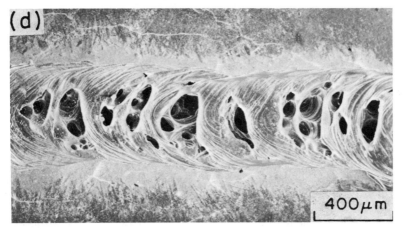

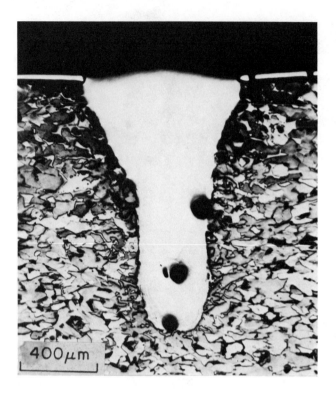

FIGURE 22   Keyhole porosity in deep-penetration alloyed casing.


and the corresponding chromium concentration profile plots from
the top surface of the fusion zone to the fusion/base interface.
The variation in chromium content is more pronounced in an hourglass-
shaped fusion zone [Fig. 23(a)], intermediate in a triangular fusion
zone [Fig. 23(b)], and relatively none in a semicircular fusion zone
[Fig. 23(c)].

     For uniform composition to occur in the fusion zone, two mecha-
nisms have been postulated. One is based on the diffusion of solute
in the melt pool, the other on the surface tension effects caused
by the temperature gradients in the melt pool and surrounding
area. The diffusion model involves the laser beam interaction time
(t) and the diffusivity of solute in the liquid ($D_L$) as given by the
equation $X = (D_L t)^{0.5}$, where X is the diffusion distance over which
the concentration homogenization can occur. Anthony and Cline
(1978) reported that with interaction times of 50 $\mu$s to 2 ms the
diffusion distances in the melt pool are too small to obtain uniform

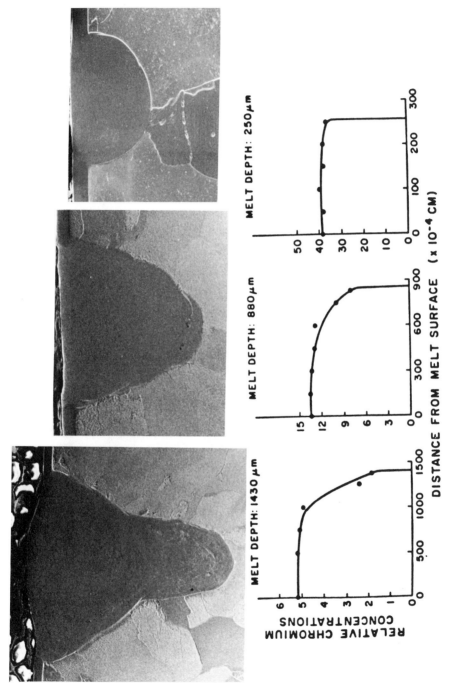

FIGURE 23   Concentration profiles of chromium as a function of fusion zone geometry.

mixing in ferrous systems. The surface tension gradient mechanism
was also not expected to be operative within this time frame, since
the convection currents produced in the melt pool by surface tension
gradients had been shown to lag the advance of the solidification
front and, hence, could not cause enough stirring. For a moment,
let us assume that the fluid flow in molten iron, induced by surface
tension gradients, would produce a more uniform alloyed region
than would the diffusion of solute in the melt pool. Then the time
that the alloy was in the liquid state (t') and the fluid flow velocity
[U(y, t')] required to homogenize the liquid were estimated (see
Table 2) by employing the following equations developed by Cline
and Anthony [6]:

$$U(y, \ t') = \frac{1}{\mu} \frac{d\sigma}{dt} \ \Delta T \ (D_u t')^{0.5} - y$$

where

   $y$ = penetration depth, cm
   $t'$ = time that the alloy was in molten state, s
   $\mu$ = surface energy per unit area, dyne-s/cm$^2$
$d\sigma/dt$ = surface tension temperature gradient, erg/cm$^2$ K
   $\Delta T$ = temperature difference between the liquid below the laser
        beam and the liquid at the solid-liquid interface of the
        melt pool, K
   $D_u$ = $\mu/\rho$, "diffusivity" of velocity profile, dyne-s/gm

Contrary to these mechanisms, Weinman et al. (1979), using electron
microprobe analysis of laser-surface-alloyed 1018 steel, have shown
that complete mixing occurred in 50 μs of interaction time. They
postulated a different mechanism, namely, that the mixing effects
caused by superheated gases above the melt pool were sufficient
to stir the melt pool. However, in the present investigation, the
diffusion model has been followed to explain the uniform concentration
of chromium in semicircular fusion zones. However, the uniform
compositions obtained in hour glass and triangular fusion zones
(see Table 2) suggest that the surface tension gradient and superheated gas mechanisms are also operative.
    The most serious limitation of laser alloying of chromium on
medium- and high-carbon steels is the frequent occurrence of cracks.
The fusion zone, the heat-affected zone, and the coating adjacent
to the melt pass are the locations usually observed to contain cracks
(Fig. 24). The centerline cracks in the fusion zone are identified
as shrinkage cracks and are due to inadequate feeding of the molten
metal. The susceptibility to this type of cracking is a function of

TABLE 2  Calculations of Concentration of Chromium Based on Theoretical Models

| Laser beam parameters | Hourglass | Fusion zone shape | |
|---|---|---|---|
| | | Triangular | Semicircular |
| Power density, W/cm$^2$ | $3.5 \times 10^7$ | $2.4 \times 10^6$ | $1.0 \times 10^6$ |
| Interaction time, s | $3.12 \times 10^{-3}$ | $1.2 \times 10^{-2}$ | $1.9 \times 10^{-2}$ |
| Penetration depth, μm | 1430 | 880 | 250 |
| Depth of uniform concentration of chromium (diffusion model), μm* | 190 | 380 | 470 |
| Time that the alloy was in molten state, s* | $5.8 \times 10^{-2}$ | $4.6 \times 10^{-2}$ | $1.7 \times 10^{-2}$ |
| Fluid flow velocity required to achieve homogenization (surface tension model), cm/s* | 4500 | 2550 | 500 |

*Estimated based on diffusion and surface tension models.
*Source*: Ayers et al. (1981) and Belmondo and Castagna (1979).

both fusion zone geometry and composition. The fusion zones containing 0.2% C and 1.0% C were produced under identical laser alloying conditions. This illustrates the effect of carbon in promoting crack formation. Unlike carbon, the chromium content of the fusion zone does not appear to increase the cracking tendency. The aspect ratio (melt depth/melt width) is another factor that increases cracking in the fusion zone.

Nonuniform, discontinuous fusion zones are occasionally observed (Fig. 25). This is believed to be due to the instability in laser power during processing.

In summary, the major flaws associated with laser surface alloying include porosity, cracks, and nonuniform composition. Controlling laser parameters and coating thickness is necessary to reduce, and if possible eliminate, these defects. It is generally observed that the hourglass fusion zones, characterized by deep penetrations

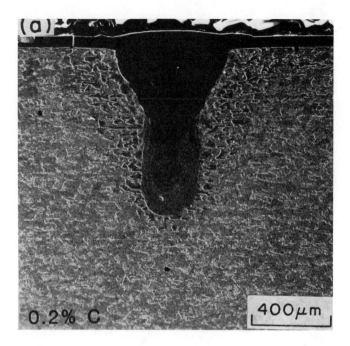

FIGURE 24   Influence of carbon content of the substrate on crack formation.

(melt depth > 700 μm), are accompanied by loss of material through vaporization, low cooling rates, shrinkage cracks, keyhole porosity, and highly nonuniform compositions. On the contrary, semicircular fusion zones have the advantages of very high cooling rates, uniform composition, and minimum defects, but they are limited to shallow melt depths (10-300 μm). The triangular fusion zones have intermediate characteristics.

## Cooling Rates in Laser Surface Alloying

Because of the experimental difficulties involved in measuring the temperature and cooling rates in this process (due to high energy density and short interaction time), several heat flow models based on theoretical concepts were addressed. The details of the mathematical descriptions and the numerical computation techniques used are available in many publications dealing with one-, two- or three-dimensional heat flow analyses. Two such models were followed to predict the cooling rates during solidification. Both models used the following assumptions: (a) semi-infinite solid, (b) constant

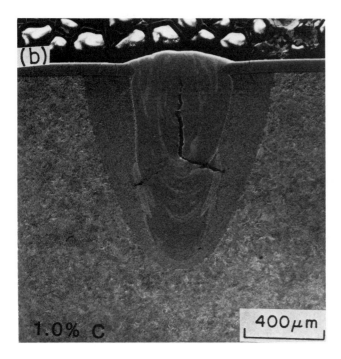

heat input, (c) conduction heat transfer, (d) homogeneous material
with constant thermal properties, (e) no latent heat effect, and
(f) temperature-independent absorptivity. A one-dimensional model,
described by Greenwald, Breinan, and Kear (1978), was employed
when the aspect ratio of the fusion zone was less than 1. The moving
gaussian energy source (three-dimensional) models (Anthony and
Cline, 1977) were applied to fusion zones having an aspect ratio
greater than 1. These models were usually applied to laser melting
substrates. However, the laser surface alloying involved three
materials, namely the coating, substrate, and alloy formed during
solidification. Nevertheless, only the thermal properties of initial
coating were considered in applying these models.

An alternative technique to determine the solidification cooling
rates is to measure the dimensions of microstructural features such
as grain size, dendrite or cell spacing, and inclusion size. The
fineness of cellular-dendritic solidification structures was related
to the cooling rates during solidification. Figs. 26(a)-(c) are the
representative scanning electron micrographs showing the cellular
dendrite structures observed at the top views of laser-alloyed zones.
The dendrite spacing ranged from 0.1 to 4 μm, depending on pene-

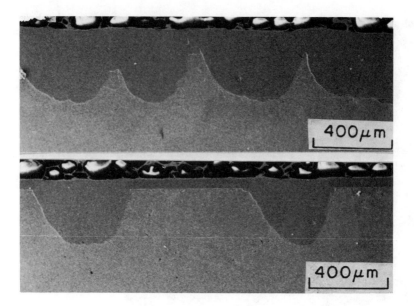

FIGURE 25   Laser-alloyed surfaces using a continuous wave laser, showing nonuniform melting.

tration depths, indicating the high cooling rates attained in this process. The appropriate equation employed here to establish the cooling rates based on dendrite spacing was given by Brower, Strachan, and Flemings (1970) as $d = 60r^{-0.41}$, where d is the dendrite arm spacing ($\mu$m) and r is the cooling rate ($^\circ$C/s). This equation was experimentally determined for Fe-1.0%C-15% Cr alloy and for cooling rates in the range $10^0$ to $10^5$ $^\circ$C/s. Figure 27 shows the cooling rates as a function of dendrite arm spacing, which in turn depends on penetration depth and alloy content of the fusion zones.

Generally, the nature of the surface alloying process is such that it is difficult to isolate the effects of composition from penetration depth. Assuming that the composition effects are minimal in altering the cell spacing, Fig. 28 provides the cooling rate as a function of penetration depth. This plot also presents the predicted values of cooling rates by heat flow models as described earlier.

*Microstructures*

Laser surface alloying is capable of producing novel nonequilibrium and ultrafine microstructures. An example of nonequilibrium structure

is shown in Fig. 29 for a surface alloy of the nominal composition
Fe-0.2%C-20%Cr. The microstructure consisted of austenite (nonequilibrium phase) needles in a ferrite (equilibrium phase) matrix.
Conventional processes produce a structure of ferrite and carbides
for this alloy composition. Laser surface alloying thus eliminated
the carbide formation and improved corrosion resistance.

## 4.2 Surface Alloying of Plasma-Sprayed Coatings

Table 3 shows the coating materials selected for laser surface alloying
of AISI 6150 steel substrate. These coating materials were first
plasma-sprayed and then subjected to melting and alloying using
a 1.2-kW $CO_2$ gas laser. Cross sections of the laser-alloyed steel,
presented in Figs. 30 and 31, reveal nonporous, structurally integral
alloy casings. Other notable features include

1. Retention of WC particles in laser-alloyed 36C coating due to
   larger-sized particles.
2. Uniform mixing of thermally deposited coatings into the substrate.
3. Physical continuity across the fusion zone-heat-affected zone
   interface.
4. Relatively smooth outside surface.
5. Some cracking in the fusion zone.
6. X-ray analysis of laser surface alloys indicated a uniform composition throughout the melt.
7. Microhardness readings of several of these coatings showed
   that the hardness exceeded 1000 HK.

## 4.3 Surface Alloying of Boron and Boron Nitride on Cutting Tools

The cutting tool industries have traditionally sought to extend the
life of tools through the use of surface treatments. The feasibility
of generating amorphous coatings on cutting tools using high-power
$CO_2$ gas lasers and boron as an alloying element was examined.
Ferrous amorphous (metallic glasses) materials offer promise as
wear- and corrosion-resistant coatings on steel substrates. The
additional benefits include high hardness, soft magnetic characteristics, high electrical resistivity, and fracture strength.

The cutting tool materials used were AISI T1 and M2 high-speed
steels and cobalt-base cemented carbides VC-2 and VC-5. The tools
were precoated with several alloying elements (prior to laser irradiation) in order to fulfill the composition requirements for producing
amorphous coatings. A list of precoating methods that were employed
either individually or in combination follows:

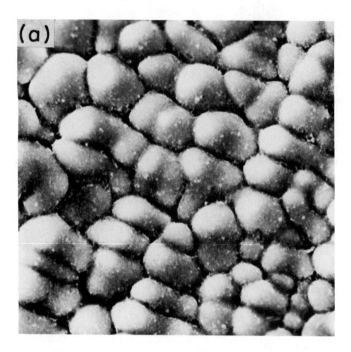

FIGURE 26   Solidification structures of laser-surface-alloyed steels
with penetration depths (a) 1000 μm, (b) 700 μm, (c) 350 μm.

1.   Electroplating to deposit nickel and phosphorus.
2.   Pack cementation process to boronize the surfaces.
3.   Ion nitriding to apply iron-nitride surface coatings.
4.   Injection of hexagonal boron nitride powders into the laser
     melt pool.

     B, Ni, P, and N are selected as alloying elements to aid in
amorphization during laser rapid solidification. Surface melting
and subsequent rapid solidification were accomplished by three
CW $CO_2$ gas lasers having power levels of 1.2 kW (Photon Sources
model V1003), 2.5 kW (Photon Sources model T3000), and 5 kW
(Spectra-Physics model 975). Laser experiments were performed
with a multitude of process variables.
     A transverse section of boronized tool steel is shown in Fig. 32.
The boride layer consists of $M_2B$ and MB compounds along with
metal carbides. Exponential variation of melt depth and melt width
with scan rate is given in Fig. 33. It was obtained with a 2.5-kW
laser. Figure 33 shows that the width-versus-speed curve levels

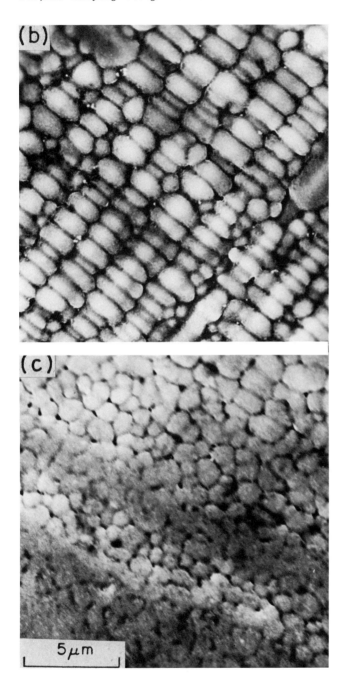

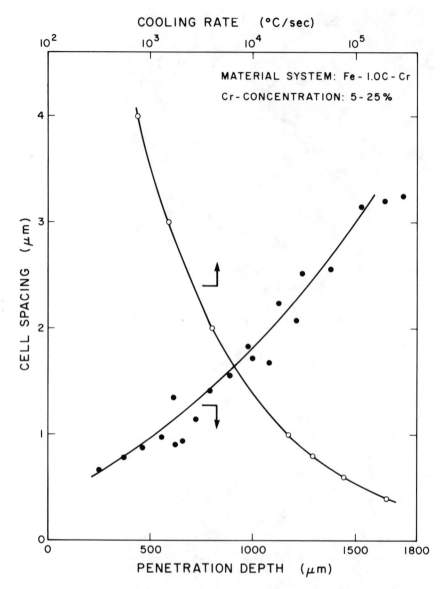

FIGURE 27   Penetration depth-cooling rate-cell spacing relations
for laser-alloyed Fe-C-Cr layers.

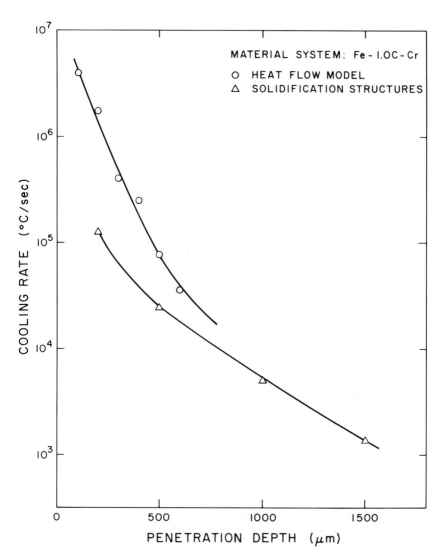

FIGURE 28   A comparison of heat flow models and solidification structures to predict the cooling rates.

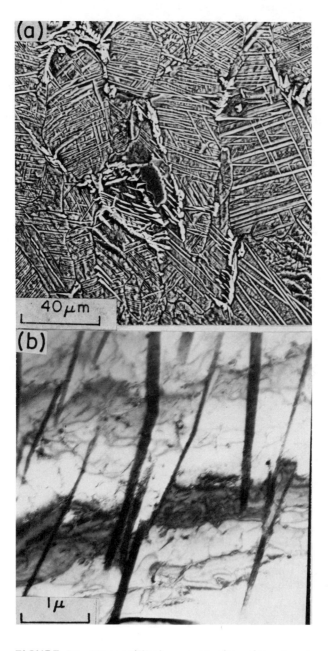

FIGURE 29  Nonequilibrium austenite microstructures in laser-alloyed casings of Fe-0.2%C-20%Cr. (a) Scanning electron micrograph, (b) transmission electron micrograph.

TABLE 3  Coating Materials Used in Plasma Spraying of AISI 6150 Steel Prior to Laser Alloying

| Metco powder | Description | Constituent | Percent | Typical size range |
|---|---|---|---|---|
| 19E | S/F nickel-chromium alloy | Chromium | 16.0 | -140 +325 mesh |
| | | Silicon | 4.0 | -106 +45 μm |
| | | Boron | 4.0 | |
| | | Iron | 4.0 | |
| | | Copper | 2.4 | |
| | | Molybdenum | 2.4 | |
| | | Tungsten | 2.4 | |
| | | Carbon | .5 | |
| | | Nickel | Balance | |
| 36C | S/F tungsten carbide alloy | WC-8% nickel composite | 35.0 | -100 +325 mesh |
| | | Chromium | 11.0 | -150 +45 μm |
| | | Boron | 2.5 | |
| | | Iron | 2.5 | |
| | | Silicon | 2.5 | |
| | | Carbon | .5 | |
| | | Nickel | Balance | |
| 81VF-NS | Chromium carbide-nickel chromium | Chromium carbide | 75.0 | -325 mesh + 5 μm |
| | | Nickel | 20.0 | -45 +5 μm |
| | | Chromium | | |
| 201B-NS-1 | Zirconium oxide-ceramic composite | Zirconium oxide | 92.0 | -140 mesh +10 μm |
| | | Calcium carbonate | 8.0 | -106 +10 μm |
| — | Molybdenum | — | 99.0 | -140 +325 mesh |
| | | | | -100 +45 μm |

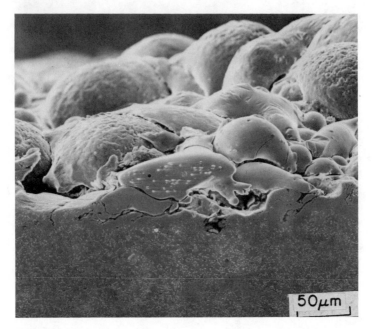

(a)

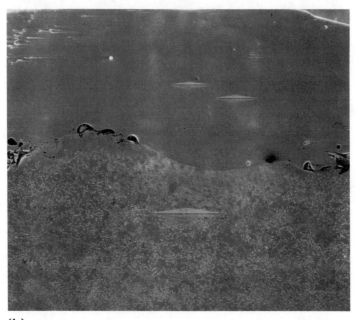

(b)

FIGURE 30  Transverse sections of plasma-sprayed and laser-alloyed coatings: (a), (c), (e), (g) plasma-sprayed; (b), (d), (f), (h) laser-alloyed. Substrate: 6150 steel.

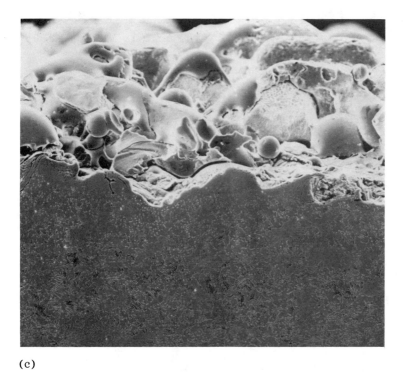

(c)

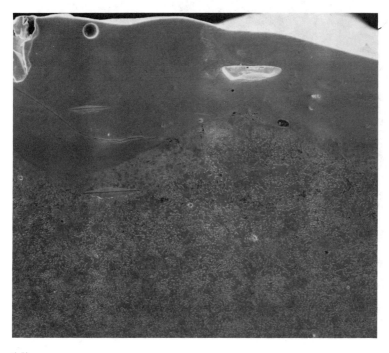

(d)

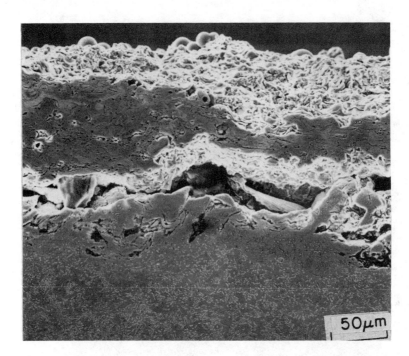

(e)

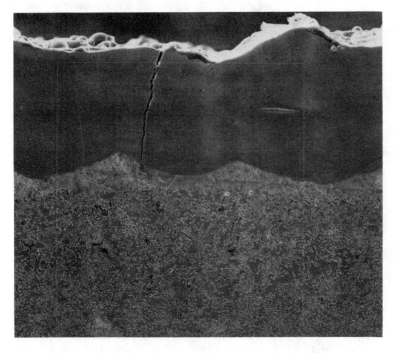

(f)

FIGURE 30 (continued)

(g)

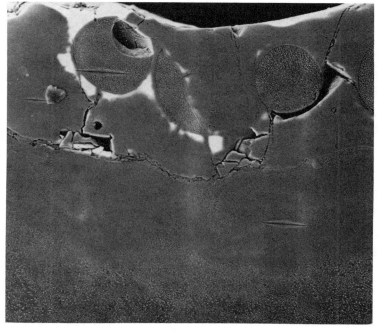

(h)

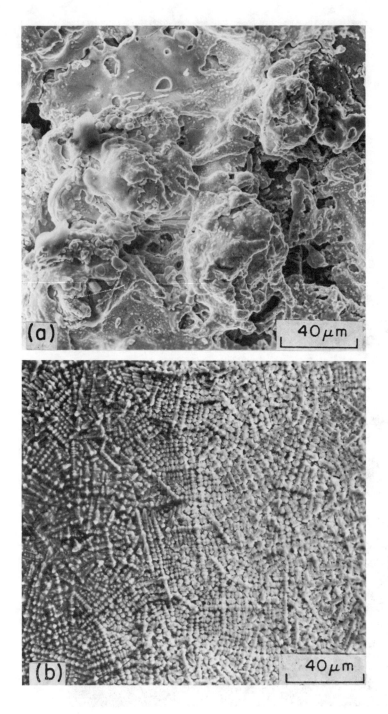

FIGURE 31  Top view of (a) plasma-sprayed and (b) laser-alloyed
molybdenum coating.

FIGURE 32   A transverse section of boronized tool steel.

off at about 250 μm. This trend was observed in all experiments conducted with differing laser parameters. It is also apparent that the boronizing time did not appreciably influence the melt zone size. A transverse section of laser-melted layer with a depth of 70 μm is shown in Fig. 34.

The most significant result in this study was that laser-processed coatings were nearly three times harder than those of tool substrates (Fig. 35) and exhibited large hardness gradients along the melt depths. Hardness was increased, whereas the hardness gradient was decreased, with a corresponding increase in travel speed. Figure 35 shows that the hardness of laser-processed, high-speed steels was comparable to that of cemented carbides and ceramics; the hardness of laser-processed cemented carbides was equivalent to that of $B_4C$.

Laser-processed coatings were substantially thicker (by as much as 1 mm) and had excellent adherence to the substrate. The coatings possessed good resistance to spalling under high mechanical loads. The additional benefits of the laser coating method were negligible distortion and a small heat-affected zone in the tool substrates. The presence of the hardness gradient in the coating will

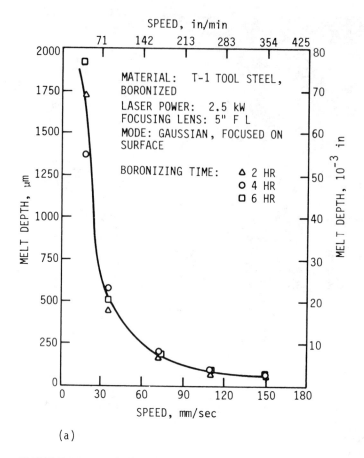

(a)

FIGURE 33 Variation of laser-alloyed depth and width of boronized steel.

minimize wear as well as fatigue of cutting tools by plastic deformation rather than by brittle fracture.

Results on melt depth, melt width, hardness, and microstructures
are given in Table 4. The microstructures obtained by laser processing are substantially different from those that emerge under conventional methods. Extremely fine cellular dendritic solidification structures
were observed in laser-processed coatings. Complex composition
precipitates were identified at the cell boundaries on an ultrafine
scale. There was not sufficient evidence of amorphization during
laser processing in spite of the fact that rapid solidification conditions
and suitable alloy compositions for glass formation were achieved in

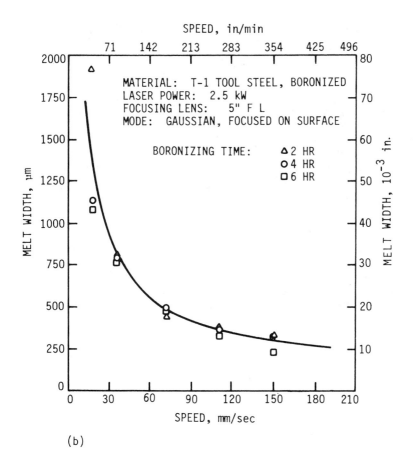

(b)

the melt zones. Laser parameters were varied to a large extent but failed to generate an amorphous structure. It is concluded that amorphization is much more difficult to achieve in laser processing because of the intimate contact between the substrate and the melt layer. The substrate allows for heterogeneous nucleation of the melt pool to occur before amorphization is completed. The end result is that only a portion of the laser-processed coating contains an amorphous structure.

Among the several coatings attempted, the most promising one was the addition of hexagonal BN powders to the cutting tools surfaces through a laser melt-particle injection process. This technique yielded high-quality coatings with superhardness (Table 5).

The deleterious effect of laser processing on the coatings was cracking. But cracks were kept to a minimum by a judicious choice

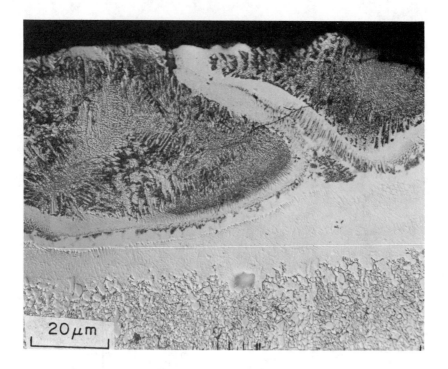

FIGURE 34   Transverse section of laser-alloyed casing of the boronized steel.

of laser parameters, coating materials, and laser techniques. For example, the hexagonal BN coating of tools via the laser melt-particle injection process resulted in few cracks.

### 4.4   Corrosion, Wear, Hardness, and Microstructures of Laser-Alloyed Surfaces

*Corrosion*

Moore and McCafferty (1981) carried out anodic polarization corrosion tests (potentiodynamically at a scan rate of 10 mV/min) on three different Fe-Cr laser surface alloys in deaerated 0.1 M $Na_2SO_4$. They observed that the critical current density for passivation and the current density in the passive region decreased with an increase in the chromium content of the alloy. Further, the laser surface alloy with 80% Cr did not exhibit any active-passive transition. However, they noted that for laser surface alloys the current densities in the passive region were higher than the corresponding bulk

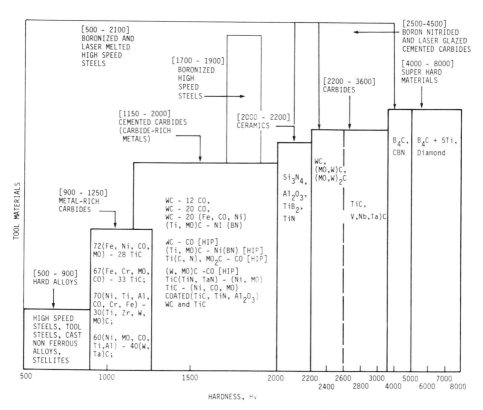

**FIGURE 35** Comparison of hardness of conventionally processed and laser-alloyed cutting tools.

alloys. For example, the passive current density for a 20% Cr surface alloy was about five times greater than that observed for bulk Fe-18%Cr and about 50 times that observed for Fe-25%Cr in 1 N $H_2SO_4$. According to McCafferty et al. (1982), these differences were not serious, and the passivation behavior of laser surface alloys was similar to that of the 400-type bulk stainless steels. McCafferty et al. (1982) also studied the effect of laser surface melting on the anodic behavior of 304 stainless steel under potentiodynamic conditions. The resistance to pitting corrosion was found to be improved and was superior to 316 stainless steel, which is usually employed for this purpose. Furthermore, the improvement in pitting potential due to laser surface melting was independent of the anodic sweep rate. They attributed these results to the modification of

TABLE 4 Summary of Results Obtained in Laser-Alloyed Tool Materials

| Tool material | Melt depth (μm) | Melt width (μm) | Nominal B-concentration (wt.%) | Maximum hardness (HV) | Micro-structure | Solidification Conventional method | Solidification Laser processing |
|---|---|---|---|---|---|---|---|
| Plain tool steels | 50–1400 | 300–1150 | — | 900 | $\delta$, $\gamma$, carbides | $L \to \delta + L$ $\to \gamma + M_2C + M_{23}C_6$ $\to \alpha + M_2C + M_{23}C_6$ | $L \to \delta + L$ $\to \delta + \gamma + M_2C + M_{23}C_6$ |
| Boronized Armco iron | 50–1500 | 500–1900 | 1–9 | 2300 | $Fe_2B$, eutectic | $L \to Fe_2B + L$ $\to \alpha + Fe_2B$ | $L \to Fe_2B + L$ $\to Fe_2B + $eutectic |
| Boronized tool steels | 40–2000 | 500–1000 | 1–7 | 2100 | $M_2B$, $\gamma$, $M_{23}(C,B)_6$, eutectic | $L \to \gamma + L \to \gamma + M_2B$ $\to \alpha + M_2B$ $+ M_{23}(B,C)_6$ | $L \to \gamma + L + \gamma + M_2B$ $\to M_2B + \gamma + $eutectic $+ M_{23}(B,C)_6$ |
| Boronized and ion-nitrided tool steels | 250–700 | 4600–5100 | 1–5 | 1750 | $\gamma$, $M_2B$, $M_4N$, carbides | $L \to L + \gamma \to \gamma + M_2B$ $\to \alpha + M_4N + M_2B$ $+ M_{23}(B,C)_6$ | $L \to L + \gamma \to \gamma + M_2B$ $\to \gamma + M_2B + M_4N$ $+ M_{23}(B,C)_6$ |
| Boronized and Ni-P-plated tool steels | 175–300 | 1600–3200 | 1–5 | 1950 | $\gamma$, $M_2B$ complex boronitrides, steadite | $L \to L + \gamma \to \gamma + M_2B$ $\to \alpha + M_2B$ $+ M_{23}(B,C)_6$ $+ $steadite | $L \to L + \gamma \to \gamma + M_2B$ $\to \gamma + M_2B + $metal boronitride $+ $steadite |
| BN-injected tool steels | 200–850 | 4500 | — | 1950 | $M_2B$, $\gamma$, complex metal boronitrides | — | — |
| BN-injected cemented carbides | 10–100 | 150–300 | — | 4500 | — | — | — |

TABLE 5   Data from Laser Melt-Particle Injection Process

| Speci-men | Laser treatment | Vickers' hardness range (HV) | | Melt depth (μm) | | Process defects |
|---|---|---|---|---|---|---|
| | | T1 steel | M2 steel | T1 steel | M2 steel | |
| 1 | Substrate melting (one pass) | 650-900 | 650-900 | 230 | 240 | None |
| 2 | BN injected and laser melted once | 1140-1360 | 1050-1190 | 200 | 220 | Few cracks and pores |
| 3 | BN injected and laser melted four times | 1570-1840 | 1090-1240 | 310 | 320 | Some pores |
| 4 | BN injected and laser melted 10 times | 1200-1940 | 1190-1840 | 840 | 670 | Porosity |
| 5 | Laser remelting of specimen 4 | 1750-2150 | 1700-1930 | 840 | 670 | Very few cracks and pores |

segregation patterns and the uniform distribution of fine carbides which were caused by rapid solidification. They also cited an example to illustrate the effectiveness of laser surface melting in minimizing the pitting resistance. Laser melting of a plasma-sprayed Ti coating on steel reduced the resistance to pitting in seawater and was thought to be due to the consolidation of pores in the plasma coating.

In some preliminary studies of aqueous corrosion of electron beam-melted M2 and M7 tool steels, Lewis and Strutt (1982) observed a greater reduction in corrosion current in electron beam-glazed specimens than in untreated specimens. Lumsden and Gnanamuthu (1982) presented the potentiodynamic polarization curves (scan rate = 20 mV/min) of laser surface alloyed 4140 steels and compared the data with bulk 304 stainless steel. The laser-surface alloys were 29%Cr-13%Ni, 19%Cr-8%Ni and 11%Cr-4%Ni. Electrochemical measurements in deaerated 1 N $H_2CO_4$ demonstrated that the surface alloy 19%Cr-8%Ni and bulk 304 stainless steel had identical corrosion behavior, while the surface alloy 29%Cr-13%Ni had a corrosion resistance

superior to 304 stainless steel. Draper et al. (1981) reported that
laser surface alloying of Pd on Ti was an effective method to passi-
vate Ti against corrosion by acid environments. They also added
that the corrosion performance of laser surface alloys was equivalent
to that of bulk alloys. This illustrates the conservation of precious
metal by the laser surface alloying technique. Draper et al. also
found that laser processing was better than thermal diffusion and
ion implantation techniques for reducing noble metal consumption.

*Wear*

Glaeser and Fairand (1979) investigated the effects of laser melting
of plasma-sprayed $TiB_2$ and $B_4C$ coatings (on A6 tool steel substrate)
on the abrasion wear resistance. Feeding a slurry of silicon carbide
in water into the sliding contact area, they observed that the abrasion resistance of laser-alloyed steel was 10 times greater than
that of untreated steel. No pitting, spalling, or chipping was found
in wear tests. Some surface cracking occurred in laser treatment,
but the occasional crack did not seem to influence the abrasive
wear properties. Belmondo and Castagna (1979) deposited and laser-
melted a mixture of chromium carbides, chromium, nickel, and
molybdenum coatings onto mild steel, stainless steel, cast iron,
and superalloys substrates. Through microstructural characterization,
they inferred that there was an excellent bonding of the melt zone
to the substrate and a uniform distribution of elements across the
melted layer. They also showed that laser melting enhanced the
sliding wear resistance at high contact pressures. Schmidt (1969)
reported that laser infusing of a WC coating into a high-speed steel
tool increased the tool life and reduced wear.

Ayers et al. (1981) injected TiC and WC powders into the laser
melt pool of Ti-6A1-4V, 304 stainless steel, Inconel X-750, and
5052 aluminum substrates and reported improvements in abrasion
wear resistance. They also expected the resistance to other types
of wear to be superior. Ayers and Schaefer (1979) observed the
improvement in corrosion resistance and adhesion of plasma-sprayed
Ti and 316L stainless steel coatings by laser melting of Cr and
Cr-Ni coatings on 1018 steel-generated alloys of high corrosion
resistance and with mechanically stable coatings. Seaman and Gnanamuthu (1975) reported that laser melting of small amounts of chromium
powder to the valve seat of an internal combustion engine enhanced
its service life.

Mordike and Bergmann (1982) used a CW $CO_2$ gas laser to
inject SiC, WC, and TiC powders into the laser melt pool of tool
steel and observed the hardness to be 1700 HV, 1100 HV, and 1800
HV, respectively. They claimed that the melt layers were free from
cracks and that subsequent conventional furnace heat treatments

did not cause cracking. The absence of cracking in the as-melted condition is rather surprising, if not impossible. Ayers and Tucker (1980) produced WC- or TiC-rich surface layers on AISI 304 stainless steel via a combined laser melt and carbide injection process. Both WC and TiC powders were added to the melt pool through a pneumatic powder delivery method. The Rockwell C hardness of the melt layer was 58 HRC, which is substantially higher than the hardness that can be obtained by conventional treatment. A deleterious effect observed by these investigators was carbide cracking. Matthews (1983) placed a paste of Hastellite (50 wt.% WC and 50 wt.% Ni-Cr-Si-B) on a steel coupon and laser-melted it using a 1.2-kW $CO_2$ laser. He noted cracks and pores in the melt zones, but believed that such a hard-facing technique is highly promising for many applications.

Snezhnoi, Zhukov, and Kokora (1980) reported the formation of an amorphous phase on the surface of chilled cast iron (C = 3.20%, Si = 2.67%, Mn = 0.64%, S = 0.014%, and P = 0.060%) by surface-melting with a pulsed Nd:glass laser. Etching, x-ray, and electron diffraction techniques were utilized to identify the amorphous layers. The hardness of amorphous surfaces was found to be 1200 $H_V$ which was significantly greater than the base iron hardness of 800-1000 $H_V$. These researchers attempted unsuccessfully to obtain an amorphous structure in Fe - 6.67%C alloy. They attributed this to the low thermal conductivity of cementite. This is the only known source that discusses the formation of amorphous cast iron. However, this investigation did not explore the potential applications and mechanical properties other than hardness. Bergmann and Mordike (1981) observed amorphous phases in laser-melted chromium steel and niobium alloy. In their experiments, an Fe-B coating was deposited on a 2.0%C-12% Cr steel substrate to a thickness of 40 μm and laser-melted to a total depth of 60 μm. The $CO_2$ gas laser parameters were 3 kW power, 0.5 mm beam size, and 20 cm/s travel speed. The hardness of amorphous (Fe, $Cr_{12})_{80}(C,B)_{20}$ was determined to be 1625 HV, whereas it was 400 HV for the base steel. Bend and compression tests performed to test the mechanical adhesion between the glass layer and substrate resulted in cracks when the compressive strain exceeded 5%.

Draper and Poate (1985) believe that amorphization is much more difficult to achieve in laser processing, even with a large heat transfer coefficient, thermal gradient, and quench rate, because of the intimate contact between the substrate and the melt layer that ensures nucleation. Kear, Breinan and Greenwald (1978) question whether the favorable nucleation sites for crystallization would prevent the undercooling necessary to obtain amorphous structures. Bergmann and Mordike (1981) suggest that epitaxial growth is common in many Fe-based alloys during rapid solidification by laser techniques and should be prevented to achieve glassy structures.

## 5  LASER CLADDING

The laser cladding process is schematically described in Fig. 36. Cladding can be accomplished either by prior placing of the coating in the form of loose powder or by pneumatically injecting the powder into the melt pool during laser processing. Between these two methods, the pneumatically delivered powder process was established to be more energy efficient (Li and Mazumder, 1985) because the beam-material interaction zone is covered with a melting powder layer that enhances the absorption of laser power. Furthermore, the dilution is controlled by the powder flow rate rather than by power density (Weersinghe and Steen). Laser cladding can also be carried out by applying the clad material as a wire, sheet, plasma-spray coat, or electroplate coat. The powder delivery system, as illustrated in Fig. 37, consists of a small hopper with a metering

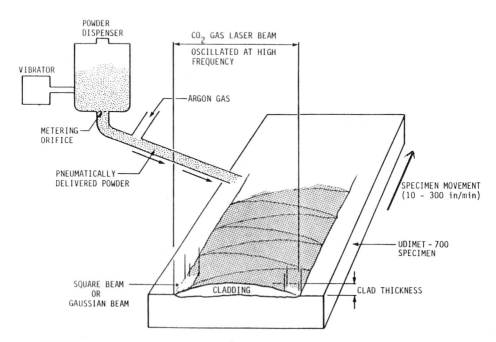

FIGURE 36  A schematic of laser cladding process.

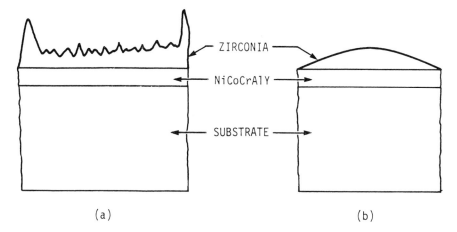

(a)                                                    (b)

FIGURE 37  Typical clad profiles of ceramic coatings. (a) Nonuniform clad profile with a rough surface. (b) Convex clad profile
with a smooth surface.

orifice in its base. The powder is fed through this hopper into
a tube connected to an argon gas cylinder. The powder is carried
by the stream of argon. The purpose of the vibrator attached to
the hopper is to produce an even powder flow. The powder flow
rate may be varied by controlling the metering orifice and argon
gas flow rate. Powder flow rate is one of the key factors affecting
the shape of cladding, porosity, dilution, and adherence of the
coating (Weersinghe and Steen).

Laser cladding offers numerous advantages, including the potential to apply cladding alloys of high-melting-point on low-melting-
point workpieces, controlled dilution, localized application of coating,
good fusion bond, fine microstructures, minimal heat-affected zone
(HAZ), and homogeneous and flawless coatings. These benefits
reduce machining costs, save clad material, and reduce scrap,
because of improved quality. Gnanamuthu (1979) applied WC-Fe,
Tribaloy T-800, and stellite coatings on steels via laser cladding
and obtained claddings with a dilution of less than 5%, good metallurgical bond, and uniform composition and microstructure. He also
attempted to clad 0.3-$\mu$m $Al_2O_3$ on 2219-aluminum substrate. Weersinghe and Steen (1983) extensively studied this process for cladding
stainless steels and observed no evidence of macrosegregation,
very little dilution, uniform microstructure, and negligible distortion.
They observed no delamination or interface cracks during the 180°
bend test of the claddings. They also suggested that process

parameters should be controlled to eliminate the defects and to produce a deposition rate comparable to that of other cladding techniques.

A more spectacular, commercial application of this process is cladding of Ni-base-alloy turbine blades of aeroengines for the Rolls Royce Company since 1982. Laser cladding was substituted for a TIG arc welding method that was time consuming, dependent on the operator skill, and nonreproducible in dilution. Laser cladding produced clad deposits having optimal wear characteristics: they were free from HAZ cracking, consistent in quality, and cheaper in material costs by 50%. Macintyre (1983) claims that laser cladding was matched only by the most rigorous execution of conventional technique. Other examples of laser cladding for some commercial applications are given in Table 6 (Eboo and Lindemanis, 1985).

### 5.1 Laser Cladding of Ceramic Coatings for High-Temperature Applications

Currently, thermal spraying methods such as flame spraying, plasma spraying, and detonation-gun, are employed to apply a wide range of ceramic coatings. None of these processes is fully satisfactory because they produce coatings with excessive porosity, slag inclusions, microcracking, coating segmentation, and poor adhesion. This leads to premature coating degradation and failure by internal sulfidation, spallation, reduced mechanical strain, pitting corrosion, condensed salt penetration, and oxygen penetration at high temperatures.

The author conducted numerous laser experiments to clad thermal-barrier, zirconia coatings on Udimet-700 alloy and AISI 4140 steel. Laser cladding parameters are shown in Table 7.

Zirconia coating thicknesses of representative specimens, given in Table 8, illustrate the effect of power density. The two substrates, Udimet-700 and 4140 steel, did not exhibit any noticeable difference in the cladding of ceramic coatings. The reason may be that both substrates were initially clad with the bond coat Ni-Co-Cr-Al-Y. To elucidate the differences, cladding of ceramic coatings directly on the substrates was not conducted. The effect of the thermal properties of the metal substrate on the cladding of the ceramic coating was minimized by the deposition of a thicker bond coat.

Laser-clad specimens can be conveniently classified into two distinct categories by the geometry of bead profiles (Fig. 37). The bead profile shown in Fig. 37(a) was nonuniform with an excessive buildup of material on the edges of cladding and rough surface. This type of profile was characteristic of all the zirconia-clad specimens processed with a 1.2-kW laser with a 5-in focal length lens.

TABLE 6   Commercial Applications of Laser Cladding

| Company | Component | Cladding material and method |
|---------|-----------|------------------------------|
| Rolls Royce | Turbine blade shroud interlock | Tribaloy/nimonics powder feed |
| Pratt & Whitney | Turbine blade shroud interlock | PWA 694/nimonics pre-placed chip |
| General Electric | Proprietary | Reverse machining with Ti power feed |
| Pilot demonstration stage | | |
| Combustion Engineering | Offshore drilling & production parts, valve components, boiler firewall | Stellites, colmonoys, and other alloys, including carbides, powder feed |
| Fiat | Valve stem, valve seat, aluminum block | $CrC_2$, Cr, Ni, Mo/cast Fe preplaced powder |
| General Motors | Automotive | Cast iron systems |
| Rockwell | Aerospace | T-800, stellites, powder feed |
| Westinghouse | Turbine blades | Stellites, colmonoys, preplaced beds, gravity feed |
| NRL | Proprietary | Multiple alloys, powder feed, preplaced beds |

*Source*: After Eboo and Lindiminis, 1985.

In specimens processed at lower power densities, the zirconia clad had a convex shape with a smooth surface. The difference in the bead profiles may be explained on the basis of surface tension of the melt pool, which in turn depends on the power density of the laser beam and composition of the cladding.

Representative convex bead profiles are shown in Figs. 38 and 39. Figure 38 presents two photomicrographs of the transverse sections of a specimen showing the bond and zirconia coatings. Delamination cracking occurred near the boundary between zirconia and the bond coat (or the substrate). It may be seen that, contrary to zirconia, the Ni-Co-Cr-Al-Y has an integral bonding to

TABLE 7   Laser Cladding Parameters for Zirconia Coatings

| Laser | Approximate beam size | Beam shape | Approximate max. power density (W/in$^2$) | Scanning |
|---|---|---|---|---|
| Photon Sources V1200 (1200 W) | 5-in lens ~ 0.005 in dia. | Gaussian | 61 × 10$^6$ | Eliptical width - 0.160 in - 0.120 in |
| | 15-in lens ~ 0.015 in dia. | Gaussian | 6.8 × 10$^6$ | Elliptical width - 0.160 in 0.120 in |
| | Integrator ~ 0.20 × 0.20 in | Square | 30 × 10$^3$ | — |
| Spectra Physics 975 | 18-in focal length ~ 0.15 in × 0.03 in | Multimode elliptical | 1.4 × 10$^6$ | Linear width - 0.20 in |
| | Integrator ~ 0.20 in × 0.20 in | Square | 125 × 10$^3$ | — |

TABLE 8   Clad Thickness Data
(bond coat thickness = 0.010-0.020 in.)

| Power density (W/in.$^2$) | Zirconia clad thickness (in., nominal value) |
| --- | --- |
| $1.4 \times 10^6$ | 0.0004 |
| $6.8 \times 10^6$ | 0.040 |
| $61 \times 10^6$ | 0.050 |

the substrate. Figure 38 confirms that the deposition of zirconia over the bond coat is no better than cladding zirconia over the 4140 steel substrate.

Figure 39 shows the transverse section of a specimen that was processed with the 5-kW laser. The zirconia claddings were thin, smooth, and glassy. The claddings had integral bonding to the substrate with no apparent cracks. Occasionally the cladding was found to be discontinuous over the substrate surface. The conclusion that can be drawn is that the power density is the dominant factor for controlling coating thickness, adhesion, and cracking of the zirconia coating.

In general, thicker claddings of zirconia contained cracks and delaminated from the substrate. Cracking is undoubtedly due to the thermal stresses developed during cladding. Some cracks might have occurred during sectioning or polishing. Adhesion of zirconia to the substrate was poor for thicker coatings and resulted in chipping of the material by thermal stresses and propagation of cracks into the zirconia.

In order for the laser cladding of ceramic coatings to be feasible, methods to improve the adhesion characteristics and to reduce the thermal stresses (which cause cracking) should be emphasized. The following methods are suggested for future work in overcoming the obstacles described and for producing a thicker coating.

1.  Apply graded composition coatings: mixtures of zirconia and bond coat should be applied in various proportions. A typical sequence would be 90% bond coat and 10% zirconia, 80% bond

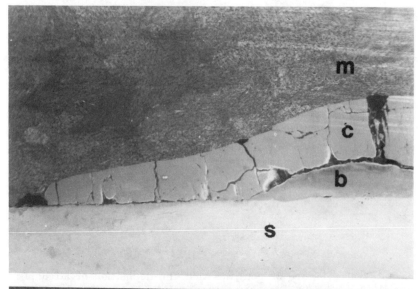

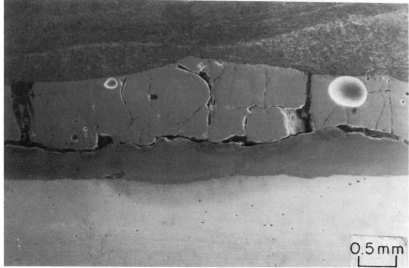

FIGURE 38   Claddings of zirconia and Ni-Co-Cr-Al-Y on AISI 4140
steel (C = zirconia, b = Ni-Co-Cr-Al-Y, S = substrate).

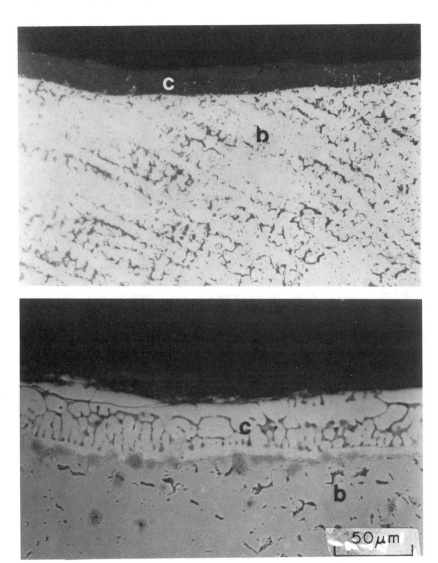

FIGURE 39 Claddings of zirconia and Ni-Co-Cr-Al-Y on Udimet-700 alloy (c = zirconia, b = Ni-Co-Cr-Al-Y).

coat and 20% zirconia, 50% bond coat and 50% zirconia, and so on. This will reduce the thermal stresses and provide a crack-free coating.

2.  Optimize power density: power density controls the vaporization, melting volume, and cooling rate and thereby offers a method to enhance the quality of claddings.

3.  Preheat the substrate and powders: this will reduce the thermal stresses accompanying the process and eliminate spallation.

4.  Optimize flow rate: powder flow rate will assist in buildup of coating once optimal conditions are established.

5.  Vary powder size: a relationship between powder size and power density appears to exist, so the powder size should be optimized.


## 6  CONCLUSIONS

The applications of lasers for surface modification brought several possibilities for tailoring the surface properties and to meet the challenging demands of tribological and high-temperature requirements. Laser alloying generates novel nonequilibrium microstructures, solid solutions, and amorphous alloys with exceptionally high wear, fatigue, and corrosion resistance. Laser surface alloying is still young, although a significant number of applications in metal working industries utilize this method. Further development is essential to identify possible production applications. This can be done only by devoting continued efforts to optimize, evaluate, and understand laser processing capabilities. Detailed technical and cost evaluations of laser alloying and comparisons with other methods are not fully available to appreciate the cost effectiveness of laser alloying. However, fast production rate, cleanliness, flexibility, simplicity, and versatility of laser surface alloying are the chief advantages.


## REFERENCES

Anthony, T. R. and Cline, H. E. (1977). *J. Appl. Phys.*, *48*:3895.
Anthony, T. R. and Cline, H. E. (1978). *J. Appl. Phys.*, *49*(3):1298.
Ayers, J. D. and Schaefer, R. J. (1979). Proceedings of Society of Photo-Instrumentation Engineers, vol. 198, p. 57.
Ayers, J. D., Schaefer, R. J., and Robey, W. P. (1981). *J. Metals*, 19.
Ayers, J. D. and Tucker, T. R. (1980). *J. Thin Solid Films*, *73*:201.
Belmondo, A. and Castagna, M. (1979). *J. Thin Solid Films*, *64*:249.
Bergmann, H. W. and Mordike, B. L. (1981). *J. Mater. Sci.*, *16*:863.
Breinan, E. M. and Banas, C. M. (1975). Technical Report, United Technologies Research Center.

Brower, W. E., Strachan, R., and Flemings, M. C. (1970). *AFS Cast Metals Res. J.*, *12*:176.

Draper, C. W., Mayer, L. S., Jacobson, D. C., Buene, L., and Poate, J. M. (1981). *J. Thin Solid Films*, *75*:237.

Draper, C. N. and Poate, J. M. (1985). *International Metal Reviews*, *30*:85.

Eboo, G. M. and Lindemanis, A. E. (1985). *Proceedings of Society of Photo-Instrumentation Engineers on Applications of Lasers*, 527, p. 86.

Engel, S. L. (1976). *Laser Focus*, *2*:44.

Glaeser, W. A. and Fairand, B. P. (1979). *Proceedings on Wear of Materials* (K. D. Ludema, ed.), American Society for Mechanical Engineers, p. 304.

Gnanamuthu, D. S. (1979). *Applications of Lasers in Materials Processing*, American Society for Metals, Metals Park, Ohio, p. 177.

Greenwald, L. E., Breinan, E. M., and Kear, B. H. (1978). *Proceedings of Laser-Solid Interactions and Laser Processing* (S. D. Ferris, ed.), p. 129.

Kear, B. H., Breinan, E. H., and Greenwald, L. E. (1978). *Metals Technol.*, *6*:121.

Lewis, B. G. and Strutt, P. R. (1982). *Corrosion of Metals Processed by Directed Energy Beams* (C. R. Clayton and C. M. Preece, eds.), Met. Soc. AIME, Warrendale, Pennsylvania, p. 119.

Li, L. J. and Mazumder, J. (1985). *Proceedings, Symposium on Laser Processing of Material* (K. Mukherjee and J. Mazumder, eds.), The Metallurgical Society of AIME, Warrendale, Pennsylvania, p. 35.

Locke, E. V. and Hella, R. A. (1974). *IEEE J. Quantum Electron.*, *2*:179.

Lumsden, J. B. and Gnanamuthu, D. S. (1982). *Corrosion of Metals Processed by Directed Energy Beams* (C. R. Clayton and C. M. Preece, eds.), Met. Soc. AIME, Warrendale, Pennsylvania, p. 129.

Macintyre, R. M. (1983). *Applications of Lasers in Materials Processing*, American Society for Metals, Metals Park, Ohio, p. 230.

Matthews, S. J. (1983). *Applications of Lasers in Materials Processing*, American Society for Metals, Metals Park, Ohio, p. 138.

McCafferty, E., Moore, P. G., Ayers, J. D., and Hubler, G. (1982). *Corrosion of Metals Processed by Directed Energy Beams* (C. R. Clayton and C. M. Preece, eds.), Metallurgical Society fo AIME, Warrendale, Pennsylvania, p. 1.

Moore, P. G. and McCafferty, E. (1981). *J. Electrochem. Soc.*, 1391.

Mordike, B. L. and Bergmann, H. W. (1982). *Japan Inst. Metals*, *1*:197.

Schmidt, A. O. (1969). *J. Eng. Industry*, 549.

Seaman, F. D. and Gnanamuthu (1975). *Metal Prog.*, 108:67.

Snezhnoi, R. L., Zhukov, A. A., and Kokora, A. N. (1980). *Metal Sci. Heat Treatment*, 22:900.

Weersinghe, V. M. and Steen, W. M., (1983). *Applications of Lasers in Materials Processing*, American Society for Metals, Metals Park, Ohio, p. 166.

# 7
# Electron Beam Coating

SIEGFRIED SCHILLER, ULLRICH HEISIG, and PETER FRACH
*Manfred von Ardenne Research Institute, Dresden, German Democratic Republic*

Evaporation in a vacuum is a significant process for the production of thin films [1,2,3,4]. A thorough knowledge of physical and applicative features of the films [5,6] and the results of considerable technological efforts in the field of evaporation techniques have increased industrial application of physical vapor deposition (PVD) in many fields. The advent of electron beam (EB) evaporation in vacuum coating has therefore exerted a strong impetus on this development. In this mode of evaporation the electron beam as an energy carrier serves to heat the evaporant in the crucible directly.

## 1 THE ELECTRON BEAM AS ENERGY CARRIER

For EB coating an electron beam is generated in the electron gun and, by suitable focusing and deflection, is passed through the work chamber to the evaporant [7,8,9]. Because generation and unimpeded propagation of electron beams are possible only in a high vacuum, both the work chamber and the beam generating system must be evacuated. Interactions between the beam electrons and gas molecules which take place on the path from the electron gun to the evaporant at increased pressures from $10^{-2}$ to 1 Pa, influence the evaporation and condensation processes.

After passage through an acceleration voltage $U_B$, the kinetic energy of an electron becomes $E = eU_B$. When the beam hits the

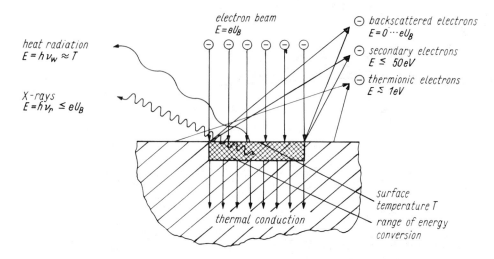

FIGURE 1   Beam action upon impingement on matter (E = energy
of the electron beam and the secondary effects, respectively).

evaporant, various elementary processes (Fig. 1) transform this
kinetic energy. The decisive portion of the beam power is converted
into heat when the beam impinges on the evaporant, a portion that
represents the useful energy of evaporation. Above all, the losses
encountered in the coating process are backscattered electrons,
secondary electrons, thermionic electrons, and x-radiation. Concerning energetic significance, however, only the backscattered electrons
count. X-radiation may represent a health hazard and must therefore
be duly considered. Moreover, heat conduction from the region
of energy conversion and heat radiation from the hot surface represent other substantial losses in the coating process.

## 1.1   Electron Backscattering

Backscattered electrons [10] exhibit an energy spectrum ranging
up to the energy of the primary beam electrons. Above all, the
portion of backscattered beam electrons, their energy spectrum,
and their directional distribution are determined by the atomic number
and by the angle between the normal to the target surface and the
direction of the incident beam [11]. The portion of backscattered
electrons is almost independent of the energy of the incident beam,
and grows with increasing atomic numbers of the material hit [12]
(up to 50% for materials with high atomic numbers, such as tantalum

and tungsten). The energy lost during the actual process due to electron backscattering may be considerable (Fig. 2). Electron backscattering increases with the angle between the direction of beam incidence and the normal to the surface. Depending on the material, it increases at glancing incidence by a factor of 2-3, compared with normal incidence of the beam [13]. The intensity distribution of backscattered electrons over the solid angle is lobe-shaped and has an intensity maximum of which the direction can be approximated from the law of reflection (the angle of incidence is equal to the angle of emergence) [13].

## 1.2 Thermionic and Secondary Electron Emission

Secondary electron emission at the point of beam action occurs with an energy less than 50 eV [9]. When EB processes are performed with high temperatures of the corresponding material, there is also a noticeable emission of thermionic electrons [9]. However, the emission of secondary and thermionic electrons is practically meaningless as far as the energy balance of the EB process is concerned.

## 1.3 X-Radiation

The emitted x-rays consist of two components: characteristic radiation and bremsstrahlung [9]. The total energy of x-radiation depends on the energy of the beam electrons and on the atomic number of the evaporant hit by the beam [9]. For the beam electron energies

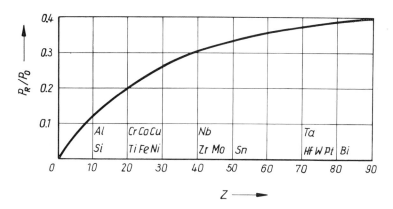

FIGURE 2    Ratio of backscattered power $P_R$ to irradiated power $P_0$ as a function of the atomic charge Z for normal beam incidence on a plane target.

under consideration, the x-radiation is about 1% or less for all
materials concerned. Even whole-body irradiation, the x-ray dose
rate can be reduced to the human tolerance dose by shielding.
The attenuation of x-ray intensity takes place exponentially with
increasing thickness. For EB evaporators with acceleration voltages
of less than 60 kV, the wall thicknesses of the work chamber and
electron gun usually suffice for proper x-ray shielding. In addition,
gaps or materials with low atomic numbers must be avoided (leaded
glass viewports must be used) [14].

## 1.4  Energy Transfer

One objective of process technology is to transfer the highest possible
portion of the beam energy to the vapor-emitting surface and to
use it to generate the vapor stream. In addition to the losses mentioned in previous chapters it is necessary to take into account
losses in the gun, losses due to residual gas scattering, and, at
high pressures above the vapor-emitting surface, losses caused
by the interaction of the beam with the vapor cloud [15,16]. Regarding the beam power $P_0$ it is generally necessary to consider such
substantial losses as electron backscattering [17,18], heat radiation,
and conduction from the hot evaporant. The useful power $P_V$ of
the evaporation process is the portion of the power that causes
an increase in the inherent energy of the evaporant needed for
the generation of the vapor stream. $P_V$ comprises the melting power,
the power which heats the evaporant to evaporation temperature,
and the evaporating power.

Let us assume for simplicity that the vapor-emitting surface
and the radiant surface are alike, that there is a steady state
temperature distribution, and that the specific heat is independent
of the temperature. Then $P_V$ exhibits an exponential temperature
dependence [7]. The power losses due to heat radiation and
conduction, however, are only proportional to the temperature in
the fourth and first powers, respectively. For low evaporation temperatures the heat losses due to conduction and convection dominate.
When the temperature of the vapor-emitting surface is high enough,
the losses due to heat conduction can, under certain conditions,
be neglected compared with the heat losses due to radiation. If
the temperature of the vapor-emitting surface continues to rise
by an increase in the power density on the evaporant, the vapor
stream and, thus, $P_V$ also increase. At a particular temperature $T_{VS}$
[19] the latent thermal power of the vapor stream reaches the magnitude of the radiation losses. If the evaporant is heated far beyond
$T_{VS}$, even the heat loss due to radiation loses its significance compared with the latent thermal power of the vapor stream. Figure 3
demonstrates this for aluminum evaporation.

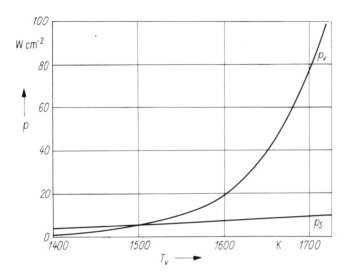

FIGURE 3 Power density $p_V$ of the latent heat flow of the vapor compared to the power density $p_S$ of the radiation as a function of the temperature $T_V$ of the vapor-emitting surface for the evaporation of aluminum (calculated with formulas derived in Ref. 7, Chapter 3.6, pp. 200).

The ratio of $P_V$ to $P_0$ characterizes the thermal efficiency of the evaporation process. With water-cooled crucibles the efficiency is only a few percent. But the high-power coating of large areas becomes profitable only at an adequate thermal efficiency. When, for instance, evaporation takes place from ceramic crucibles at a high temperature of the vapor-emitting surface, the thermal efficiency may approach 50% [20].

## 2 EVAPORATION

### 2.1 Principle of Electron Beam Evaporation

Evaporation (i.e., PVD) is a vacuum coating process in which a directed vapor stream propagates from the evaporator to the substrate. Since generation and guidance of the beam must also take place in a vacuum, evaporation generally requires no additional expenditure for vacuum equipment. Figure 4 shows the principle of EB evaporation. In effect, a plant for EB evaporation consists of a work chamber with a vacuum pumping system, a crucible for the evaporant, an electron gun, a shutter, and a substrate with

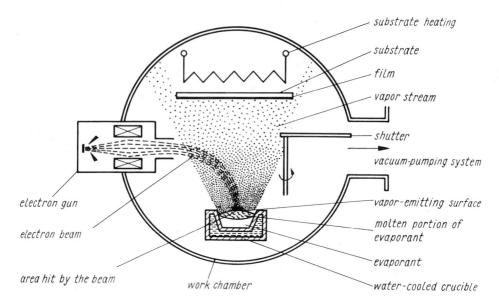

**FIGURE 4**  Electron beam evaporation equipment (schematic).

its fixtures and heating appliance. In contrast to conventional heating modes, the evaporant is heated by the beam that impinges
directly onto its surface, the greatest portion of the kinetic energy
in the beam being converted to heat. The surface is therefore
brought to such a high temperature that it becomes the source
of a vapor stream. The substrate to be coated is arranged in this
vapor stream, and part of the vapor condenses on it in the form
of a thin film.

Direct heating of the evaporant with high energy density of
the electron beam therefore allows the evaporation of materials from
water-cooled crucibles, an application often found in EB evaporation.
Water-cooled crucibles are needed for evaporating reactive and,
in particular, reactive refractory materials. Moreover, evaporation
from cooled crucibles allows the production of high-purity films
because reactions with the crucible walls are avoided almost completely; that is, crucible materials or their reaction products are
practically excluded from evaporation. Heat-insulating crucible inserts
can reduce the heat losses at the crucible so as to obtain extremely
high evaporation rates as a consequence of enhanced energy utilization.

Measuring and monitoring equipment complements the evaporator arrangement. This instrumentation serves mainly for the control of the beam parameters during the evaporation process. Beam control with respect to time and location further influences the energy flow to the evaporant, the evaporation rate, and the vapor stream distribution. This is another favorable characteristic of industrial EB evaporation.

An important prerequisite to the introduction of EB evaporation was the development of industrial coating units, so evaporators with transverse guns [21] and evaporators with axial guns and 90° beam deflection [22] were soon highly developed. A significant step toward industrial coaters was the creation of full-fledged devices for high-power EB evaporation [23,24,25]. Evaporators with guns of 5 to 600 kW are currently being used, which yield evaporation rates on the order of 100 g·h$^{-1}$ to 100 kg·h$^{-1}$. These performance parameters have opened up a wide range of industrial applications.

## 2.2 Interactions Between Residual Gas, Vapor, and Electrons

In the work chamber an adequately low pressure must be maintained to ensure that both the vapor stream and beam propagation take place unimpeded. This condition exists when the particle density of the residual gas is kept low enough so that the effects of collisions of vapor particles and electrons with gas particles can be neglected. In the region of the evaporator the high vapor density produces interactions between the electron flow and the vapor stream so that collisions cause the electrons to depart from their initial energy and trajectory. In addition, this process is accompanied by the excitation and ionization of vapor and residual gas, respectively.

From the kinetic theory of gases, one obtains the cross sections and the mean free path for collisions among gas particles and between electrons and residual gas or vapor, respectively [26]. The mean free path is the path within which the probability of collision of the particle under consideration is 1-1/e. For a residual gas pressure of 10$^{-2}$ Pa the mean free path of residual gas particles between two collisions is approximately 500 mm. Under these conditions the effect of the gas on both the vapor stream and electron flow can be neglected. At a gas pressure of 10$^{-1}$ to 1 Pa as used for reactive or ion-assisted coating, however, the interactions with the gas are significant.

Usually the vapor pressure above the vapor-emitting surface is higher than 10$^{-1}$ Pa, so the interactions between electrons and the vapor cannot be disregarded [27]. At vapor pressures of 1 Pa and 10 Pa the mean free path of the electrons, under the simplifying

assumption of the same mean free path in gas and vapor, would
be as low as 30 mm and 3 mm, respectively. But since the cross
section for collisions between electrons and vapor particles reduces
with the square root of electron energy [28,29], the evaporation
under high vapor pressures calls for higher acceleration voltages.

The number of collisions per unit volume is decisive in the
interactions between vapor particles or electrons and the gas. The
interactions of these collision partners on the substrate and wall
surfaces essentially depend on the number of collisions per unit
of surface area. At a pressure below $10^{-3}$ Pa the number of collisions
with the wall is greater than the number of collisions in the volume
[29]. But as far as the effect of the gas on condensation is concerned, it is not the number of collisions with the wall ($\nu_G$) itself
but the ratio of the collision rates of gas particles to those of the
vapor particles on the substrate that counts.

With the aid of the kinetic theory of gases it is possible to
determine the ratio $\nu_G/\nu_D$ of gas particle and vapor particle collision
rates on the substrate surface. Assuming that the gas consists of
air at 293 K, this yields

$$\frac{\nu_G}{\nu_D} \approx 55.5 \, \frac{M_D \cdot p_G}{\rho \cdot a_K} \tag{1}$$

where $\nu_G$ (in $cm^{-2} \cdot s^{-1}$) is the specific collision rate of the gas
particles, that is, the number of gas particles (air) that impinge
on a unit area per unit time; $\nu_D$ (in $cm^{-2} \cdot s^{-1}$) is the specific collision
rate of the vapor particles, that is, the number of vapor particles
that impinge on a unit area per unit time; $M_D$ is the atomic or
molecular weight of the vapor particles; $\rho$ (in $g \cdot cm^{-3}$) is the density
of the evaporant; $p_G$ (in Pa) is the gas pressure in the work chamber
during vacuum deposition; and $a_K$ (in $\cdot nm \, s^{-1}$) is the condensation
rate, given here as the growth rate of the film. In the case of
aluminum, for instance, Eq. (1) yields

$$\frac{\nu_G}{\nu_D} \approx 5.6 \cdot 10^2 \, \frac{p_G}{a_K} \tag{2}$$

The condensation rate $a_K$ for the evaporation of aluminum is
often from 1 to 100 $nm \cdot s^{-1}$. With a pressure of $10^{-2}$ Pa the ratio
of collision rates is then

$$\frac{\nu_G}{\nu_D} \approx 10 - 10^{-1}$$

From this example it can be seen that the number of collisions $\nu_G$ of residual gas particles may approach that of the vapor particles $\nu_D$ even at a relatively low residual gas pressure. But the influence of the gas on the film properties during vapor condensation cannot be assessed by means of the ratio of collision numbers alone. According to the kind of gas, substrate, and vapor, and such condensation conditions as the temperature and rate, the interaction at the surface results in binding states of different strength that may range from physisorption to chemisorption. Aside from the ratio of collision rates, it is the nature of the partners that determines the probability of condensation of vapor and gas particles [29]. For chemically inactive gas components the probability of condensation is often so slim that during condensation of the vapor the number of collisions $\nu_G$ and $\nu_D$ may reach the same order of magnitude without gas incorporation in the film. The probability of condensation found for such gas components as oxygen, water vapor, or vapors of the working fluid in the pumps, however, is by no means small with respect to unity; so in these cases the ratio of collision rates must be very low in order to avoid any effect of such gases. For this it is necessary to strongly reduce the residual gas pressure or to have a very high vapor condensation rate. The most significant measure for obtaining a drop in residual gas pressure, particularly the partial pressure of contaminants, is to select a suitable vacuum pumping system and a well-matched evaporator design.

Keeping down the influence of residual gas by using a high condensation rate has led to rates of up to 10 $\mu$m$\cdot$s$^{-1}$ as far as the evaporation of aluminum is concerned. With these extremely high condensation rates and a gas pressure of only $10^{-1}$ Pa, the ratio of collision rates, $\nu_G/\nu_D$, is still lower than $10^{-2}$. So the possibility of yielding high rates by EB evaporation is a significant prerequisite to carry out coating processes without detrimental effects of the residual gas and at rather limited vacuum requirements. The cost of vacuum pumping systems can be kept low, and there are a variety of particular coating jobs that can be solved in a profitable manner by EB evaporation alone.

## 3 ON THE EVAPORATION OF VARIOUS MATERIALS

### 3.1 Elements

For vaporization of a substance in a high vacuum, the specific evaporation rate—that is, the amount evaporated per unit time per unit area—according to Langmuir [30] is

$$a_{v1} = \alpha(4.4 \times 10^{-4})p_S \left( \frac{M_D}{T} \right)^{1/2} \tag{3}$$

where $a_{V1}$ (in $g \cdot cm^{-2} \cdot s^{-1}$) is the specific evaporation rate, $\alpha$ is
the evaporation coefficient (for idealized evaporation $\alpha = 1$), $p_S$
(in Pa) is the saturated vapor pressure at a temperature T, $M_D$ is
the molecular weight of the evaporant, and T (in K) is the absolute
temperature of the evaporant.

Equation (3) holds under the condition that, once vaporized,
none of the particles will be backscattered on the vapor-emitting
surface by the gas or a vapor cloud above the evaporator; otherwise
it is the specific evaporation rate reduced by a factor $\tau$, that is,
$a_{V2}$, that is being measured. Hence,

$$a_{V2} = \tau a_{V1} \tag{4}$$

where $\tau$ is the transmission coefficient which, depending on the
rate $a_{V1}$ and the gas pressure $p_G$, ranges from zero to unity
(Fig. 5). This reveals that a gas pressure of 1 Pa already exerts
a noticeable effect on the specific evaporation rate $a_{V2}$.

To calculate the evaporation rates, one must use empirically
determined values of the vapor pressure [19]. The saturated vapor
pressure is related to the temperature and approximately

$$p_S = K_1 \exp\left(\frac{-K_2}{T}\right) \tag{5}$$

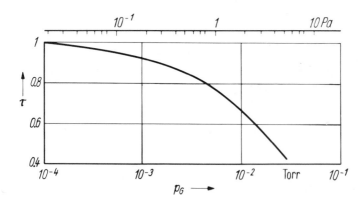

FIGURE 5 Transmission coefficient $\tau$ as a function of the gas pressure $p_G$ for the evaporation of copper. Residual gas: air, $p_R < 10^{-2}$
Pa; admitted gas: argon; vapor pressure above melt $p_D = 1 \cdots 3$ Pa;
$a_{V1} = 3.2 \times 10^{-4}$ $g \cdot cm^{-2} \cdot s^{-1}$ at $p_R < 10^{-2}$ Pa; $a_{V1}$ and $a_{V2}$ determined
by difference weighing; curve averaged over measured values.
(From Ref. 7)

where $K_1$ and $K_2$ are matter constants. With aluminum, for instance, the saturated vapor pressure rises from 1 to 10 Pa when the temperature is increased from about 1400 to 1500 K. Thus the range of saturated vapor pressures for aluminum specified for industrial use is attained only at temperatures of 500-600 K above the melting point. Tungsten, on the other hand, has a similar saturated vapor pressure at a temperature of 3600-3900 K, that is, at temperatures just above the melting point. Substituting Eq. (5) in Eq. (3) gives

$$a_{V1} = \alpha(4.4 \times 10^{-4})K_1 \left(\frac{M_D}{T}\right)^{1/2} \exp\left(\frac{-K_2}{T}\right) \qquad (6)$$

Just as the saturated vapor pressure depends on the temperature, the evaporation rate also increases almost exponentially with the evaporation temperature (Fig. 6).

Industrial coating calls for evaporation rates from $10^{-5}$ to $10^{-2}$ g·cm$^{-2}$·s$^{-1}$. From Fig. 6 it is evident that, in general, evaporation must take place from the molten state in order to obtain the required rates. Only in some special cases can evaporation from the solid state, that is, by sublimation, also be employed. Contaminants such as oxides and carbides—the densities of which are lower than that of the evaporant—rise to the vapor-emitting surface during melting, so they will partly cover this surface. The evaporation rate then reduces accordingly. But trouble of this kind does not become important where the vapor pressure of contaminants is high, where they are subject to thermal decomposition, or where the vapor is able to diffuse through the contaminating layer. The high surface temperature caused by direct EB heating stimulates the thermal decomposition of annoying surface layers on the melt. Using high-purity evaporants and evaporation from water-cooled crucibles suppresses the effect of contamination on the evaporation rate.

## 3.2 Alloys

If alloys are to be deposited, a uniform composition of the film must be obtained over the total substrate surface and film thickness. In effect, two basic principles are used for the deposition of alloys: deposition from single EB evaporation sources and deposition from multiple EB evaporation sources (Fig. 7) [3].

For multiple EB evaporation sources [22,31-34] the constituents are separately evaporated from several crucibles, the number of which corresponds to that of the alloying elements, and jointly condensed on the substrate. With single EB evaporation sources a vapor stream of the required composition is produced and condensed. One variant of single-source evaporation is partial evaporation

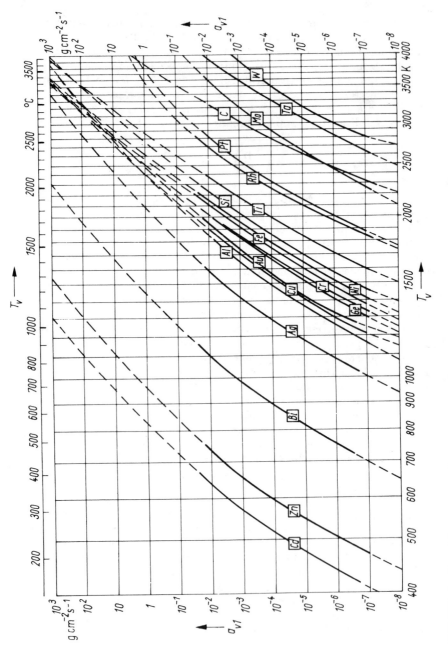

FIGURE 6  Specific evaporation rate $a_{v1}$ of some elements in the 400–4000 K range. (From Refs. 3, 19)

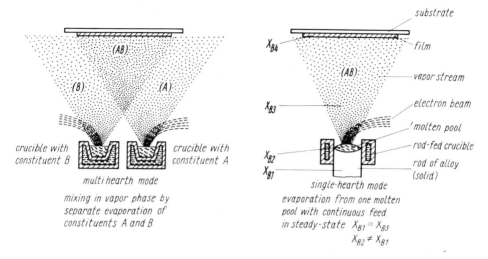

FIGURE 7 Principles of alloy deposition with defined composition in the film (example of a binary alloy AB, $X_{Bi}$ ... portions of constituent B in Wt.%).

from a large molten stock in compliance with specifically timed regimes [35]. A more significant version, however, is alloy deposition from a continuously fed single crucible where the composition of the alloyed evaporant complies, with the requirements placed on the film to be deposited [22,36].

*Multiple-Source Evaporation*

Codeposition from several crucibles shall now be explained by using as an example a binary alloy AB, that is, an alloy of constituents A and B. Separate vapor streams with evaporation rates as given by Eqs. (3) or (6); that is, the rates $a_{vA} = a_{vA}(M_A, T_A)$ and $a_{vB} = a_{vB}(M_B, T_B)$, respectively, are obtained above both crucibles. $M_A$ and $M_B$ are the molecular weights of the constituents A and B, whereas $T_A$ and $T_B$ denote the temperatures of the corresponding vapor-emitting surfaces. When the crucibles are separated by a distance l, which is short compared with the distance $h_v$ between the substrate and the crucibles, one obtains an extended range where the vapor stream contains both alloying constituents. Owing to the directional dependence of the vapor streams, however, adequate alloying constancy can be obtained only within a restricted substrate area. The influence of the geometric array on the uniform composition of the film depends on the ratio $l/h_v$.

When the condensation coefficient of both constituents approaches unity (i.e., when the amount $X_{B_4}$ of constituent B in the film is equal to the amount $X_{B_3}$ in the vapor), the relationship between the evaporation parameters and the portion $X_{B_3}$ becomes

$$\frac{a_{vB}(M_B, T_B)F_B}{a_{vA}(M_A, T_A)F_A} = \frac{X_{B_3}}{100 - X_{B_3}} \tag{7}$$

The amounts of the constituents are given in weight percent, which yields $X_{B_3} + X_{A_3} = 100$. The symbols $F_A$ and $F_B$ denote the vapor-emitting surfaces. Let us assume that the specific evaporation rates are uniform over all evaporator surfaces, that $1 \ll h_v$, and that $F_B/F_A$ is constant; then a defined portion of the constituent $X_{B_4}$ in the film can be adjusted by means of the temperatures $T_A$ and $T_B$ of both crucibles if codeposition takes place by multiple-source evaporation.

Since the evaporation rate shows a pronounced dependence on the temperature, the accuracy of the alloy composition is limited. From Eq. (7) and the values plotted in Fig. 6 the fluctuation in the amount of one constituent, for example, nickel in a Fe-Ni film ($T_{Ni} = 1900$ K), can be determined as follows:

$$\frac{\Delta X_{Ni_3}}{X_{Ni_3}} \approx 20 \frac{\Delta T_{Ni}}{T_{Ni}}$$

For a temperature fluctuation of only 10 K the amount of nickel in the vapor already changes by 10%. Thus a highly constant evaporator temperature is a necessary condition for obtaining uniform evaporation rates, and represents a basic requirement for producing films of an adequate alloy constancy when using coevaporation from several crucibles. Multiple-source evaporation allows one to obtain an alloy constancy in a range of some percent.

Such a codeposition can be performed with the aid of various EB evaporators. Another possibility is to use the beam of one gun to heat several crucibles.

Multiple-source evaporation is employed in the manufacture of alloy films whenever the evaporant cannot, or only with difficulty, be produced with the required composition, so single-source evaporation proves to be impossible. Mixing in the vapor phase is practical in cases where the vapor pressures of the constituents are vastly different, for example, differing by four or more orders of magnitude. In addition, its use is indicated if only very small portions of a constituent are to be added to an alloy.

## Single-Source Evaporation

The law of partial vapor pressures and the Langmuir equation [see Eq. (3)] with due consideration of the thermodynamic properties of an alloy [37] are the fundamentals for determining the composition of alloy deposits applied by means of single-source evaporation. In that manner the composition in the film can be calculated from the composition in the melt [37,38].

First, let us consider the amount of a constituent contained in the melt and in the deposit. If, for instance, a binary alloy AB with a portion $X_{B2}$ of constituent B in the melt is evaporated, the correlation of the portion $X_{B4}$ in the deposit to the portion $X_{B2}$ must be determined. As before, let us assume that $X_{B4} = X_{B3}$; that is, with both constituents the condensation coefficient shall approach unity so that the composition of the film is the same as the composition of the vapor.

When an alloy is evaporating, the vapor pressure $p_{SAB}$ above the melt consists of the partial vapor pressures

$$p_{SAB} = p_{SA}^* + p_{SB}^* \tag{8}$$

where $p_{SA}^*$ and $p_{SB}^*$ are the vapor pressures of the constituents above the melt. Then the vapor pressure $p_{Si}^*$ of a constituent and its saturated vapor pressure $p_{Si}$ are related by

$$p_{Si}^* = \gamma_i N_{i2} p_{Si} \tag{9}$$

where $N_{i2}$ is the molar concentration of constituent i in the melt and $\gamma_i$ is the activity coefficient of the said constituent [39]. The values of Eq. (5) refer to $p_{Si}$, and the ratio of the evaporation rates above the melt therefore becomes

$$\frac{a_{vB}}{a_{vA}} = \frac{\gamma_B}{\gamma_A} \frac{N_{B2}}{N_{A2}} \frac{p_{SB}}{p_{SA}} \left( \frac{M_B}{M_A} \right)^{1/2} \tag{10}$$

In due time, the melt generally shows a depletion in the more volatile constituent, that is, the constituent having a higher vapor pressure at the same temperature. Therefore the composition of the deposit generally does not coincide with that of the melt, since it changes with time [37,7].

For constancy of the composition of the film needed for industrial applications, coating can take place only after the steady-state condition has been attained. This state requires that the amount of material fed to the melt per unit time be precisely the same as that evaporated per unit time. Moreover, the composition of the

feed stock must correspond accurately to that of the film to be
deposited. It is attained when the more volatile constituent in the
melt has been depleted to such an extent that its composition satisfies
the equilibrium condition for evaporating at the vapor composition
required. The time needed to establish the steady-state condition,
that is, the composition required in the melting bath, is called
transient time. It depends mainly on the properties of the alloy
constituent, the volume of the melting bath, the vapor-emitting
surface, and the temperature of the latter [37,7].

But transient times are downtimes in the evaporation process
and should be reduced to a minimum. Short transient times can
be obtained by using a high evaporation temperature and, therefore,
a high vapor pressure. When evaporation takes place from crucibles
having a volume $V_2$ that is great compared with the vapor-emitting
surface F, that is, one has high values of the ratio $V_2/F$, the
transient time is rather long.

To obtain short transient times despite large ratios of $V_2/F$,
one uses an initial charge in the crucible—a charge having a volume
of the melting bath and an alloy composition that corresponds to
that of the melt in the steady-state condition. The alloy composition
of the feed material must be the same as that of the deposited film
and therefore differs from the composition of the initial charge.

Considerations on the transient time further contribute to estimating the influence of contaminations during evaporation. For less
volatile contaminants, that is, contaminants with a low vapor pressure,
the transient time is long and, in most cases, much larger than
the actual coating time. During evaporation the portion of contaminants in the film will therefore be lower than in the charge. But
this advantageous effect is counteracted by an enrichment of less
volatile contaminants in the melt which, in turn, reduces the evaporation rate.

During the time needed to obtain steady-state conditions, the
total evaporation rate and the evaporation rates of the participating
constituents also show some characteristic changes (see Fig. 8).
From these curves it can be seen that the initial slope of the evapora-
stion rate is very steep during the heat-up process of the evaporant.
The very high rate of the more volatile constituents encountered
at the beginning of the evaporation results in a high total evaporation
rate. At the very beginning the overall rate will be higher than
that obtained under steady-state conditions. Up to the end of the
transient time interval the evaporation rate of the more volatile
constituents shows a continuous decrease, whereas that of the less
volatile constituents increases.

An example of evaporating a binary alloy from a single crucible
is NiCr, which is frequently used in the manufacture of resistive

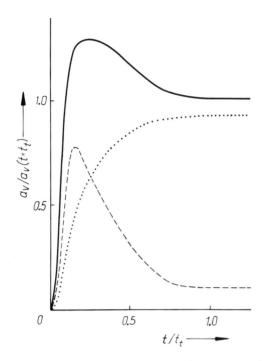

FIGURE 8   Relative evaporation rate $a_V/a_V(t=t_t)$ of a binary alloy (——), the more volatile constituent (----), and the less volatile constituent (....), where $t_t$ = transient time.

films and corrosion-protective layers. When an alloy with a composition of 80/20 has to be evaporated at a temperature of about 2000 K the chromium content in the melt must be reduced to some percents during the transient time because the vapor pressure of chromium at $T_V$ = 2000 K is 100 times as high. For a crucible with a ratio of volume to vapor-emitting surface of 0.7 cm, the transient time is about 50 min [38]. Single-source evaporation from a rod-fed crucible yields a constancy of the alloy composition in the deposit of about ±2% over several hours [22].

Single-source evaporation is also used for the manufacture of Fe-Cr and Cu-Al films. Moreover, it is suitable for depositing stainless steel ternary alloys [40,41]. Because evaporation works like a refining process with respect to some particular contaminants, this may further yield films of higher purity and therefore enhanced properties.

But if materials exhibit very different vapor pressures for the individual constituents, the use of evaporation of alloys from

a continuously fed single crucible becomes somewhat restricted.
Pronounced differences in the vapor pressures result in extensive
critical transient times. Instabilities in the composition of the film
may also be due to time-dependent changes in the beam power
supplied as well as to variations of heat losses in the crucible or
of the temperature gradients over the evaporator surface [42].
For instance, a change in the volume of the melting bath of only
$10^{-3}$ already changes the percentage of constituent B in the film
by 10% if the ratio of both concentrations differs by 100. Hence,
alloy deposition by evaporation from a continuously fed melting
bath reaches its limits for materials that differ widely with respect
to their vapor pressure [43].

## 3.3 Compounds

Most compounds dissociate during thermal evaporation either completely or in part. So thin films with stoichiometric composition
cannot be produced from these compounds by plain evaporation.
But there are a series of compounds, such as chlorides, sulfides,
selenides, or tellurides [1], as well as a limited number of oxides
and polymers for which thermal evaporation is possible, owing to
weak dissociation or recombination of the constituents during condensation.

 However, it is not only thermal dissociation but also reactions
with the crucible material and the resulting changes in the composition
of the films obtained that restrict the vaporizability of compounds.
In part these limitations can be overcome by using EB evaporation.
Evaporation from water-cooled crucibles eliminates any possible
reaction with the crucible wall and thus renders some reactive com-
pounts vaporizable. Since EB heating produces a high temperature
at the vapor-emitting surface, it is also possible to evaporate barely
vaporizable and refractory compounds such as aluminum oxide [44,46],
silicon dioxide [45], and various glasses [45] and carbides [31].

 But the use of electron beams may also entail some drawbacks
with respect to the evaporation of compounds, for example, enhanced
dissociation due to the increased surface temperature of the evaporant
and collision dissociation of the already vaporized material [46].

 In order to produce films of some particular compounds with
an almost stoichiometric composition in spite of thermal dissociation
and collision dissociation, various methods are used for thermal
evaporation and, in particular, for EB evaporation.

 The most significant method is reactive evaporation, that is,
the formation of the chemical compound during evaporation from
a vaporized constituent and a gaseous constituent admitted in the
work chamber [47]. Generally, it is the aim of reactive evaporation

to stoichiometrically deposit the compound. It first established itself in the production of oxides by evaporation of metals or suboxides [48], but is now being used also for the deposition of films consisting of compounds such as carbides and nitrides [49].

Another possible way of producing compound coatings is by the evaporation of, say, two constituents that react during condensation on the substrate in a defined ratio. Among other things, it is possible to produce thin $Nb_3Sn$ films by separate EB evaporation of niobium and tin [50].

A variant of this method is to deposit the constituents in the required ratio and establish the chemical compound by subsequent annealing. In this way it is possible to produce Ti-C films, just to cite an example [51].

For reactive evaporation the pressure or partial pressure of the gaseous component in the work chamber required for the reaction (generally $10^{-2}$ to $10^{-1}$ Pa) is adjusted by means of a gas admittance valve. If the pressure of the reaction gas is too high, there is the possibility of gaseous inclusions and energy losses of the vaporized constituent due to frequent collisions between the gas and the vapor, which has a negative effect on the condensation of the compound [48].

In the above-mentioned pressure range the collisions in the volume and therefore the volume reactions can be neglected. This means that the reaction must take place between the condensing atoms of the vapor and the atoms of the reaction gas absorbed on the substrate surface [48]. The portion of the stoichiometrically deposited substance depends on such thermodynamic parameters as temperature, activation energy, and condensation coefficients, and on the ratio of collision numbers of the constituents [29]. The ratio of collision numbers is given by the partial pressure of the reaction gas and the condensation rate of the vapor. For a reactive evaporation it must be generally high with respect to unity in order to obtain stoichiometric compounds [48]. This requirement results in a general restriction of conventional reactive evaporation, because the increase in reaction gas pressure is accompanied with a drastic reduction in the energy of the collision partners and therefore in the reaction probability.

One method of increasing the reaction probability for a particular ratio of collision numbers—that is, without an increase in the pressure of the reaction gas—is called activated reactive evaporation (ARE). To this end, either the reaction gas or the vapor (or both) must be activated. This can be accomplished by microwaves, electrons, or an electric gas discharge (that is, in a plasma) [49]. During ARE a plasma is produced between the evaporant and the substrate which, aside from vapor and reaction gas particles, also contains ions, excited particles, electrons, and sometimes even the dissociation products of the involved constituents. The collision cross sections of the charged or excited particles differ widely (up to an order of magnitude when compared with neutral particles) due to the Coulomb fields. This increases the collision probability of the reaction partners. It is known from plasma chemistry that the reaction probability is much higher for the ionized or excited state. Because the substrate is also hit by the charged particles, the reaction probability increases at the substrate as well. For instance, ARE of titanium, with $C_2H_2$ as the reaction gas, results in a joint condensation of Ti and C in the required stoichiometric ratio even at a high condensation rate of 0.2 $\mu m \cdot s^{-1}$ [49,51]. The formation of TiC then occurs according to the following reaction:

$$2 \text{ Ti (vapor)} + C_2H_2 \text{ (gas)} \rightarrow 2 \text{ TiC (film)} + H_2 \text{ (gas)}$$

At substrate temperatures around 1000°C a reaction did occur that resulted in a Ti-C compound [52].


## 4   VAPOR PROPAGATION

### 4.1   Small-Area Evaporators

The vapor stream emerging from an evaporator is characterized by the vapor stream density distribution $\Phi(\alpha)$. With the aid of this function (also called the evaporator characteristic) and a given arrangement of the substrates relative to the evaporator, the prospective film thickness distribution on the substrate can be calculated as outlined in Refs. 1, 3, and 53.

The simplest case of a point source [1,54] with a constant vapor stream density distribution $\Phi(\alpha)$ is of practically no importance in the field of EB evaporation. Only the special case of crucible-free evaporation from a molten drop (see Section 5.2) can be characterized by such a distribution function. The following is a more suitable approach.

The vapor stream emerging from a plane surface element dF is described by the formula [1]

$$\Phi(\alpha) = \Phi_0 \cos \alpha \qquad\qquad (11)$$

where $\Phi(\alpha)$ is the vapor stream density in a direction describing an angle $\alpha$ from the normal to the vapor-emitting surface, and $\Phi_0$ is the vapor stream density for $\alpha = 0$.

It is characteristic of small-area evaporators that the dimensions of the vapor-emitting surface are small compared with the distance from the substrate. Equation (11) approximately describes the vapor stream distribution of such a small-area evaporator. For larger evaporator surfaces or extended source distributions (see Section 4.2) the vapor stream density is determined by a summation of either the evaporator characteristics of the individual small-area evaporators or the surface elements. EB evaporators with crucibles of diameter $D_V$ and distance $h_V$ away from the substrate are to be considered as small-area evaporators when $D_V \ll h_V$.

But if evaporation takes place with the aid of an EB evaporator that is classified according to the above definition, as among the small-area evaporators, a series of effects is likely to change the vapor stream distributions; this is the reason why the simple cosine distribution model generally proves to be inadequate in practice. Figure 9 illustrates the effects that cause the evaporator characteristic of the EB evaporator to depart from a simple cosine distribution.

If the molten evaporant has a high surface tension, so that the crucible walls are not wetted, the plane vapor-emitting surface may become convex. The beam may produce a very high vapor pressure above the melt so that the vapor-emitting surface becomes

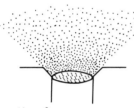

*formation of a convex vapor-emitting surface due to the surface tension of the evaporant*

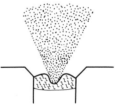

*formation of a concave vapor-emitting surface due to a local increase in vapor pressure*

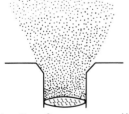

*obstruction of vapor propagation by the crucible wall due to inadequate feeding of the crucible*

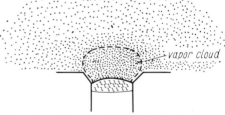

*vapor cloud*

*formation of a vapor cloud which, instead of the vapor-emitting surface, acts as a virtual source of the vapor stream*

FIGURE 9  Effects of the vapor-emitting surface of actual small-area evaporators on the distribution of vapor stream density.

deformed (see Section 6). A concave vapor-emitting surface results in a more pronounced drop in the vapor stream density with the angle $\alpha$ than that predicted by Eq. (11). Vapor sinks in the neighborhood of the vapor-emitting surface will act similarly: if the propagation of the vapor stream is impeded by cold sections of the crucible wall, its effect on the vapor stream density distribution is called the funnel effect.

High evaporation rates may result in the creation of a vapor cloud above the vapor-emitting surface (see Section 4.3). Collisions between the vapor particles prevent a portion of the particles from passing from the vapor-emitting surface directly to the substrate, but their virtual source is located somewhere in the vapor cloud. If the range of high vapor pressure is restricted to the surroundings of the vapor-emitting surface, the propagation of the vapor particles outside the cloud is straight-lined, and calculations have to be based on a corrected evaporator characteristic [55,56].

One approach to describing the vapor stream density distribution of real small-area evaporators is via a cosine function of higher order:

$$\Phi(\alpha) = \Phi_0 \cos^n \alpha \tag{12}$$

where $n > 1$. It has been shown that the description of the vapor stream density distribution of EB evaporators according to Eq. (12) is fully adequate in an angular range of up to about 30° if the evaporation rate is not too high. With growing evaporation rates— that is, with increasing deformation of the melting bath surface— a more pronounced directional dependence is to be expected. In other words, the exponent $n$ in Eq. (12) depends on the evaporation rate. Whereas values between 2 and 3 are given for the exponent $n$ with condensation rates of about 5 nm·s$^{-1}$ [22,55], rates of about 100 nm·s$^{-1}$ may result in values of $n = 6$ [55,57]. The effect of a vapor cloud can be described by an additive isotopic portion to Eq. (12). For a 12-kW evaporator [55] this portion has been determined to 14% $\Phi_0$.

Now the film thickness distribution on any desired substrate arrangement will be calculated from the evaporator characteristic by means of an example [1]. One wants to know the film thickness distribution on a plane substrate or substrate arrangement parallel to the surface of a small-area evaporator. When the vapor emitted by such an evaporator is condensed on a spherical collector surface, one obtains

$$d(\alpha) \sim \Phi(\alpha) \tag{13}$$

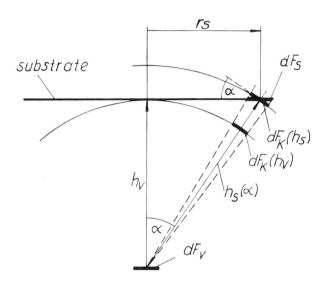

FIGURE 10   Plot for determining the film thickness $d_S$ on a plane
substrate from the vapor stream density distribution: $dF_V$ = vapor-
emitting surface element; $dF_K(h_V)$ = surface element on a sphere
around the evaporator having radius $h_V$; $dF_K(h_S)$ = surface element
on a sphere having a radius $h_S$; $dF_S$ = surface element in the substrate plane. Geometric relations: $\cos \alpha = h_V/h_S = dF_K(h_S)/dF_S$.
The distance formula for the vapor stream density is $\Phi_K(h_S)/\Phi_K(h_V) = h_V^2/h_S^2 = \cos^2 \alpha$.

From the relationships given in Fig. 10, the film thickness
on a plane substrate surface $d_S(\alpha)$ then becomes

$$d_S(\alpha) \sim (\cos^3 \alpha) \, \Phi(\alpha) \tag{14}$$

If Eq. (12) applies for the vapor stream density distribution $\Phi(\alpha)$,
this yields

$$d_S(\alpha) = d_0 \cos^{n+3} \alpha \tag{15}$$

where $d_0$ is the film thickness for $\alpha = 0$. With $\tan \alpha = r_S/h_V$, it
follows from Eq. (12) that

$$\frac{d_S}{d_0} = \left[ 1 + \left( \frac{r_S}{h_V} \right)^2 \right]^{-(n+3)/2} \tag{16}$$

where $h_v$ is the distance of the substrate plane to the evaporator, and $r_S$ is the distance between the normal to the evaporator and the substrate point under consideration.

Instead of computing the film thickness distribution from the vapor stream density distribution described by Eq. (12) and an exponent $n > 1$, one can also calculate it from a simple cosine distribution (i.e., $n = 1$) when a virtual vapor source is being introduced [58]. For a highly directional evaporator the virtual vapor source lies above the actual vapor-emitting surface at a distance $\Delta h$.

The example of coating a plane substrate by means of a small-area evaporator with a $\cos^n \alpha$ characteristic of the vapor stream density shall further be used to discuss the relative change $\Delta d_S/d_S$ in the film thickness on the substrate and the utilization factor $\eta_m$ of the evaporated material [3].

On a plane substrate the relative change in film thickness according to Eq. (15) is

$$\frac{\Delta d_S}{d_S} = (n + 3)\tan \alpha \, \Delta\alpha \qquad (17)$$

Thus the relative change in film thickness increases as the exponent $n$ increases. Because $n$ increases with the evaporation rate of the EB evaporator, the uniformity of the film thickness on a given substrate arrangement deteriorates with higher evaporation rates.

The ratio of the material deposited on the substrate to the total weight of evaporated material is called the utilization factor $\eta_m$ of the evaporated material. When the substrates are arranged within an angle $\alpha$ from the normal of the vapor source, the utilization factor becomes

$$\eta_m = 1 - \cos^{n+1} \alpha \approx \frac{n+1}{2} \alpha^2, \qquad \text{if } \alpha \ll 1 \qquad (18)$$

With a single small-area evaporator a film thickness of adequate uniformity over extended substrate surfaces is not possible, and $\eta_m$ of the evaporated material is small. Based on theoretical and experimental investigations, various principles have been postulated in an attempt to achieve both uniform film thickness and high material utilization, even on extended substrate arrangements [1,59,60]. The most significant [3] are the following:

Distribution of the vapor source intensity over an extended vapor-emitting surface
Arrangement of the substrates over areas of equal vapor stream density

Matching the residence time of the substrates in the vapor stream
  to the vapor stream density distribution by relative motion
Correction of the film thickness distribution by scattering at increased gas pressures
Correction of the vapor stream density distribution by shutters

There are a great many substrate arrangements where, independent
of the use of EB evaporation, either one of these principles or
a combination thereof are used in.

To deposit films of uniform thickness over large areas, extended
source distributions are often used, especially if a high condensation
rate is necessary.

## 4.2 Evaporator Arrangements with Extended
## Source Distribution

One approach to the implementation of an extended source distribution
is to use a given number of small-area evaporators. By optimizing
the number, the mutual arrangement, and the evaporation rates,
one can obtain the required vapor stream density distribution,
with due consideration of the characteristics of the individual evaporators. The resulting film thickness distribution can then be determined by either summation or integration of the film thickness
contributions of the individual evaporators. Several small-area
evaporators are often arranged in a line. When the substrates are
moved normal to this line, it is possible to obtain a film thickness
of high uniformity (Fig. 11). This principle is used for the coating
of strip-type materials and substrates arranged either on plane
fixtures or a rotary cage.

A particular source distribution may also be obtained by passing
the beam over the surface of the evaporant in a programmed mode.
For constant beam power the temperature distribution and thus
the specific evaporation rate are functions of the residence time
of the beam at each point of the surface. In addition, the evaporation
rate can also be changed via power control.

When the beam is deflected over a longitudinal crucible in one
direction only, one obtains a so-called line evaporator. For a sinusoidal deflection the residence times along the beam track will differ
accordingly. At the inversion points of the beam track on the evaporant the residence time $t_w$ reaches a maximum. From Fig. 12 one
can see how the energy density distribution $e(x)$ along the beam
track can be changed by a simple variation of the beam diameter
$d_F$ and of the maximum deflection. To ensure a uniform film thickness over the whole substrate at a limited extension of the vapor-
emitting surface, one must increase the vapor stream density near

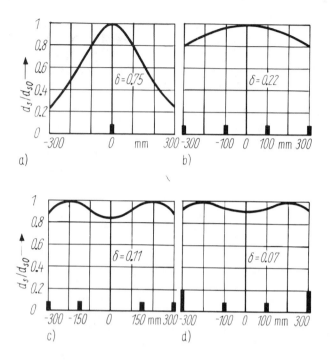

FIGURE 11    Improved film thickness distribution $d_s/d_{s0}$ for coating
of a plane substrate with several small-area evaporators arranged
in a line (calculated): (a) one small-area evaporator; (b) four small-
area evaporators (same distance between evaporators and same evaporation rate); (c) four small-area evaporators (different distance,
between evaporators and same evaporation rate); (d) four small-area
evaporators (same distance between evaporators and different evaporation rates). Distance between the evaporator and the substrate,
$h_V = 300$ mm; angular aperture of vapor stream, $2\alpha = 180°$; vapor
stream density distribution, $\Phi(\alpha) = \Phi_0 \cos \alpha$; $\delta = (d_{s,max} - d_{s,min})/d_{s0}$.

the substrate edge. Sinusoidal deflection of the beam increases the
energy density in the region of the inversion points; this facilitates
correction of the film thickness distribution to a degree that is fully
adequate for many applications. Although strong "peaking" of the vapor
stream density at the crucible rims yields a uniform film thickness on
the substrate, this reduces the utilization factor $\eta_m$ of the evaporant
also. Thus it may often be necessary to select a deflection program
that offers more than two energy density maxima, and thus more

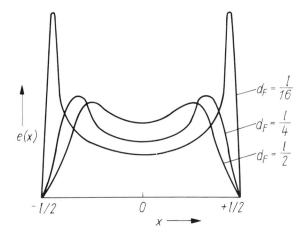

FIGURE 12 Energy density distribution e(x) on the evaporant over the deflection trace 1: sinusoidal beam deflection; parameter, beam diameter $d_F$.

than two source density maxima over the beam track (Fig. 13). In this case programmed beam deflection is obtained by combining a dynamic deflection field with a static beam-turning field (see Section 5.1). Even though electronic deflection programs with ramp-type deflection currents are readily generated, they have been abandoned. The beam has to be controlled in two coordinates to obtain an extended distribution of the vapor stream over a large-area evaporator.

### 4.3 Vapor Propagation for Short Free Path Lengths

In the previous discussion of vapor stream generation and propagation it has been assumed that, according to Section 2.2, the propagation of vapor particles is basically straight-lined. It was only because of the need to explain the departure of the vapor stream density distribution from the cosine distribution that collisions between the vapor particles had to be taken into consideration when within reach of the vapor-emitting surface. Interactions causing, for instance, the creation of a vapor cloud and a vapor pressure acting on the vapor-emitting surface, however, were restricted to the immediate vicinity of the evaporator.

For increasing vapor pressures or a high gas pressure, interactions occur within the entire space between the evaporator and the substrate. In these cases vapor propagation departs from a

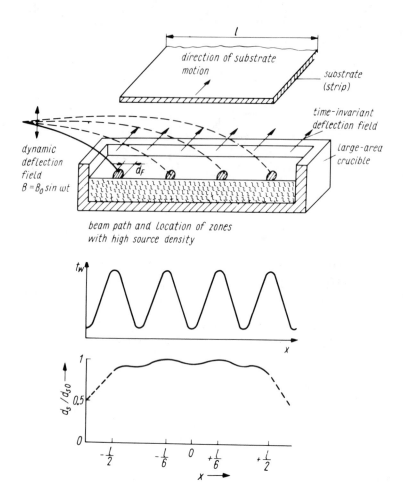

FIGURE 13  Implementation of an extended source distribution
through programmed beam deflection in the x direction on a large-
area crucible. Evaporation characteristic $\Phi = \Phi_0 \cos^3 \alpha$; strip
width, l; crucible width, (4/3) l; distance of zones with high
source density, (1/3) l; distance between the evaporator and the
substrate, (2/5) l; $t_w$ = residence time of beam; $d_s/d_{s0}$ = relative
film thickness distribution on the strip in the x direction; and
$d_F$ = beam diameter on the evaporant.

straight-lined mode, which makes the relations specified in Sections 4.1 and 4.2 practically meaningless.

It is characteristic of the vapor propagation in the region of pronounced interactions among vapor particles, and between the vapor and gas, that the mean free paths between the vapor particles, between vapor and gas particles, and between the gas particles are small compared with the distance between the evaporator and the substrate. These conditions for vapor propagation exist at a gas pressure of about 1 Pa, and if coating takes place at an extremely high evaporation rate. Because under these circumstances each vapor particle experiences several collisions along its path to the substrate, the initial direction and energy of the particles are meaningless by the time the particles impinge on the substrate. Instead of a directed vapor stream there is a vapor cloud in the entire space between the evaporator and the substrate that may also surround the latter. Figure 14 illustrates the change in the directional distribution of the vapor particles due to scattering in a vapor cloud. Whereas in a directed vapor stream the vapor particles impinging on the substrate are shown to have only minor directional differences that correspond to the diameter of the vapor-emitting surface and the distance between the evaporator and the substrate, any space point P located within the vapor cloud can now be a source of vapor particles that propagate in any direction. For extremely frequent interactions the vapor propagation can be described with the aid of the vapor pressure. For EB evaporation at about 100 kW [20] this pressure ranges from about $10^{-1}$ to about 10 Pa.

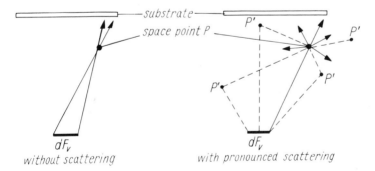

FIGURE 14 Effect of scattering on the directional distribution of the vapor particles at the space point P in the vicinity of the substrate from where the vapor particles virtually emerge: $dF_v$ = vapor-emitting surface element; P' = scatter center.

With coating under enhanced gas pressure [16,61], that is, reactive evaporation and ion plating, vapor scattering is caused by the interactions between the vapor and gas particles. The vapor stream density distribution according to Eq. (15) will change toward lower exponents and, with pronounced interactions, may even reach complete isotropy.

In the region of the vapor-emitting surface, interactions with the gas causes the vapor to backscatter. The resulting drop in the evaporation rate can be described by means of the transmission coefficient $\tau$ as outlined in Section 3.1. As a consequence of the interaction, the condensation rate drops because of the reduced evaporation rate and the additional scatter along the path to the substrate. Scattering of the vapor with the residual gas also impairs the utilization factor $\eta_m$ of the evaporant. On the other hand, however, residual gas scattering has the advantage of reducing the shadowing and penumbral effects of surface irregularities and the edges of the substrate. Therefore, pressure plating improves the coverage, especially in the case of rough surfaces.

## 5  EVAPORATOR UNITS

### 5.1  Electron Beam Evaporators

*Principles*

Electron beam evaporators consist of a gun and a crucible. In many cases such a unit is supplemented by a feeding device for the evaporant. In most designs the devices for beam generation and guidance as well as the crucible are united into one assembly.

Compiled in Fig. 15 are EB evaporators that operate on different principles. This illustration further reveals the various steps toward full-fledged engineering solutions.

In simple arrays with electron impact heating a cathode is arranged near the evaporant or crucible. The acceleration voltage across the cathode and crucible then directs the electron flow onto the evaporant. A control electrode is connected to the cathode potential and serves to bundle the electron flow to a given diameter upon impingement on the evaporant. When ring cathodes are used, the control electrode also has the shape of a concentric ring about the crucible. Because the cathode is arranged close to the evaporant, the evaporated cathode material may contaminate the coating layer. Through suitable shaping of the control electrodes, however, vaporized cathode material can be prevented from directly reaching the substrate. The arrangement of high-voltage components in the immediate vicinity of the vapor source is likely to cause flashovers and glow discharges between the cathode and the crucible. This kind of risk increases

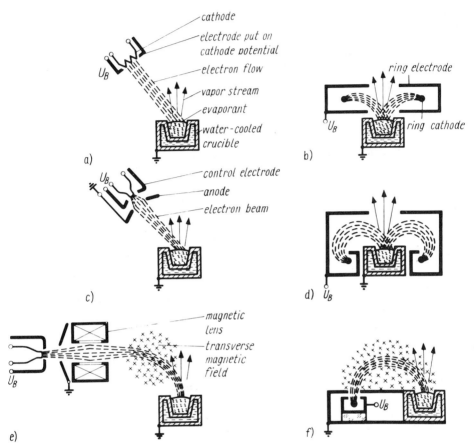

FIGURE 15 Principles of EB evaporators: (a) evaporator with electron impact heating, linear cathode, and electrostatic focusing; (b) evaporator with electron impact heating, ring cathode, and electrostatic focusing; (c) evaporator with axial gun and electrostatic telefocusing; (d) evaporator with electron impact heating, ring cathode, electrostatic focusing, and beam turning; (e) evaporator with axial gun, magnetic focusing, and magnetic beam turning by 90°; (f) evaporator with transverse gun and magnetic beam turning by 180°.

with the evaporation rate and acceleration voltage. With such ring sources the vapor pressure near the evaporant should not exceed $10^{-1}$ to $10^{-2}$ Pa. Usually the acceleration voltage is kept well below 10 kV; in most cases it is only 3-5 kV. Because of these restricted parameters the use of ring-type sources is limited to power levels of up to about 3 kW. The maximum evaporation rate depends on the evaporant, the power, and the evaporation geometry. The condensation rates attainable with such evaporators are from 1 to 10 $nm \cdot s^{-1}$ if the distance between the evaporator and the substrate is 250 mm. Hence the application of ring-type sources is somewhat restricted. Specialized shaping of the control electrode allows proper focusing and turning of the electron bundle. In this way the cathode can be completely shifted into the shadow zone of the vapor. This electrode configuration has the ability to reduce contamination and improves the dielectric strength.

When an additional anode is arranged in the space between the cathode with control electrode and the crucible, an electron beam can be formed. The advantage of this principle is that a field-free region exists between the gun and the vapor source; that is, beam generation and evaporation are separated. In the simplest case the beam is directed onto the evaporant from above, either vertically or at some angle. Telefocus guns are used to obtain small beam diameters and thus an adequate power density on the evaporant, even where the distance between the gun and the crucible is long.

Another approach is to use a magnetic lens for beam focusing. The arrangement of the gun above the crucible has the disadvantage of being located directly in the vapor stream. In this case it is not only the gun coating that is a problem, but, due to the shadow zone of the gun, the useful range within which the substrates can be arranged is restricted. These drawbacks can be avoided with a horizontal gun arrangement where the beam is turned electrostatically or, more commonly, magnetically before it hits the evaporant.

Separation of beam generation and crucible, magnetic deflection of the beam prior to impingement on the evaporant, and the use of linear cathodes have been implemented in the development of transverse evaporators. Usually the beam is deflected by an angle of not less than 180° and frequently as high as 270° [21]. The EB evaporator with axial gun and magnetic deflection, and the evaporator with transverse gun have found, owing to their advantageous properties, the widest field of industrial application.

*Evaporators with Axial Guns*

Evaporators with axial guns are implemented in the power range of 5-600 kW [20,22,62,63] and are standard for powers higher than

15 kW. Built-in versions have been developed for lower power levels. Evaporators with an axial gun, magnetic deflection, and a separate crucible proved to be sound as add-on systems. Separation of the gun and crucible allows the separate optimization of both assemblies.

Vacuum decoupling between the beam generating and evaporation chambers can be readily accomplished with add-on systems by incorporating flow resistances between the gun and the crucible, or by pumping down the gun separately. Such decoupling proves to be advantageous in reactive evaporation, pressure plating processes, and ion plating for pressures of $10^{-1}$ Pa or more. It is required whenever high-rate evaporation is to be performed with guns having power levels above 50 kW where a residual gas pressure of about $10^{-1}$ Pa can be tolerated in the process, or where a high-pressure vapor cloud will be formed in front of the gun. Figure 16 shows an axial gun used for large-area evaporation. The schematic of an evaporator with a high-power axial gun of add-on design and magnetic beam turning is shown in Fig. 17 [20,64,65,66]. The gun is mounted with a flange to the evaporation chamber. Both the crucible and the magnetic deflection system are arranged inside the chamber. Another evaporator with two axial guns and beam deflections by 180° is shown in Fig. 18. Both guns are arranged symmetrically to the crucible [67].

FIGURE 16 View of a 600-kW EB evaporator EH 600/40 with an axial gun. (Courtesy of the M. von Ardenne Institute) 40-kV acceleration voltage gun with magnetic lenses and beam deflection system.

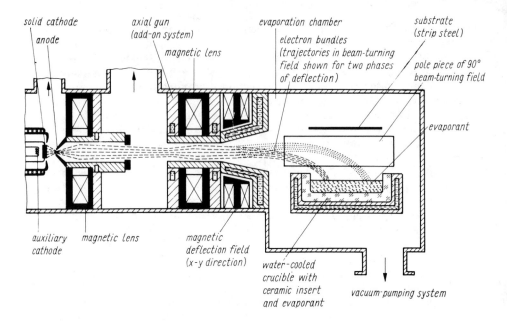

FIGURE 17   Schematic of an evaporator unit with a 250-kW and
a 600-kW axial gun, respectively. (Courtesy of the M. von Ardenne
Institute) Acceleration voltage 30 kV for the 250-kW gun and 40
kV for the 600-kW gun, respectively; gun with magnetic lenses
and beam deflection system.

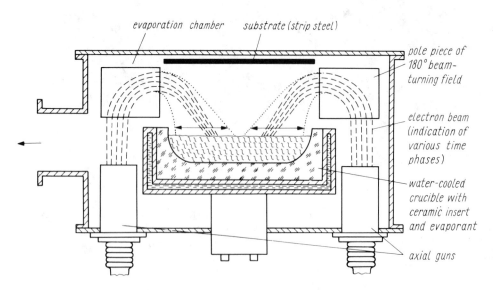

FIGURE 18   Schematic of an evaporator unit with two 50-kW axial
guns. (Courtesy of Leybold-Heraeus; from Ref. 18)

*Transverse Evaporators*

The principle of the transverse gun [21] allows one to design an evaporator unit where the devices for beam generation and guidance as well as the crucible are united in one assembly without giving up the advantages of local separation between beam generation and vapor generation. In this way, construction is less expensive, and high performance is retained.

The schematic of an evaporator with a transverse gun is shown in Fig. 19. Its beam-generating system consists of a linear, directly heated cathode, a control electrode, and an anode. The cathode is made of a helical wire. Initial beam formation is obtained by a suitable arrangement of the cathode relative to the control electrode, which is usually connected to the cathode potential, and to the anode and by the shaping of the electrodes. Similarly to the axial systems, the electric field accelerates and focuses the electrons emitted at the cathode, which, owing to the linear beam-generating system, yields a strip beam. Emerging from the beam-generating

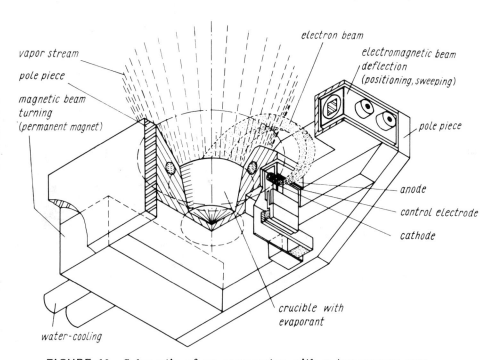

FIGURE 19  Schematic of an evaporator with a transverse gun.

system, the strip beam is further deflected by an extended transverse magnetic field and thus hits the crucible. Specialized shaping
of the pole pieces—that is, a defined inhomogeneity of the beam-
deflection field—allows one to influence the width and length of
the strip beam cross section and, thus, the power density on the
vapor-emitting surface, in compliance with the coating job to be
performed.

The angle between the beam direction upon entry into the beam-
deflecting field and the normal to the plane of the crucible is called
the angle of turn. It significantly determines the design of the
evaporator. Angles of turn from 180° to 270° are commonly used.

For a great many materials about 20% of the electrons are backscattered during evaporation, even for normal beam incidence. Hence,
backscattered electrons must be prevented from reaching the substrate so that uncontrolled and usually negative changes in the
film properties are avoided. Extending beyond both sides of the
crucible, a transverse magnetic field acts as an electron trap. In
this way scattered electrons will hit the usually water-cooled collector
plates in a predetermined manner.

In the crucible the strip beam of a transverse gun produces
an oblong focal spot, the smaller axis of which is about 1-5 mm.
The length of this spot depends essentially on the beam power
and, at 15 kW, is about 10-15 mm. For the beam power to be distributed on the evaporant, the beam should be deflected mainly
in the direction of the smaller axis of the focal spot. Frequently,
however, one also uses an additional dynamic deflection normal

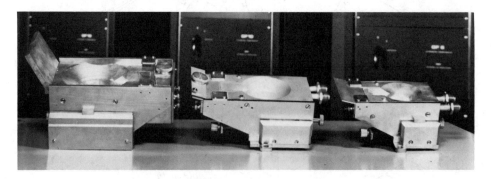

FIGURE 20   Models ES 100, ES 40, and ES 8 EB evaporators with
transverse gun; beam power 16 kW, 10 kW, and 6 kW, respectively;
acceleration voltage 10 kV; crucible volume 100 cm$^3$, 40 cm$^3$, and
8 cm$^3$, respectively; turning angle 270°. (Courtesy of VEB Hochvakuum Dresden)

to this axis; that is, the beam is wobbled. Such sweeping devices are generally engineered for deflection frequencies of about 50 Hz, or even up to 400 Hz.

Transverse evaporators of 4 to 15 kW are standard in industrial applications. Figure 20 shows the actual design of transverse evaporators of different beam power.

## 5.2 Crucible Designs

Crucibles are used to accommodate the evaporant. Material selection and the design of the crucibles depend on principal features such as EB heating, energy transfer through the surface of the evaporant, and high power density.

Even at high temperatures the crucible material must have a low vapor pressure and low susceptibility to reactions with the evaporant and the gas in the work chamber. Severe erosion and excessive evaporation of the crucible material contaminate the evaporant or deposit and limit the crucible life and profitability of evaporation.

Depending on the coating jobs to be performed, either water-cooled copper crucibles or hot crucibles (i.e., water-cooled supports with ceramic crucible inserts) are used (Fig. 21). Crucible-free evaporation also plays a certain role in various specialized applications.

Water-cooled copper crucibles have found the most widespread application. The high thermal conductivity of copper may result in extremely large temperature differences at the crucible wall-evaporant interface. Crucibles of this type may be used for the evaporation of refractory materials (such as tungsten) at temperatures of 3000-4000 K, as well as for highly reactive materials (such as titanium). The temperature gradient at the interface between the copper crucible and the evaporant is so large that one can speak of a jump in temperature. The zone where the molten evaporant contacts the crucible wall is essential for the heat transfer to the crucible. Wetting properties, surface condition of the crucible material, and particularly the formation of an oxide layer on the copper surface determine the height of the temperature jump and, therefore, the heat transfer. Among other things, the widespread use of water-cooled copper crucibles can be attributed to the fact that a variety of material can be evaporated from one and the same crucible. But there are some applications where the high heat losses represent a drawback in the use of water-cooled copper crucibles. When high power losses cannot be tolerated or when, at a given power level, the evaporation rate and, thus, the thermal efficiency have to be enhanced, crucible inserts must be employed that act as a heat barrier.

Heat losses through the walls are determined mainly by the thermal conductivity of the material used for the crucible insert.

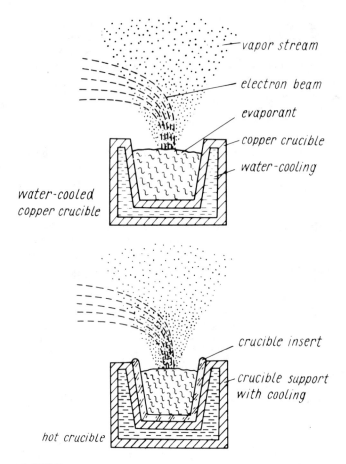

FIGURE 21   Different types of crucibles for EB evaporators.

The use of such an insert yields a more uniform temperature distribution over the molten pool and a greater depth of this pool.
But material selection depends on its chemical resistivity to the
hot evaporant as well as on its thermal conductivity. Repeated heating and cooling impose high mechanical stresses on these crucibles;
therefore a high resistance to thermal shocks is required of the
material employed for the crucible insert.

$Al_2O_3$-based ceramic crucibles are of particular importance.
Graphite, titanium, diboride, or boron nitride are often used for
small-sized crucible inserts. Their service life is about 100 h. Crucible dimensions are determined by the parameters of the evaporation

process (that is, the type and quantity of the evaporant, the
evaporation rate, and requirements with respect to energy utilization)
and by the necessary vapor stream density distribution. Guide
values for the crucible dimensions for a given evaporator power
can be derived from the following requirements: the evaporant has
to be kept in a molten state up to the crucible rim, and the power
density shall not exceed the limit where splash formation starts
to occur. With evaporators in the power ranges 5-15 kW and 15-600
kW it is common usage to employ crucible diameters of 15-75 mm
and 50-1000 mm, respectively.

With small-area evaporators the rotationally symmetrical crucible
dominates. The crucibles used in line evaporators and large-area
evaporators, on the other hand, have to be matched to the size
and shape of the substrates or substrate arrangements and are
therefore of a rectangular design.

Compiled in Fig. 22 are the characteristic crucible shapes found
in small-area evaporators.

Whenever various materials are to be evaporated with a single
EB evaporator in one cycle, it is common to employ multihearth

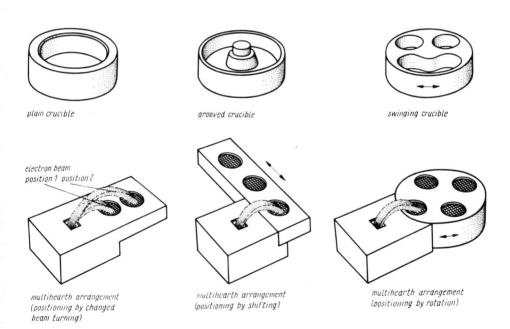

*plain crucible*     *grooved crucible*     *swinging crucible*

*electron beam position 1 position 2*

*multihearth arrangement (positioning by changed beam turning)*     *multihearth arrangement (positioning by shifting)*     *multihearth arrangement (positioning by rotation)*

FIGURE 22   Characteristic crucible shapes or arrangements of small-
area evaporators.

arrangements. Here the individual crucibles are sequentially brought to evaporation position by either rotation or longitudinal shifting. Another possibility is to use a programmed beam deflection (beam shifting device) for evaporation from several crucibles either sequentially or concurrently.

The need to completely eliminate any contamination of the evaporant by the crucible material or to vaporize evaporants of a particular initial shape has led to crucible-free evaporation. The simplest method is evaporation from a limited molten pool in an extended ingot. Evaporation from rods or plates that are to be moved relative to the beam is suitable for sublimable materials, especially if the beam is additionally swept. A well-coordinated mechanical shift of the evaporant and beam sweep almost completely avoids pit formation in the sublimable material. Crucible-free evaporation from a drop of feeding material is also possible but proves to be suitable only for very low evaporation rates.

## 5.3  Feeding Modes

When constant evaporation conditions have to be maintained for long periods, the crucibles of the evaporators must be fed during operation. This is particularly important during operation at extremely high evaporation rates or when great quantities have to be evaporated within one coating cycle. The evaporation of alloys calls for a feeding system to ensure a definite and constant vapor stream composition (see Section 3.2). This can be accomplished in a continuous mode or in batches. The simple batchwise mode requires a large crucible volume. Degassing of the predegasified feed stock can be made in the coating plant prior to evaporation and with the shutters closed.

For continuously fed crucibles only a material quantity corresponding to the evaporation rate has to be supplied per unit time. Continuous feeding places higher demands on the device used but ensures constant evaporation parameters over long operating periods. Depending on the material in question, the shape of the evaporant differs widely; that is, the evaporant can be in the form of wires, rods, or granules. Figure 23 shows some of the principles of continuous feeding.

Wire-fed crucibles are the most frequently used [68]. Adequate selection of the wire diameter and rate of feed allows one to accurately determine the quantity of material feed per unit time.

Moreover, a specialized crucible, the annual crucible, was developed to use rod-shaped evaporants [22]. Here the rod forms the bottom, and a water-cooled copper ring forms the wall of the crucible. The molten pool is then formed in a zone of the annular crucible.

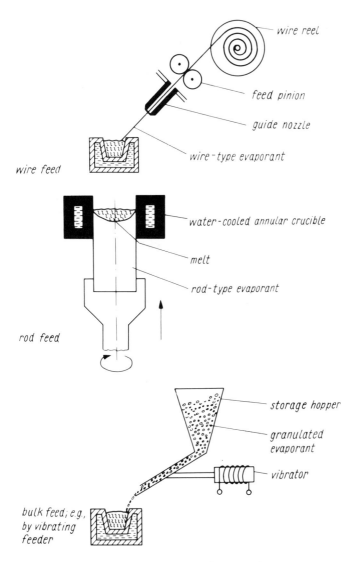

**FIGURE 23**  Feeding principles for crucibles.

Bulk feeding is used for either lumpy, pelleted, or granulated stock. Here it is the vibrating feeder that offers a favorable industrial solution [69].

Another specialized feeding method for low-melting evaporants such as tin or zinc is the liquid mode [67]. The feed stock is first liquefied in a premelting crucible and, after being fed to the evaporator crucible, is heated to evaporation temperature by the beam.

## 6  RATES OF THE EVAPORATION PROCESS

### 6.1  Power Density and Evaporation Rates

Owing to the exponential dependence of the evaporation rate on temperature (see Section 3.1), the rate increases with the power density on the evaporant. At a given power $P_0$ a higher power density can be achieved by reducing the beam diameter $d_F$ on the evaporant, that is, by improving the focus. Figure 24 shows the evaporation rate as a function of the current flow in the magnetic lens for an evaporator with axial gun. With optimized focusing the evaporation rate reaches a pronounced maximum.

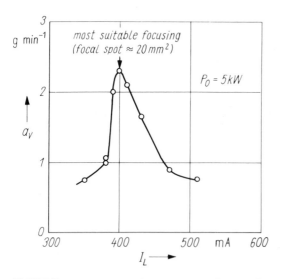

FIGURE 24  Evaporation rate $a_V$ for various focusing states, i.e., different focal spot diameters (change in power density due to an alteration of the lens current $I_L$); water-cooled crucible with a diameter of 20 mm; evaporant CoFe 30. (From Ref. 22)

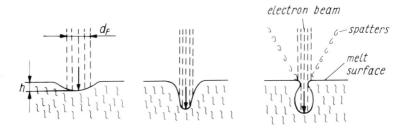

FIGURE 25   Distortion of melt surface and spatter formation with increasing beam power densities: h = depth, $d_F$ = beam diameter.

In optimized focusing the evaporation rate increases with $P_0$ until the losses in the vapor cloud become effective. Usually, however, another limitation, arising from melt surface distortions and spatter formation, imposes itself first. These effects are due to recoil phenomena of the vapor stream, that is, the action of the vapor pressure on the melting bath (Fig. 25). The trend toward spattering depends not only on the power density but, equally, on the purity of the evaporant and the temperature gradient in the crucible. For a great many materials the onset of spattering has been observed from $10^8$ to $10^9$ W·m$^{-2}$.

To obtain high evaporation rates without spattering for a given power and specific crucible dimensions, one commonly adjusts a small beam diameter $d_F$ (to achieve a high power density) and pass the beam over the vapor-emitting surface via programmed deflection. Such a sweep mode allows the use of power densities in the focal spot that would cause spatters when working in the steady-state mode.

When the beam is passed over the vapor-emitting surface in a programmed mode and at higher power densities, the beam is followed by a vapor pressure wave (blast) that is liable to produce undulatory motions of the bath surface.

Irrespective of the evaporator power rating (i.e., throughout the power range 3-600 kW), power densities of $10^8$ to $10^9$ W·m$^{-2}$ are used for evaporation. Today, values up to $5 \cdot 10^8$ W·m$^{-2}$ are employed in transverse guns, with axial guns also operating at values in the upper limit of the above-mentioned power range. The beam power is distributed via programmed beam deflection over an extended area and, in the limiting case, over the total crucible surface [20].

The mean power density with programmed deflection often lies between $2 \cdot 10^6$ and $2 \cdot 10^7$ W·m$^{-2}$, where the lower values refer to operating with a crucible insert. The crucible size and deflection program have to be chosen in accordance with the necessary evaporation rate and vapor stream density distribution. For small, water-cooled crucibles the heat losses have a substantial bearing on the evaporation rate. The temperature gradient between the vapor-emitting surface and the crucible wall and, thus, the heat losses decrease with increasing crucible diameter. Under otherwise identical conditions a larger crucible therefore yields a higher evaporation rate.

Since heat losses increase with lower crucible filling levels, the evaporation rate is at its maximum when the crucible is full and drops as the evaporation progresses. In order to ensure a high long-term constancy of the evaporation rate, one must employ feeding devices. Compared with evaporation from water-cooled crucibles, the use of a crucible insert allows the evaporation rate to be increased by a multiple owing to the reduced heat losses.

## 6.2 Evaporation Parameters and Condensation Rates

The performance of an evaporator is generally characterized by the evaporation rate, that is, the material evaporated per unit time, or the condensation rate referenced to a particular distance above the crucible (usually 250 mm). Typical condensation rates for 15-kW transverse evaporators with water-cooled crucibles, determined at the above-mentioned distance, are 2-10 $\mu m \cdot min^{-1}$. With high-power evaporators of 50 kW or more, condensation rates of up to 3000 $\mu m \cdot min^{-1}$ have been implemented for the evaporation of aluminum from ceramic crucibles [64]. As far as high-power evaporators are concerned, it is common to specify the evaporation rate. The power dependence of the evaporation rate for aluminum is illustrated in Fig. 26 for a 600-kW evaporator.

## 6.3 Control of the Evaporation Process

Control and monitoring of the EB coating process, especially of the condensation rate, is done with the aid of generally known methods [3,4,70-74]. According to the various factors that determine the rate of an EB evaporator, many parameters have to be kept constant by a suitable plant design or by electronic circuits. An essential condition for obtaining a steady rate is to keep the beam power constant. Generally, the beam current or power is used as a control variable for the evaporation rate. Control circuits for obtaining a steady evaporation rate with EB evaporators of up to 15 kW are often based on condensation rate measurements and implemented with the aid of quartz monitors. Moreover, evaporation rate controls have been integrated in transverse evaporators that use the measured ion current generated by the beam in the vapor stream

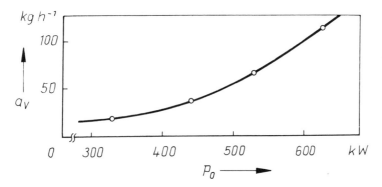

FIGURE 26   Evaporation rate $a_V$ as a function of evaporator power $P_0$ for aluminum (EH 600/40 evaporator from the M. von Ardenne Institute, evaporator with axial gun).

to control the evaporation rate of the evaporators [75]. If both the evaporation rate and the vapor stream density distribution have to be kept constant for several hours, monitoring and control of evaporators with feeding devices includes not only the beam current but also the height of the melting bath, which is maintained at the specified level by the feed control.


## 7   PLASMA-ASSISTED ELECTRON BEAM EVAPORATION

The conventional EB evaporation uses the vapor stream generated in a high vacuum, and its condensation on the substrate. In order to widen the field of use of EB evaporation and to improve the properties of the film obtained, one must influence the condensation conditions directly. At present the most important possibility is the plasma-assisted regime [4,76,77,78]. To this end, a plasma must be produced in a working gas of a pressure ranging from $10^{-2}$ to 1 Pa, which is to be admitted into the space between EB evaporator and substrate. In this plasma both the vapor produced and the working gas are excited, ionized, and recharged. Aside from the neutral vapor, excited and ionized particles as well as uncharged fast neutral particles are rendered active on the substrate. However, the coating at increased pressure without a plasma—that is, a transition from the unimpeded vapor propagation to the conditions of gas scattering—will first deteriorate most of the film properties. Therefore the plasma-assisted process becomes useful in its various applications only after this negative action has been overcompensated [79].

The plasma-assisted coating techniques developed and their descriptions common in the relevant literature differ with respect to the effects attained, the elementary processes used, and the engineering approaches employed. The most important basic techniques of plasma-assisted coating are ion plating and activated reactive evaporation.

## 7.1  Ion Plating

Ion plating is a plasma-assisted coating process which uses mainly the actions of ions and fast neutral particles in both the vapor and the carrier gas [79-85].

### *Elementary Processes and Energy Balance*

The impingement of ions on the substrate surface or the deposited layer is combined with a transfer of energy and momentum. Within a certain radius of action about the point of impingement, both phenomena evoke a number of elementary processes. After all, these processes are the cause of the altered film properties obtained under ion action during condensation. With ion plating the following elementary processes take place [76,79,86,87].

Impingement of high-energy particles first causes desorption or sputtering of either the adsorption layers and contamination layers or the substrate surface [4]. The substrate becomes activated due to an increase in the energy of the particles on the surface as well as interfacial defect generation. As a rule, this is linked with a temperature increase of the substrate. At the onset of vapor condensation further sputtering of substrate material accompanies partial resputtering of the condensing material, but as the condensation advances, only partial resputtering of the deposit takes place. The condensation of the film occurs under the condition of a steady activation. Part of the ions are incorporated in the film after they have given off their energy and charge [88]. When the carrier gas used for the generation of the plasma contains reactive gases, this also results in the generation of ions and excited particles of the reactive constituent. In this way, compounds are deposited in the condensation process. The reactive ion plating process implemented in this manner is identical to the so-called plasma-activated reactive evaporation performed under the action of ion draw-out voltages (see Section 7.2).

The extent to which the energy balance is subject to change during condensation is significant in determining the influence of ions on film properties [79].

Linked with the condensing vapor stream is an energy flow $E_D$ that is proportional to the number of condensing vapor particles

$n_D$ per unit time and unit area and to their evaporation temperature $T_V$. The energy transport via impinging ions is proportional to the number of ions per unit time and unit area and to the mean ion acceleration voltage $\overline{U}_I$.

One measure of the energetic activation of the film surface is the energy coefficient [79,89], which becomes

$$\varepsilon_I = \frac{E_I + E_D}{E_D} \qquad (19)$$

This coefficient gives the factor by which the condensation energy increases, owing to ion action, compared with the condensation energy of mere evaporation. Provided that the ion action does not yet cause a temperature increase of the whole substrate due to transient phenomena, the energetic activation of the surface corresponds to a higher effective condensation temperature $T_K$ as it is given by the mean substrate temperature $T_S$ [86,90]. Among other things, this yields a greater mobility of the condensing particles and thus an increased diffusion rate [86].

Since $E_D$ is generally much lower than $E_I$, the energy coefficient approximately corresponds to the ratio $n_I/n_D$ of ions and vapor particles on the substrate surface and to the mean ion acceleration voltage. Hence,

$$\varepsilon_I \approx 6 \times 10^3 \frac{\overline{U}_I}{T_V} \frac{n_I}{n_D} \qquad (20)$$

where $\overline{U}_I$ is measured in volts and $T_V$ is given in kelvins.

A value of about 0.2 eV is characteristic of the mean energy of a vapor particle obtained at an evaporation temperature of about 2000 K. Typical values of the mean energy of the ions during ion plating are 50-5000 eV. According to the selection of evaporation rate and ion current density on the substrate, and therefore according to the ratio $n_I/n_D$ chosen, it is possible to obtain energy coefficients of 1-1000. At a ratio of ion density to vapor particle density of $10^{-2}$, even an ion energy of 100 eV will suffice to obtain an energy coefficient of 7. This energetic excitation approximately corresponds to that obtained for sputtering without bias potential, due to the increased energy of the sputtered particles (some eV).

Because of the varied dependence of elementary actions on the ion energy, ion plating effects may generally differ even for like values of $\varepsilon_I$. Figure 27 shows that the predominant ion actions are associated with certain ranges of the ion energy [7,76]. To achieve high values of $\varepsilon_I$, one must use ionizing devices for ion plating, which allow realization of high values for $n_I/n_D$ [89,91].

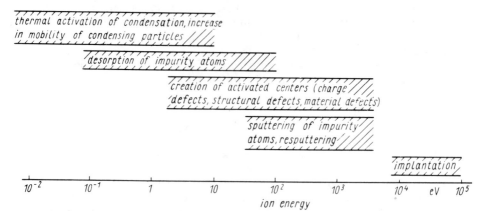

FIGURE 27   Predominant ion actions within different energy ranges.

However, restrictions will arise from the increase in resputtering
and in the thermal load imposed on the substrate.

A significant prerequisite to ion plating is that the condensation
rate $a_K$ must exceed the resputtering rate $a_S$. Here $a_K$ is the condensation rate obtained when coating takes place without ion action.
Under certain conditions the energy coefficient can also be expressed
by

$$\varepsilon_I \approx (\overline{U}_I)^{1/2} \frac{a_S}{a_K} \tag{21}$$

so that it reduces to quantities that are easy to measure [7,89].

*Elementary Processes and Film Properties*

From the elementary processes of ion plating described, it follows
that the condensation conditions are much more complex than those
of mere evaporation. By influencing adhesion mechanisms, nucleation,
and growth, ion plating allows one to change the morphology of
the film, especially with respect to its orientation and structure,
and the stoichiometry and surface topography.

The sputtering of adsorption layers on the substrate surface,
the formation of mixed interface layers composed of sputtered substrate and film materials, the diffusion of both materials, the higher
binding energy of activated centers [86], and the altered sticking
coefficients of the ionized vapor result in a good adhesion [92]
of ion-plated films. There are combinations of the substrate and
film materials for which evaporation alone fails to provide for ade-

quate adhesion. Strong adhesion, however, is necessary for application of very thick films (10-100 μm).

Ion bombardment during the condensation of the complete film causes in situ cleaning, which in turn may noticeably increase the purity of the deposited layer.

Resputtering and gas adsorption can result in an improved nucleation density and, therefore, a more fine-grained film structure [93]. Energetic excitation, resputtering, and gas adsorption may also affect the kind of film growth: for example, by changes in crystal orientation [90] and transition from columnar to isotropic [91,94,95] or equiaxial [96] growth at substrate temperatures, at which such a growth normally proves to be impossible [97,98]. Under the action of ions, recrystallization has been observed even at room temperature [99].

The enhanced binding energy of activated centers and altered sticking coefficients of the ionized gas are the cause of increased gas content and the directed incorporation of alloy constituents. In this way it is possible to incorporate insoluble constituents, for example, helium in gold [100] or phosphorus and lithium in zinc selenide [101]. The incorporation of constituents via ion bombardment during condensation of the vapor of another constituent can be called in situ ion implantation.

Gas scattering and differentiated resputtering, especially at peaks and edges as a consequence of the distribution of the electric field, yield a higher degree of coverage [79,81,102] combined with a low porosity [89,91]. Because energetic activation through ion plating yields an effective condensation temperature $T_K$ higher than the substrate temperature $T_S$, there are many new methods of successfully coating temperature-sensitive substrates with refractory materials that would otherwise call for substrate temperatures which are much too high.

But ion plating has also been successfully used to change other film properties: for instance, density [91,94], electrical conductivity [102], ductility and hardness [90], tensile strength [92], wear resistance [103,104], and optical reflectivity [89,102,105]. In addition, corrosion protective coatings have also been produced with this technique [86,106,107].

From the point of view of production engineering the reduction of the condensation rate due to scattering and resputtering in the ion plating process is a drawback with respect to the rate of mere evaporation; also, the material utilization $\eta_m$ usually proves to be somewhat lower [79,108].

*Interaction of the Vapor Stream and Ion Current*

In its initially suggested, simplest version [86] ion plating is performed with the substrate at a negative potential of a few thousand

volts. After inlet of a carrier gas (e.g., argon) the voltage across
the grounded evaporator and the substrate causes a gas discharge
at a pressure of about 1 Pa. Thus a vapor stream and an ion current
will flow to the substrate. In this conventional mode of ion plating,
and for rather small substrates, it can be assumed that the densities
of the vapor stream and ion current are constant; that is, $n_D =$
$const_1$ and $n_I = const_2$ over the substrate surface with respect to
time. In the general case of industrial coaters, the vapor stream
and ion current are cyclic functions of time for each point of a
large substrate or substrate arrangement; that is, $n_D = f_1(t)$ and
$n_I = f_2(t)$. If ion plating takes place under these conditions, it is
called alternating ion plating [89] (Fig. 28).

The principle of AIP is readily applied by locally separating
the areas subjected to vapor action and ion action, moving the
substrates relative to these areas. To this end, the vapor source
and ionizing device have to be arranged on different sites, so the
vapor stream and ion current are rendered active simultaneously
but at different points of the substrate. The relative motion of
the substrates must be carried out such that in one cycle each
point to be coated passes once in the zone of vapor action and
once in the zone of ion action. During each cycle the film in the
vapor zone grows by a partial layer. When being moved through
the zone of ion action, the deposited film in each case, especially

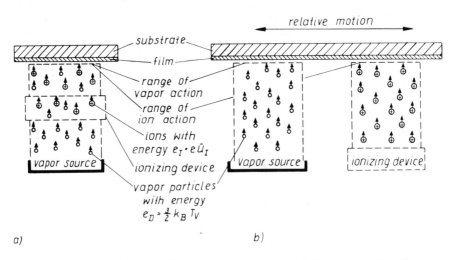

FIGURE 28 Principles of the vapor stream and ion current action
in (a) conventional ion plating (IP), (b) alternating ion plating
(AIP).

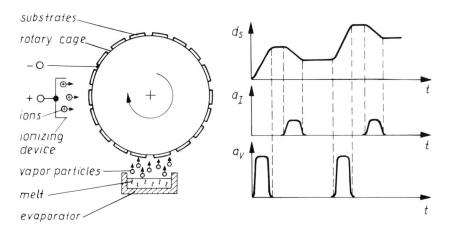

FIGURE 29 Example of a possible arrangement for alternating ion plating (AIP) as well as time response of the evaporation rate $a_V$, the resputtering rate $a_I$, and the growth of film thickness $d_S$ on the substrate (schematic).

the partial layer, is subjected to energetic excitation and partial resputtering. At a correspondingly high ion energy, this will result in a certain removal of the film (Fig. 29).

It has been found that AIP exerts effects on the film properties identical, or similar, to conventional ion plating if accomplished in such a manner that the thickness of a partial layer corresponds to only a few atomic layers [89]. Therefore AIP proves to be advantageous for the high-rate coating of large areas because the restricting condition of keeping the densities of the vapor stream $n_D$ and ion current $n_I$ constant can be dropped. Evaporator and ionizing devices can be separately selected, optimized, and arranged.

*Ion Plating Equipment*

In compliance with the various possible ways of producing vapor and ions and according to the strongly diverging requirements to be placed on the performance parameters, a great many units have been suggested. The simplest instrumental solution is obtained when the beam is used as a power source for both production and ionization of the vapor stream. Generated by the primary beam within the crucible region, the vapor ions are drawn out of this region and, by application of a negative potential of a few thousand volts to the substrate, accelerated toward the latter. However, the ion current density attainable on the substrate for such an arrange-

ment is too low for industrial use. Ion current is low because of
the slim probability of ionizing the vapor particles by fast electrons
with energies of 10-30 keV [9]. Nevertheless, this solution is advantageous in that the process can be carried out in a high vacuum
(pressures below $10^{-2}$ Pa).

One possible way of ion plating in a high vacuum is to use the
EB evaporator to produce the vapor stream only, while a separate
ion source produces and directs an ion flow onto the substrate
[103,109]. Although such an arrangement allows separate control
of the ion current and vapor stream, ion beam generation is rather
expensive. Because of its features, this arrangement will be restricted
mainly to research work.

The most widespread application is in a device that uses an
EB evaporator to generate the vapor stream and a dc gas discharge
between the evaporator and substrate to produce a current consisting
of vapor ions and ions of the carrier gas. Both the beam and the
plasma formed between the electrodes of the discharge are ionization
sources. Figure 30 shows the principal arrangements for ion plating
with an EB evaporator and a glow discharge between evaporator
and substrate. The substantial difference to conventional ion plating
devices is the pressure stage between the gun and the discharge
space. For transverse evaporators the necessary pressure difference
is obtained by means of shutters in the work chamber; for axial
guns, pressure stage tubes between evaporator and work chamber
are used. If ion plating is to be made at a carrier gas pressure
in the Pa range the pressure drop must amount to 100:1. In self-
maintained gas discharges with current densities of approximately
1 mA·cm$^{-2}$, however, high evaporation rates allow only small $n_I/n_D$
ratios.

Especially with respect to vapor ions, the ion current density
may be enhanced by arranging the crucible in a magnetic axial
field [110]. In this way the slow secondary electrons may be utilized
for additional ionization. But ionization of the vapor or carrier
gas can also be obtained by rf discharges [41,101]. Such an arrangement allows ion plating of nonconducting substrates. Also, increases
in the ion current can be obtained via discharges that are not self-
maintained (for example, from a heated auxiliary cathode). Ion
current densities of more than 100 mA·cm$^{-2}$ can be achieved with
a ring-gap discharge facility as the ionizing source, which operates
on the magnetron principle [89]. Because of the pronounced inhomogeneity of the discharge, however, this device is related to AIP almost
exclusively. Of great significance in the dimensioning of ion plating
equipment with electron beams is proper matching of $n_I/n_D$ ratio
and the energetic activation to the high evaporation rate inherent
in EB evaporators. The selection of the ionizing source also fixes
the pressure range and the kinetic energy of the ions.

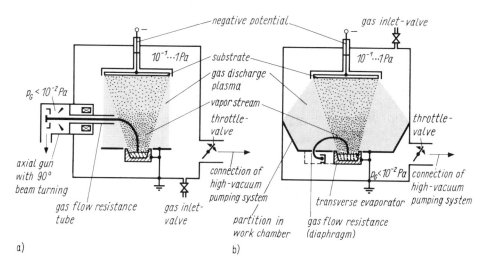

**FIGURE 30** Schematic ion plating equipment with EB evaporator and dc gas discharge between evaporator and substrate (a) with an axial gun and (b) with a transverse gun.

The timed regime (that is, either operation in alternating or nonalternating mode) confers on the applicative and instrumental development another degree of freedom.

### 7.2 Activated Reactive Evaporation

Plasma-activated reactive EB evaporation is a coating technique in which the activation of a vaporous component and of a gaseous component is used mainly for depositing compounds [47,49,111]. The simplest example of such a reactive coating process is the deposition of TiN. Used as reactants are the Ti metal atoms of the vapor stream from the crucible and the $N_2$ gas molecules in the space between the crucible and substrate. Excitation of the reactants by the plasma also takes place mainly in the space between the crucible and the substrate. Generally, the reaction occurs on the surface of the substrate on which the compound has to be deposited. Under very high work pressures, part of the reactions also takes place in the plasma. Low-energy electrons (10-100 eV) exhibit a large cross section of action for ionization and for excitation.

An earlier suggestion about the use of an EB facility for ion plating is to utilize the secondary electrons produced on the melting bath surface by the primary high-energy beam, which, after acceleration through an annular electrode, generates a plasma that in turn activates the reactants (Fig. 31) [49].

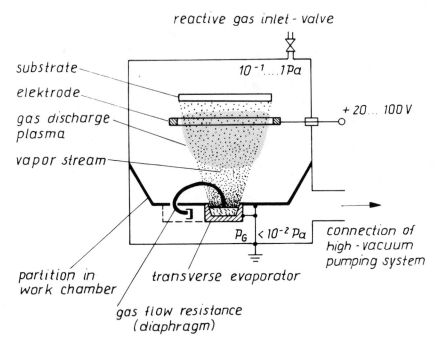

reactive gas inlet - valve

substrate

elektrode

gas discharge plasma

vapor stream

$10^{-1} \ldots 1\,Pa$

$+20 \ldots 100\,V$

$P_G$ $< 10^{-2}\,Pa$

connection of high - vacuum pumping system

partition in work chamber

transverse evaporator

gas flow resistance (diaphragm)

FIGURE 31   Equipment for plasma-activated reactive deposition with transverse electron beam gun and ring electrode (schematic).

As with ion plating, the purpose is to create improved facilities for plasma-activated reactive evaporations which yield high condensation rates of the compounds, that is, to match the activation of the reactants to the high metal vapor streams. The introduction of an rf discharge by means of an rf coil within the space of vapor propagation [112] or of a non-self-maintained dc discharge (e.g., via heated auxiliary cathode [113,114]) are approaches for increasing the plasma density and thus the activation rate. A plasma is already obtained when a bias voltage is applied to the substrate fixture [115,116]. The electrons of this plasma activate the reactants. Furthermore the bias electrode serves to draw out the ions of the plasma and direct them onto the substrate. It can be seen that the solutions of plasma-activated reactive evaporation at high plasma densities and EB evaporators are almost identical to those discussed in Section 7.1 for ion plating.

TABLE 1 Extended Process Variants of Evaporating and Sputtering Techniques

| Processing variant of basic process | Objective | EB evaporation | Basic process high-rate sputtering | Low-voltage arc coating |
|---|---|---|---|---|
| High vacuum deposition | Avoidance of residual gas scattering | HV or UHV evaporation | Conditionally possible; min. work pressure $p \approx 5 \times 10^{-2}$ Pa | Conditionally possible; min. work pressure $p \approx 10^{-2}$ Pa |
| Gas-scattering deposition | Particle scattering in residual gas; step coverage | Inert gas inlet | Sputtering at selectable inert gas pressure ($p \approx 0.5$ Pa) | Evaporation at selectable inert gas pressure; already contained in basic process |
| Reactive deposition | Chemical reaction | Reactive gas inlet | Sputtering in reactive gas; basic process already plasma activated | Reactive gas inlet; basic process already plasma activated |
| Electron or plasma-activated reactive deposition | | Reactive gas and plasma generation | | |
| Ion-assisted coating | Activated condensation | Ion plating; gas intake for ion generation | Bias sputtering; ions formed in basic process | Ions formed in basic process |
| Intermediate and combined techniques | Chemical reaction; activated condensation | Plasma plating plasma pretreatment and EB evaporation | | |

547

Plasma-activated reactive coating allows deposition of a series
of oxides, carbides, nitrides, and sulfides. A survey of deposited
compounds, film properties attained, and applications is given in
Refs. 4, 78, 82, 117, and 118.

## 7.3  Comparison of Ion-Assisted Coating Techniques

Parallel to ion-assisted coating with EB evaporators, a variety of
competing techniques for physical vapor deposition have been
developed, such as high-rate sputtering with magnetron/plasmatron
devices [4,78,119-127] and low-voltage arc coating [90,95,110,128].
The device of Moll and Daxinger [129] operates with external electron
generation in a discharge chamber in which the pressure is higher
than that in the deposition chamber. The electrons emerge from
a hot cathode. In the hollow-cathode arc evaporator [130,131] of
Lunk et al. [132], the semicircular guidance of the plasma yields
a compact design similar to that of a transverse evaporator. With
the spark source of Snaper [133], the point of the arc under consideration is moved on the solid-state evaporant. This source type
allows one to choose the coating direction as required. On the other
hand, however, it has the disadvantage that the formation of big
clusters is likely to impair the quality of the film obtained.

Whereas the plasma-activated deposition by high-rate sputtering
and low-voltage arc coating is always included in the basic procedure
(depending solely on the selection of the work gas and process parameters), EB evaporation requires additional devices for plasma generation.
A survey of the possibilities of matching the various methods of vapor
generation to different process variants is given in Table 1 [76,77,121].

The main advantage of process variants with electron beams
is the high condensation rates that can be attained. Adaptability
with respect to material selection, geometric configuration, and
control of the process parameters is an advantageous characteristic
of high-rate sputtering. With this coating technique, up to 10%
of the discharge current may act on the substrate in the form of
ions [76]. However, the mean ion energy, $E_I = e\overline{U}_I$, is restricted
to a voltage level that is generally lower than the discharge voltage.
High degrees of ionization to 50%, and therefore high ion currents,
as well as the free choice of the ion acceleration voltage are characteristic of the low-voltage arc discharge.

## 8  INDUSTRIAL USE OF ELECTRON BEAM EVAPORATION

### 8.1  Use of Electron Beam Evaporators

Because of its advantages, EB evaporation has replaced a substantial
number of conventional coating techniques and opened up many new

applications. Hence a variety of requirements have to be placed on
EB evaporation. A decisive factor for the use of EB evaporators
is their good adaptability to a great many coating jobs since a variety
of materials can be evaporated from a water-cooled crucible. This
holds especially for the evaporation of reactive and refractory substances [134] because the power density attained is very high.
For certain applications the evaporation from water-cooled crucibles
is of interest wherever a high purity of the deposits is required.
The high evaporation rate attained especially with heat-insulating
crucible inserts is a precondition for the coating of large areas
as well as for the deposition of films with a thickness of 10 μm
and more [31,106,107,135,136].

In order to solve the various coating problems in a profitable
manner, both standard units and specialized equipment are made
available. With small substrates, where the basic concept of the
vacuum system proves to be significant, standard units with built-in
assemblies matched to the given requirements are used. But if a
special coating job substantially affects the plant design (for example,
by the shape or size of the substrates), specialized equipment has
to be developed. Some examples of these units are coaters for strip
steel and for foils or webs. Plants were developed for the deposition
of thick films on parts of turbines and jet engines (overlay coating)
[137] and for protection against corrosion [138] that are now used
on an industrial scale.

## 8.2  Use of Electron Beam Evaporators in Standard Plants

It was mentioned in Section 5.1 that transverse evaporators are
available as built-in devices in a power range up to 15 kW. For
most units of standard design, installation of these evaporators
is possible either parallel to or instead of conventional evaporating
devices.

Since EB evaporators can be readily matched to the various
coating jobs, it is possible in the interest of improved economy
to equip standard coaters with this evaporator type only. Such
a plant allows for depositing multilayer systems with the aid of
several EB evaporators or by multiple-source evaporation (see
Section 5.2). Often the higher costs of an EB evaporator are compensated by savings in wearing parts common to conventional evaporators. Another argument that speaks in favor of EB evaporators in
such equipment is the need to perform coating at higher rates. The
use of EB evaporators in standard-type coaters devised for the
production of aluminum films shall be cited as an example. Figure 32
shows a coater that consists of two 10-kW EB evaporators with transverse gun and spherical substrate fixtures with planetary drive.

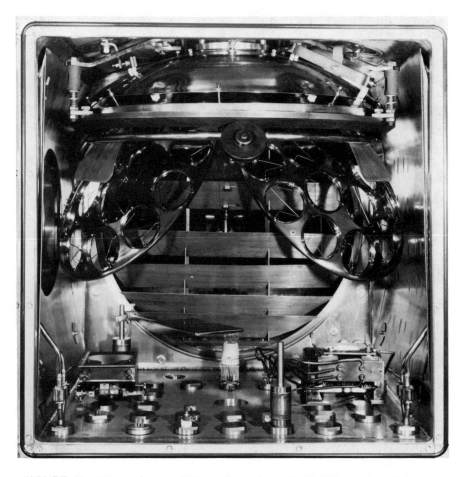

FIGURE 32  High-vacuum EB coating plant B 63 DM produced by
VEB Hochvakuum Dresden: arrangement of two 10-kW transverse
gun evaporators; substrate fixture with planetary drive for substrates of 100-mm diameter.

Today, the main application of standard equipment with EB
evaporators concerns the electronics and optical industries. Mass-
produced components and semifinished products in the machine-tool
and machine building industries can also be coated in such standard
plants. In some fields, however, standard units are needed that
have to be fitted with additional equipment for ion-assisted coating
(see also Section 7).

### 8.3 Examples of Using Electron Beam Evaporators in Specialized Plants

*Electron Beam Evaporation for Roll Coating*

Aluminum coating of plastic sheets and paper webs has been a well-established technique on an international scale for many years. Webs of 0.5-2 m width were coated with aluminum films 30-80 nm thick. With web speeds up to 10 $m \cdot s^{-1}$ the annual output is about 10 million $m^2$. Until now, aluminum was deposited mainly with the aid of conventional, resistance-heated evaporators. Some important uses of aluminum-coated plastic sheets and paper webs concern the manufacture of packaging and decorative products, capacitors, and the like.

In recent years, however, this technique has been used increasingly for the deposition of other materials, particularly the manufacture of thin-film magnetic tapes by coating a polyester sheet with cobalt or cobalt alloys. Video tapes with magnetic films for longitudinal recording are produced in roll coating plants.

The industrial implementation of roll coating equipment with EB evaporators for the above-mentioned applications becomes possible with the aid of EB line evaporators as described in Section 4.2, which ensures the required film thickness uniformity over the total web width. Figure 33 shows the arrangement of an EB line evaporator in a roll coater for aluminum. Referenced to a plant of

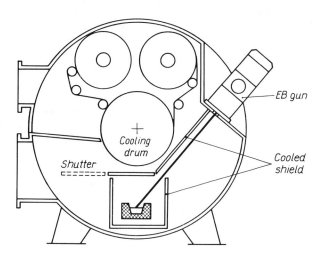

FIGURE 33 Arrangement of an EB line evaporator without magnetic trap in a roll coating plant.

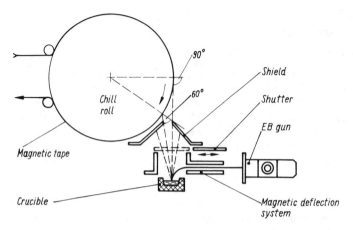

FIGURE 34   Arrangement of an EB line evaporator with magnetic
trap for the manufacture of metal thin-film magnetic tape.

1-m web width, the evaporator power required at the heat-insulated
crucible is about 65 kW, the specific power consumption is 8 kWh·kg$^{-1}$,
and the evaporation rate is 8 kg·h$^{-1}$. At a vapor utilization of 60%
and a film thickness of 50 nm, a web speed of 10 m·s$^{-1}$ can be
attained [65,4].

The schematic of an array for coating magnetic tapes by EB
evaporation of cobalt or cobalt-nickel alloys is shown in Fig. 34.
The high coercivity required is attained by oblique shadowing;
that is, masking of the vapor stream and use of a lateral offset
of the crucible relative to the chill roll. The magnetic properties
of the film will improve with growing angle of incidence. In the
interest of a reasonable material utilization and deposition rate,
it is therefore common to use an angle of incidence between 60°
and 90°. The necessary power of the EB evaporator is 100-300 kW.
When, for instance, evaporation takes place at 200 kW from a crucible
1 m wide, about 50 kg cobalt per hour can be vaporized at a specific
power consumption of 4 kWh·kg$^{-1}$ and a vapor utilization of 6%.
For a film thickness of 100 nm the web speed is about 1 m·s$^{-1}$ [65].

As with any other evaporation method, efficient cooling during
coating becomes a must so as to protect the web against overheating.
This is accomplished by passing the web over a smooth chill roll,
the temperature of which is generally -20 to -40°C. Sometimes an
intermediate aluminum layer 20 nm thick is applied as a radiation
shield for the subsequent deposition of cobalt [4].

When EB evaporation is employed for the coating of plastic
sheets or paper webs, annoying side effects may occur due to

overheating and the penetration of charge carriers into the substrate.
Since this is due mainly to backscattered electrons, a good procedure
is to use a magnetic trap, particularly where thin plastic sheets
(15 μm) or paper webs (20 μm) are to be coated. In this way the
backscattered electrons are prevented from reaching the sheet or
web. Here, some engineering solutions are available that proved
sound in practice [139,140,65]. Because in such a case a rather
dense plasma is formed above the evaporant, one can speak of a
plasma-activated regime.

## Electron Beam Coating of Metal Strips

In the manufacture of films on metal strips the use of EB evaporation
is a significant alternative to conventional coating techniques. EB
coating excels in its ability to give high evaporation rates up to
100 kg·h$^{-1}$ and more (condensation rates to 3 mm·min$^{-1}$) and yields
a much better environmental compatibility compared with other coating techniques. Films 1-10 μm thick can be obtained at high strip
speeds. Compared with electroplating, the energy expenditure required for a series of materials such as aluminum, zinc, chromium,
and titanium is less by a factor of 5 or more. In addition, EB
evaporation substantially widens the assortment of useful coating
materials. Though known for decades, it had a comeback in recent
years particularly due to its outstanding environmental compatibility,
the high engineering level of electron beam technology attained,
and entirely new demands in industry. Aluminum coating of strip
steel is backed up by industrial experience gathered over many
years. To outline the possibilities inherent in high-rate EB evaporation, we briefly review the present state of the art with respect
to processes and instrumentation.

Aluminum-coated strip steel is used in the packaging industry
and competes well with the tin plate produced by conventional electroplating. For EB deposition of aluminum coatings the energy expenditure is lower than that for electrotinning. The principal process
of strip-steel coating is EB evaporation. Used for the purpose are
the facilities for large-area deposition discussed in Sections 4.3
and 5.1. The strip steel to be coated is passed over the extended
hot crucible at the lowest possible distance (less than 300 mm).
Crucible heating takes place by means of the electron beam which
is deflected in a programmed mode. The crucible dimension in the
direction of strip motion is about 500 mm. That in the other direction
is given by the width of the strip steel to be coated. Despite the
high pressure in the vapor cloud (up to $10^3$ Pa), the beam can
be passed through the vapor at low power losses [141]. The process
parameters being decisive for profitable coating are the thermal
efficiency $\eta_t$ and the utilization coefficient $\eta_m$ of the evaporated
material.

The following reflections refer to the evaporation of aluminum by means of a 600-kW EB evaporator [64]. With due consideration of the most important power losses—heat radiation from the evaporant, electron backscattering, and absorption of some beam power in the vapor cloud—a thermal efficiency $\eta_t$ of 0.65 is attained with due consideration of the splash limit (1.3 $kW \cdot cm^{-2}$). Referenced to a strip width of 800 mm and a demanded film thickness uniformity of ±10%, a material utilization of $\eta_m$ = 0.8 is attained in practice. Hence, the total efficiency of the input power amounts to 56%. As a consequence, a beam energy of 6.4 kWh is needed for depositing 1 kg of aluminum. For implementing a strip-steel coater based on large-area EB evaporation, it is necessary to have a vacuum pumping system, the final pressure of which lies in the range of 1 to $10^{-2}$ Pa. The strip steel is passed through the plant in "air-to-air" mode. Here a pressure stage system is necessary on either end. A number of flow resistances with interposed vacuum pumping systems serves to maintain a stepwise pressure drop from atmospheric pressure to the necessary work pressure in the high-vacuum range. The gaps between roller pairs are used as flow resistances for the continuous strip locks. Between the possibilities of employing either partially contacts rolls or squeeze rolls [20], it is the latter that has gained a strong reputation because of its sturdier design, even though the flow resistance per roller pair is somewhat lower.

Aside from the actual coating in the plant [20], a series of other processing steps are needed in a production line for strip-steel coating. Above all, the strip must be pretreated at atmospheric pressure outside the plant by conventional methods. In the plant, such a pretreatment has to be supplemented by heating and degassing (see Fig. 35). The pretreatment chamber and its vacuum pumping system are to be dimensioned such that the necessary degassing action is ensured at the specified strip speed. Here EB heating proved to be very useful as an energy source for the degassing process [20]. In order to obtain well-adhered coatings, the temperature of the strip steel should be at least about 220°C [142,7].

Activation by ion bombardment has been investigated also [143, 144]. Though being energetically more favorable, it has not yet been used widely because the regime in large-scale industrial applications is rather intricate. Another possibility of surface activation is brushing in a vacuum [145]. As to the aluminum-coating of strip steel, however, this mode of activation is unsuitable because it is tied in with a too-pronounced diffusion and alloy formation. For tin coatings, on the other hand, this process yields films of adequate adhesion and quality. Moreover, coating in a vacuum must be followed by such after-treatments as heat treatment, cooling, skin-rolling, lacquering, and drying. Finally, the processing line has to be

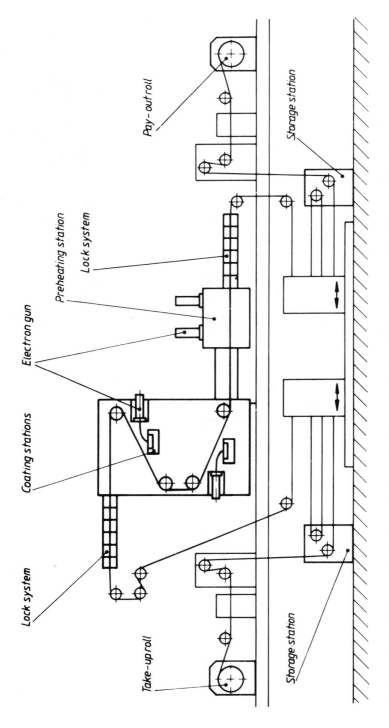

FIGURE 35   Schematic design of a production plant for strip steel coating by electron beam evaporation.

FIGURE 36   EBA 635/800 electron beam strip coater: strip width
800 mm; strip speed 200 m·min⁻¹; capacity $10^4$ m²·h⁻¹, double-sided;
evaporator power two stations per 600 kW; preheating power: two
stations per 600 kW. (Courtesy of VEB LEW Hennigsdorf and the
M. von Ardenne Institute; see Ref. 64)

complemented by such additional processes as butt welding of the
strips, charging of the rolls, and evaporant feeding to the crucibles
(see Section 5.3).

   The present state of the art is amply reflected from the EBA
635/800 electron beam strip coater engineered for handling strip
steel 800 mm wide (Fig. 36). Fitted with four high-power guns,
it is controlled and monitored by one operator only and allows for
continuous operation over one week.

## SYMBOLS

| | |
|---|---|
| $a_K$ | Condensation rate |
| $a_V$ | Evaporation rate |
| $a_{V1}$, $a_{V2}$ | Specific evaporation rates |

| | |
|---|---|
| $d_F$ | Focal spot diameter on work piece |
| $d_S$ | Film thickness on a plane substrate |
| $E$ | Energy |
| $e$ | Energy density, electron charge |
| $F$ | Area |
| $F_K$ | Surface on sphere |
| $F_S$ | Surface at substrate level |
| $F_V$ | Vapor-emitting surface |
| $h$ | Depth of melting bath, Planck's constant |
| $h_V$ | Distance between evaporator and substrate |
| $K_2$, $K_2$ | Constants |
| $k_B$ | Boltzmann constant |
| $l$ | Path |
| $M$ | Molecular weight |
| $M_D$ | Molecular weight of vapor particles |
| $N_{i2}$ | Molar concentration of ith component |
| $n_D$ | Vapor particles per units of area and time |
| $n_I$ | Ions per units of area and time |
| $P_V$ | Useful power |
| $P_0$ | Beam power |
| $p_D$ | Vapor pressure |
| $p_G$ | Gas pressure |
| $p_S$ | Saturated vapor pressure, power density of heat radiation |
| $p_0$ | Beam power density |
| $r_S$ | Distance between axis of evaporation and substrate point |
| $T$ | Temperature |
| $T_S$ | Substrate temperature, melting temperature |
| $T_V$ | Evaporation temperature |
| $T_{VS}$ | Temperature where evaporation power and radiation loss are equal |
| $t_w$ | Time of beam action |
| $U_B$ | Acceleration voltage |
| $U_I$ | Ion acceleration voltage |
| $V_2$ | Volume of melting bath |
| $X_i$ | Amount of ith alloying element (in weight percent) |
| $X_{i1}$ | Amount of ith alloying element in feed of evaporant (in weight percent) |
| $X_{i2}$ | Amount of ith alloying element in melt (in weight percent) |
| $X_{i3}$ | Amount of ith alloying element in vapor (in weight percent) |
| $X_{i4}$ | Amount of ith alloying element in film (in weight percent) |
| $Z$ | Atomic number |
| $\alpha$ | Angle; evaporation coefficient |
| $\gamma$ | Activity coefficient |

| | |
|---|---|
| $\varepsilon_I$ | Coefficient of energy |
| $\eta_m$ | Material utilization |
| $\eta_t$ | Thermal efficiency of evaporation |
| $\nu$ | Frequency |
| $\nu_D$ | Surface-referred collision rate of vapor particles |
| $\nu_G$ | Surface-referred collision rate of gas particles |
| $\rho$ | Density |
| $\tau$ | Transmission coefficient |
| $\Phi$, $\Phi_0$ | Vapor stream density |

## Important Constants

| | |
|---|---|
| Planck's constant | $h = 6.625 \times 10^{-34}$ W·s$^2$ |
| Electron charge | $e = 1.602 \times 10^{-19}$ A·s |

## Abbreviations

| | |
|---|---|
| AIP | Alternating ion plating |
| EB | Electron beam |
| IP | Ion plating |
| PVD | Physical vapor deposition |

## Conversion Tables

| | |
|---|---|
| Length | 1 m = 3.281 ft |
| | 1 m – 39.37 in |
| Weight | 1 g = 0.03527 oz |
| | 1 kg = 2.205 lb |
| Area | 1 cm$^2$ = 0.155 in$^2$ |
| | 1 m$^2$ = 10.763 ft$^2$ |
| Volume | 1 cm$^3$ = 0.061023 in$^3$ |
| | 1 cm$^3$ = 3.531 × 10$^{-5}$ ft$^3$ |
| Temperature | ( ~ °C × 9/5) + 32 = °F |
| Pressure | 1 Pa = 0.75 × 10$^{-2}$ torr |
| | 1 Pa = 0.2953 × 10$^{-3}$ inHg |
| | 1 Pa = 0.1451 × 10$^{-3}$ psi |
| | 1 Pa = 10$^{-5}$ bar |
| | 1 Pa = 0.9869 × 10$^{-5}$ atm |

## References

1. Holland, L. (1961). *Vacuum Deposition of Thin Films*, Chapman and Hall, London.
2. Maissel, L. J. and Glang, R. (1970). *Handbook of Thin Film Technology*, McGraw-Hill, New York.
3. Schiller, S. and Heisig, U. (1975). *Bedampfungstechnik*, Volkseigener Betrieb Verlag Technik, Berlin.
4. Frey, H. (1987). *Dünnschichttechnologie* (Kienel, G., ed.). Verlag des Vereins Deutscher Ingenieure, Düsseldorf.
5. Mayer, H. (1950, 1955). *Physik dünner Schichten*, vols. 1 and 2. Wissenschaftliche Verlags-Gesell, Stuttgart.
6. Chopra, K. L. (1969). *Thin Film Phenomena*, McGraw-Hill, New York, pp. 83-136.
7. Schiller, S., Heisig, U., and Panzer, S. (1982). *Electron Beam Technology*, Wiley, New York.
8. Bakish, E. (1965). "Electron and Ion Beam Science and Technology," Proceedings of First International Conference '64." Toronto, Canada, Wiley, New York.
9. von Ardenne, M. (1962). *Tabellen zur Angewandten Physik*, Vol. 1 and 2nd ed., VEB Dtsch. Verlag d. Wiss., Berlin.
10. Everhardt, T. E. (1960). *J. Appl. Phys.*, 31:1483-1490.
11. Sommerkamp, P. (1970). *Z. Angew. Phys.*, 28 (4):220-232.
12. Archard, G. D. (1961). *J. Appl. Phys.*, 32:1505-1509.
13. Kanter, H. (1957). *Ann. Phys. Leipzig*, 20 (6):144-166.
14. Jaeger, T. (1960). *Grundzüge der Strahlenschutztechnik*, Springer-Verlag, Berlin.
15. Schiller, S. and Jäsch, G. (1974). "On the electron-metal vapor interaction in high power evaporators," Proceedings of Sixth International Conference '74, San Francisco, California, Electrochemical Society, Princeton, N.J., pp. 447-456.
16. Erikson, E. D. (1974). *J. Vac. Sci. Technol.*, 11 (1):366-370.
17. Hunt, C. d'A and Hughes, J. L. (1972). "Principles of electron beam evaporation," In Silva, R. M. Second Electron Beam Processing Seminar, Frankfurt (Main), Universal Technology Corporation, Dayton, Ohio, pp. 4a, 1-4a, 18.
18. Sommerkamp, P. (1970). *Vak. Tech.*, 19 (7):161-167.
19. Dushman, S. (1962). *Scientific Foundations of Vacuum Technique*, 2nd ed. Wiley, New York.
20. Schiller, S., Lenk, P., Förster, H., Jäsch, G., and Kühn, G. (1974). "Industrial electron beam coating of strip steel," In Silva, R. M. Third Electron Beam Processing Seminar, Stratford-upon-Avon, England, Universal Technology, Dayton, Ohio, pp. 2d 1-2d 28.
21. Smith, H. R. and Hunt, C. d'A. (1966). "Advances and future of electron beam processes," In Bakish, R. Electron and Ion

Beam Science and Technology. Second International Conference, New York, AIME, New York, pp. 277-300.

22. Schiller, S., Effenberger, D., Heisig, U., Goedicke, K., and Schneider, S. (1967). *Vak. Tech.*, *16* (9):205-209.

23. Dietrich, W., Reichelt, W., and Hauff, A. (1967). *Elektrowärme* *25* (6):219-226.

24. Smith, H. R. and Hunt, C. d'A. (1967). *Vak. Tech.*, *16* (4): 80-84.

25. Schiller, S. and Förster, H. (1966). *Neue Hütte*, *11* (2):110-113.

26. Jaeckel, R. (1950). *Kleinste Drücke, ihre Messung und Erzeugung*, Springer-Verlag, Berlin.

27. Jäsch, G. (1980). Elektronenstrahlverdampfung von Aluminium in Raten größer 10 $g \cdot s^{-1}$. Thesis. Technische Hochschule Karl-Marx-Stadt.

28. Ramsauer, C. (1921). *Ann. Phys. Leipzig*, *64* (6):513-540.

29. Auwärter, M. (1971). Probleme der Vakuum- und Grenzflächenphysik, In Auwärter, M. Ergebnisse der Hochvakuumtechnik und der Physik dünner Schichten, vol. 2. Wissenschaftliche Verlags-Gesell., Stuttgart, pp. 1-15.

30. Langmuir, J. (1913). *Phys. Rev.*, *2* (5):329-342.

31. Movčan, B. A., Demčišin, A. V., and Kulak, K. D. (1974). *J. Vac. Sci. Technol.*, *11* (5):869-874.

32. Suryanarayanan, R. (1978). *Thin Solid Films*, *50*:349-355.

33. Hegner, F. and Feuerstein, A. (1978). *Thin Solid Films*, *53*:141-152.

34. Rönnefarth, B., Mattheis, R., and Dintner, H. (1982). *Phys. Stat. Sol.*, *73*:K153-K156.

35. Zinsmeister, G. (1964). *Vak. Tech.* *13* (8):233-240.

36. Kennedy, K. D. (1968). "Alloy deposition from single and multiple electron beam evaporation source," American Vacuum Society Regional Symposium.

37. Santala, T. and Adams, C. M. (1970). *J. Vac. Sci. Technol.*, *7* (6):S22-S29.

38. Andreini, R. J. and Foster, J. S. (1974). *J. Vac. Sci. Technol.*, *11* (6):1055-1059.

39. Hulgren, C. (1963). *Selected Values of Thermodynamic Properties of Metals and Alloys*, Wiley, New York, p. 732.

40. Wan, C. T., Chambers, D. L., and Carmichael, D. C. (1971). *J. Vac. Sci. Technol.*, *8* (1):312-316.

41. Harker, H. R. and Hill, R. J. (1972). *J. Vac. Sci. Technol.*, *9* (6):1395-1399.

42. Krutenat, R. C. (1974). *J. Vac. Sci. Technol.*, *11* (6):1123.

43. Stowell, W. R. (1973). *J. Vac. Sci. Technol.*, *10* (4):489-493.

44. Reichelt, W. and Müller, P. (1962). *Vak. Tech.*, *11* (8):235-239.

45. Grubb, A. P. (1969). *Res. Dev. Ind.*, *20* (6):42-46.

46. Hoffmann, D. and Leibowitz, D. (1972). *J. Vac. Sci. Technol.*, 9 (1):326-329.
47. Auwärter, M. (1960). Process for the manufacture of thin films." U.S. Patent 2,920,002.
48. Ritter, E. (1966). *J. Vac. Sci. Technol.*, 3 (4):225-226.
49. Bunshah, R. F. and Raghuram, A. C. (1972). *J. Vac. Sci. Technol.*, 9 (6):1385-1388.
50. Neugebauer, C. A. and Rairden, C. A. (1964). *J. Vac. Sci. Technol.*, 1 (2):72-73.
51. Stowell, W. R. (1974). *Thin Solid Silms*, 22 (1):111-120.
52. Raghuram, A. C. and Bunshah, R. F. (1972). *J. Vac. Sci. Technol.*, 9 (6):1389-1394.
53. Maissel, L. I. and Glang, R. (1970). *Handbook of Thin Film Technology*, McGraw-Hill, New York.
54. Ramprasad, B. S. and Radha, T. S. (1974). *Vacuum*, 24 (4):165.
55. Graper, E. B. (1973). *J. Vac. Sci. Technol.*, 10 (1):100-103.
56. Behrndt, K. H. (1972). *J. Vac. Sci. Technol.*, 9 (2):995-1007.
57. Smith, H. R. (1969). Deposition distribution and rates from electron beam heated sources, Soc. of Vacuum Coaters, Detroit, Michigan.
58. Smith, H. R. (1972). "Principles of electron beam technology," In Silva, R. M. Second Electron Beam Processing Seminar, Frankfurt (Main), Universal Technology Corporation, Dayton, Ohio, pp. 1a 1-1a 66.
59. Behrndt, K. H. and Jones, R. A. (1961). *Vacuum*, 11 (3):129-138.
60. Eschbach, H. L. (1964). *Vak. Tech.*, 13 (5):141-145.
61. Ritter, E., Vacuum coating, In Winkler, O., Bakish, R., *Vacuum Metallurgy*, Elsevier, Amsterdam, pp. 803-820.
62. Dietrich, W., Gruber, H., Sperner, F., and Stephan, H. (1968). *Vacuum*, 18 (12):657-663.
63. Schiller, S., Förster, H., Lenk, P., and Jäsch, G. (1972). "High power electron guns for evaporation," In Bakish, R., Electron and Ion Beam Science and Technology. Proceeding of Fifth International Conference '72, Houston, Texas, Electrochemical Society, Princeton, N.J., pp. 399-411.
64. Schiller, S., Beister, G., Neumann, M., and Jäsch, G. (1982). *Thin Solid Films*, 96:199-216. Int. Conf. Met. Coat., San Diego, California.
65. Schiller, S. and Neumann, M. (1985). *Vakuum-Technik*, 34 (4):99-109.
66. Schiller, S., Jäsch, G., and Neumann, M. (1983). *Thin Solid Films*, 110:149-164.
67. Sommerkamp, P. "Application of EB coating in the metal industries." In Silva, R. M. Second Electron Beam Processing Seminar,

Frankfurt (Main), Universal Technology Corporation, Dayton, Ohio, pp. S 4b 1-4b 71.

68. Schiller, S., Beister, G. and Zeißig, G. (1968). *Technik, 23* (12):775-778.

69. Flashverdampfungseinrichtung BEF 101, Brochure Balzers AG, Liechtenstein.

70. Behrndt, K. H. (1966). Film thickness and deposition rate monitoring devices, In Hass, G. and Thun, R. E., *Physics of Thin Films*, vol. 3. Academic, New York, pp. 1-59.

71. Pulker, H. and Ritter, E. (1965). *Vak. Tech., 14* (4):91-97.

72. Steckelmacher, W. (1971). *Vak. Tech., 20* (5):139-150.

73. Greaves, C. (1970). *Vacuum, 20* (8):332-340.

74. Greaves, C. (1970). Film thickness measurement by absolute methods. *Vacuum, 20* (10):437-442

75. Wulff, G. (1977). *Vak. Tech., 26* (2):39-48.

76. Schiller, S., Heisig, U., and Goedicke, K. (1977). "Ion deposition techniques for industrial application." In Dobrozemsky, R. et al., Proceedings of the Seventh International Vacuum Congress and Third International Conference on Solid Surfaces '77, Vienna, vol. 2. F. Berger and Söhne, Horn, pp. 1545-1552.

77. Weissmantel, C. Trends in thin film deposition methods. In (76), pp. 1533-1544.

78. Thornton, J. A. (1982). *Deposition Technologies for Films and Coatings, Development and Applications* (Bunshah, R. F., ed.), Noyes Publications, New Jersey.

79. Schiller, S., Heisig, U., and Goedicke, K. (1976). *Vak. Tech., 25* (3):65-72, and (4):113-120.

80. Mattox, D. M. (1964). *Electrochem. Technol., 2* (9-10):295-298.

81. Chambers, D. L. and Carmichael, D. C. (1971). *Res. Dev. Ind., 22* (5):32-35.

82. Mattox, D. M. in (78).

83. Spalvins, T. (1980). *J. Vac. Sci. Technol., 17*:315.

84. Heinz, B. and Kienel, G. (1977). "Ion plating with electron beam evaporation," Proceedings of the International Conference on Ion Plating and Allied Techniques, London, CEP Consultants, Edinburgh, pp. 73-79.

85. Pulker, H. K., Buhl, R., and Moll, E. Ion plating in industrial thin film production. In (76), pp. 1595-1598.

86. Mattox, D. M. (1973). *J. Vac. Sci. Technol., 10* (1):47-52.

87. Mattox, D. M. (1979). "Mechanism of ion plating," Proceedings of the International Conference on Ion Plating and Allied Techniques '79, London, CEP Consultants, Edinburgh, pp. 1-10.

88. Kennedy, K. D., Scheuermann, G. R., and Smith, H. R. (1971). *Res. Dev. Ind., 22* (11):40.

89. Schiller, S., Heisig, U., and Goedicke, K. (1975). *J. Vac. Sci. Technol, 12* (4):858-864.

90. Mah, G., McLeod, P. S., and Williams, D. G. (1974). *J. Vac. Sci. Technol.*, *11* (4):663-665.
91. Bland, R. D., Kominiak, G. J., and Mattox, D. M. (1974). *J. Vac. Sci. Technol.*, *11* (4):671-674.
92. McLeod, P. S. and Mah, G. (1974). *J. Vac. Sci. Technol.*, *11* (1):119-121.
93. Bunshah, R. F. and Juntz, R. S. (1972). *J. Vac. Sci. Technol.*, *9* (6):1404-1405.
94. Mattox, D. M. and Kominiak, G. J. (1972). *J. Vac. Sci. Technol.*, *9* (1):528-531.
95. Wan, C. T., Chambers, D. L., and Carmichael, D. C. (1973). "Effects of process variables on the structure of ion-plated coatings," In Vacuum Metallurgy, Fourth International Conference '73, Tokyo.
96. Boone, D. H., Strangman, T. E., and Wilson, L. W. (1974). *J. Vac. Sci. Technol.*, *11* (4):641-646.
97. Movčan, B. A. and Demčišin, A. V. (1969). *Fiz. Met. Metalloved.*, *28* (4):653-660.
98. Bunshah, R. F. (1974). *J. Vac. Sci. Technol.*, *11* (4):633-638.
99. Patten, J. W., McClanahan, E. D., and Johnston, J. W. (1971). *J. Appl. Phys.*, *42* (11):4371-4377.
100. Berg, R. S., Kominiak, G. J., and Mattox, D. M. (1974). *J. Vac. Sci. Technol.*, *11* (1):52-55.
101. Davy, J. G., Hanak, J. J. (1974). *J. Vac. Sci. Technol.*, *11* (1):43-46.
102. Carmichael, D. C. (1974). *J. Vac. Sci. Technol.*, *11* (4):639.
103. Aisenberg, S. and Chabot, R. W. (1973). *J. Vac. Sci. Technol.*, *10* (1):104-107.
104. Matthews, A. and Teer, D. G. Ion Plated Titanium Nitride Coatings for Dies and Moulds. In (87), pp. 11-20.
105. McLeod, H. A. Optical thin film deposition. In (87), pp. 74-83.
106. Steube, K. E. and McCrary, L. E. (1974). *J. Vac. Sci. Technol.*, *11* (1):362-365.
107. White, G. W. (1973). *Res. Dev. Ind.*, *24* (7):43-46.
108. Stowell, W. R. and Chambers, D. (1974). *J. Vac. Sci. Technol.*, *11* (4):653-656.
109. Wolter, A. R. (1965). *Microelectron Reliab.*, *4* (1):101-102.
110. Morley, J. R. and Smith, H. R. (1972). *J. Vac. Sci. Technol.*, *9* (8):1377-1378.
111. Bunshah, R. F. (1983). *Phys. Thin Films*, *13*.
112. Murayama, J. (1975). *J. Vac. Sci. Technol.*, *12*:818.
113. Nath, P. and Bunshah, R. F. (1980). U.S. Patent 4, 336, 277, June 22, 1982. *Thin Solid Films*, *69*:63.
114. Matthews, A. and Teer, D. (1981). *Thin Solid Films*, *80*:41.
115. Bunshah, R. F. (1976). Physical vapor deposition of metals, alloys' and compounds. New Trends in Materials Processing. American Society for Metals, Metals Park, Ohio, p. 200.

116. Kobayashi, M. and Doe, J. (1978). *Thin Solid Films*, *54*:67.
117. Bunshah, R. F. (1983). *Thin Solid Films*, *107* (1).
118. Bunshah, R. F. in (78).
119. van Vorous, T. (1976). *Solid State Technol.*, *19* (12):62–66.
120. Schiller, S., Heisig, U., and Goedicke, K. (1977). *Thin Solid Films*, *40* (1-3):327–334.
121. Schiller, S., Heisig, U., and Goedicke, K. (1978). *Vak. Tech.*, *27* (2):51–55 and (3):75–86.
122. Waits, R. K. (1978). Planar magnetron sputtering. In Vossen, J. L. and Kern, W. *Thin Film Processes*. Academic, New York, pp. 131–173.
123. Thornton, J. A., Penfold, A. S. Cylindrical magnetron sputtering. In (122).
124. Fraser, D. B. Sputter- and S-gun magnetrons. In (122).
125. Schiller, S., Heisig, U., Goedicke, K., Schade, K., Teschner, G., and Henneberger, J. (1979). *Thin Solid Films*, *64* (3):455–467.
126. Schiller, S., Heisig, U., and Goedicke, K. (1983). *Vakuum-Technik*, *32* (2):35–47.
127. Schiller, S., Heisig, U., Neumann, M., and Beister, G. (1986). *Vakuum-Technik*, *35* (4):35–54.
128. Williams, D. G. (1974). *J. Vac. Sci. Technol.*, *11* (1):374–376.
129. Moll, E. and Daxinger, H. (1977). BRD-Patent No. 282 3876.
130. Morley, J. R. (1968). U.S.-Patent No. 3,562,141.
131. Nakamura, K., Zugawa, K., Tsuruoka, K., and Komiya, S. (1977). *Thin Solid Films*, *40*:155–167.
132. Lunk, A., Schrader, F., and Wilberg, R. (1981). *Vakuuminformation*, *11*:351–355.
133. Snaper, A. A. (1971). U.S.-Patent No. 3,625,848.
134. Sherman, M. A. (1975). *J. Vac. Sci. Technol.*, *12* (3):697–703.
135. Bunshah, R. F. (1974). *J. Vac. Sci. Technol.*, *11* (4):814–819.
136. Anderson, A. (1973). Physical vapor deposition—present and future, Automobile Engineering Meeting, Detroit, Michigan.
137. Boone, D. H., Lee, D., and Shafer, J. M. The electron beam coating of turbine components and ion plating. In (84), pp. 141–148.
138. Burt, R. A. A preliminary evaluation of ion plating for the deposition of high temperature corrosion resistant alloys. In (84), pp. 135–140.
139. Schiller, S. and Jäsch, G. (1978). *Thin Solid Films*, *54* (1):9–21.
140. Brill, K. and Grothe, W. An electron beam evaporator for band coating. In (76), pp. 1587–1590.
141. Schiller, S., Jäsch, G., and von Ardenne, A. (1978). The Effects of Gas/Vapor and Plasma on High Power Electron Beam Generation and Guidance," Eighth International Conference on Electron and Ion Beam Science and Technology, Seattle, Washington.

142. Schiller, S., Förster, H., Jäsch, G. (1975). *J. Vac. Sci. Technol.*, *12* (4):800–805.
143. Schiller, S., Heisig, U., and Steinfelder, K. (1976). *Thin Solid Films*, *33*:331–339.
144. Schiller, S., Heisig, U., Steinfelder, K., and Gehm, K. (1978). *Thin Solid Films*, *51*:189–196.
145. Schiller, S., Förster, H., Hötzsch, G., and Reschke, J. (1981). "Advances in Mechanical Activation as a Pretreatment Process for Vacuum Deposition." International Conference on Metallurgical Coatings, San Francisco, California.

# 8

# Boriding and Diffusion Metallizing

RUTH CHATTERJEE-FISCHER  *Institut für Harterei Technik, Bremen, Federal Republic of Germany*

## 1   BORIDING

Boriding is a thermochemical treatment applied to an object in order to produce a surface layer of borides. This treatment can be applied to ferrous materials, cermets, and certain nonferrous materials, such as titanium, tantalum, or nickel-based alloys (Chatterjee-Fischer and Schaaber, 1976). The main purpose of this treatment is to improve wear behavior. The procedure is quite expensive, so its use has generally been restricted to engineering components which need the outstanding wear resistance of the boride layers. An increasing application range offers local boriding with pastes.

### 1.1   Boriding of Ferrous Materials

From the Fe-B phase diagram two borides can be obtained, the iron subboride $Fe_2B$ and the iron boride FeB (Fig. 1). Some physical properties of these borides are given in Table 1.

  Boriding of ferrous materials is generally performed from 840 to 1050°C. The reactive medium can be in the solid, liquid, or gaseous (plasma) state. Boriding takes place in two steps:

1.   In the first step, reactions between the reaction partners in the boriding medium and the surface of the object occur. Particles will be nucleated at the surface and increase very rapidly with boriding time till a thin compact boride layer has been obtained. The time to attain this state depends on the boriding

temperature. At 900°C this step takes about 10 minutes. The incubation time itself can be neglected because it is very short in comparison to the whole process time.

2.  The second step starts just after the formation of the thin compact layer. This step is diffusion-controlled, which means that the thickness of the boride layer grows according to a parabolic time law

$$x^2 = kt \qquad\qquad\qquad\qquad (1)$$

where x = thickness of the layer, cm; t = boriding time, s; and k = constant depending on the temperature. Values for k are given in Table 2 for some typical unalloyed steels.

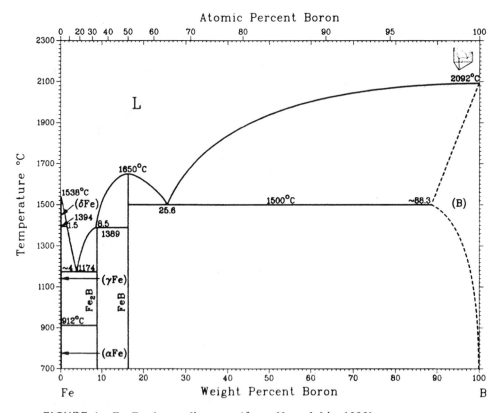

FIGURE 1   Fe-B phase diagram (from Massalski, 1986).

TABLE 1  Physical Properties for Iron Borides

1. Composition
   $Fe_2B$    8.83 wt.% B
   FeB     16.23 wt.% B

2. Lattice
   $Fe_2B$    tetragonal (a = 5.078 Å, c = 4.249 Å)
   FeB     orthorhombic (a = 4.053 Å, b = 5.495 Å, c = 2.946 Å)

3. Specific density (Kunst et al., 1967)
   $Fe_2B$    7.43 $g/cm^3$
   FeB     6.75 $g/cm^3$

3. Thermal expansion coefficients

   Due to von Matuschka (1977) between 200 and 600°C
   $Fe_2B$    7.85 × $10^{-6}$ $deg^{-1}$
   FeB      23 × $10^{-6}$ $deg^{-1}$

   Due to Kunst (1972) between 100 and 800°C
   $Fe_2B$    9.2 × $10^{-6}$ $deg^{-1}$

4. Young's modulus

   Due to Lyachova et al. (1972)
   $Fe_2B$    30,000 $kp/mm^2$
   FeB     60,000 $kp/mm^2$

   Due to Chou et al. (1977)
   $Fe_2B$    29,000 $kp/mm^2$

5. Diffusion coefficient of boron in iron

   Due to Kunst et al. (1967)
   $D_{950°C}$    1.53 × $10^{-7}$ $cm^2$ $s^{-1}$ for the diffusion zone
   $D_{950°C}$    1.82 × $10^{-8}$ $cm^2$ $s^{-1}$ for the boride layer

   Due to Krishtal (1974)
   $D_{980°C}$    1.82 × $10^{-7}$ $cm^2$ $s^{-1}$ for the diffusion zone

Iron boride layers have a characteristic appearance. The tooth-like configuration is due to the preferred diffusion direction. The layer shows a strong <002> texture (Kunst and Schaaber, 1967). This structure is especially obvious with pure iron or unalloyed low-carbon steels (Fig. 2). With increasing content in carbon and/or alloying elements this characteristic configuration diminishes, as shown in Fig. 3.

The thickness of the boride layer is reduced if the content of the alloying elements is higher. This is due to alloying elements hindering the diffusion of boron into the steel (Fig. 4).

TABLE 2   Values for k for Estimating the Thickness of the Boride
Layer for Iron and Some Unalloyed Steels

| Material | $k\ cm^2\ s^{-1} \times 10^{-8}$ | | | |
|---|---|---|---|---|
| DIN spec | 1000°C | 950°C | 900°C | 800°C |
| Iron | 3.40 | 2.16 | 1.22 | 0.330 |
| Ck15 | 2.88 | 1.87 | 1.06 | 0.319 |
| C45 | 2.45 | 1.56 | 1.02 | 0.283 |
| C60 | 2.29 | 1.44 | 0.924 | 0.262 |
| C100 | 1.69 | 0.982 | 0.497 | 0.207 |

*Source*: From Lu (1983).

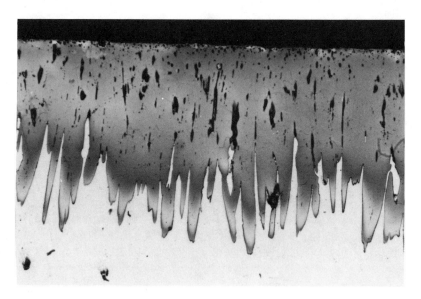

FIGURE 2   Boride layer ($Fe_2B$) on an unalloyed low carbon steel.

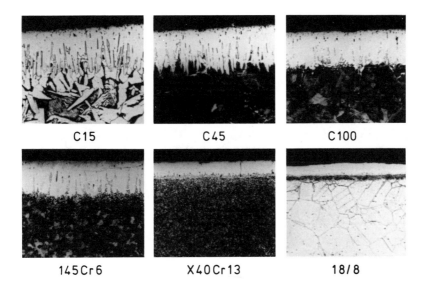

FIGURE 3  Influence of the steel composition on the configuration and thickness of the boride layer (simultaneously borided) (18/8: austenitic steel).

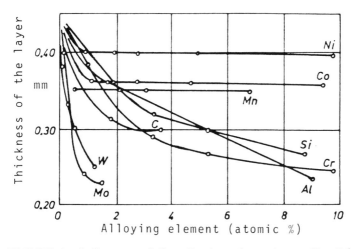

FIGURE 4  Influence of the alloying elements on the thickness of the boride layer.

Most of the alloying elements present in steel are soluble in
the boride layer (e.g., chromium, manganese). Therefore it would
be more appropriate to understand $(Fe,M)_2B$ or $(Fe,M)B$, where
M represents one or more metallic elements.

Carbon and silicon are not soluble in the boride layer. These
elements are pushed from the surface by boron and ahead of the
boride layer into the substrate, leading to an enrichment of these
elements beneath and/or between the boride teeth. An example
for the distribution of these elements in the boride layer of an
unalloyed steel is given schematically in Fig. 5.

The efficiency with which these two elements are removed from
the surface layer by boron may be demonstrated by a siliconized
sample which was subsequently borided (Fig. 6). All silicon is pushed
from the surface and concentrates beneath the boride layer. A similar
phenomenon occurs with carbon.

The enrichment in silicon beneath the boride layer may also
amount to several percent. This enrichment is responsible for forming
a soft ferritic zone beneath the boride layer when boriding low-
carbon, chromium-alloyed steels (Fiedler and Hayes, 1970). This
problem can only be overcome by reducing the silicon content of
the steel. The enrichment in carbon may lead to an oversaturation
of the austenite. Carbon will precipitate in the form of cementite
or boroncementite (e.g., $Fe_3B_{0.8}C_{0.2}$).

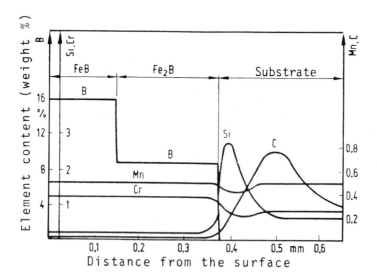

FIGURE 5   Schematic presentation of the element distribution within
and beneath the boride layer.

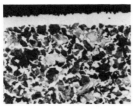

(a)

(b)

(c)

FIGURE 6 Structure and distribution of silicon of a sample that was first siliconized and subsequently borided (siliconized at 900°C, 5h, in silicon; borided at 900°C, 5h, in commercial boriding powder: (a) surface layer after siliconizing, (b) surface layer after subsequent boriding, (c) silicon distribution after boriding.

Besides the boride layer there is also a boron-containing diffusion zone beneath this layer, which may be 10 times as thick as the boride layer. Sometimes it can be observed by the change in structure of the base material due to the increase in hardenability by boron.

## 1.2 Boriding of Nonferrous Metals and Alloys

Nonferrous metals and their alloys can also be borided with great success, mostly by a special technique. Of special interest is the

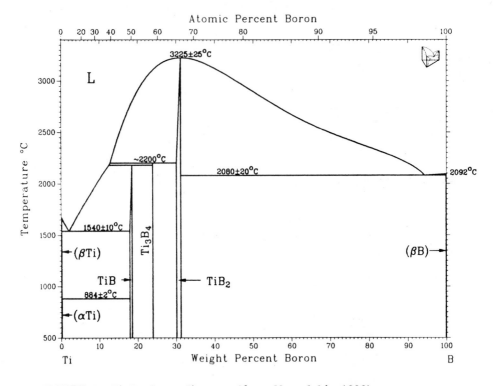

FIGURE 7   Ti-B phase diagram (from Massalski, 1986).

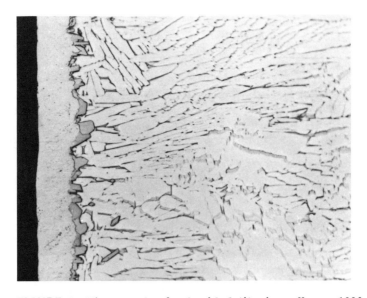

FIGURE 8   Micrograph of a borided titanium alloy. ×1000.

boriding of titanium and its alloys and boriding of nickel alloys. Good results have also been obtained with tantalum and cobalt. Principally, most of the refractory metals can be borided.

Figure 7 shows the Ti-B phase diagram. From this diagram it is clear that three borides can be obtained: TiB with 18 wt.% boron, $TiB_2$ with 30-31 wt.% boron, and $Ti_3B_4$, which has been discovered just recently.

A boride layer on a titanium alloy is presented in Fig. 8. The bright layer consists of the boron-rich compound(s); the darker teeth correspond to TiB. $TiB_2$ has a hexagonal lattice, while TiB is orthorhombic.

## 2  BORIDING PROCEDURES

### 2.1  Ferrous Materials

Ferrous alloys can be borided in solid (powder, pastes), in liquid (salt melts), and in gaseous (gas, plasma) reactive media. In the Western Hemisphere, pack boriding is the preferred procedure, while in eastern countries liquid boriding generally is applied. Gaseous reactive media are seldom used on a commercial scale.

*Pack Boriding*

Generally, pack boriding is carried out in commercial boriding powders, consisting, e.g., of 5% $B_4C$ (as boron donor) + 5% $KBF_4$ (as activator) + 90% SiC (as diluent). The components being borided will be packed into a box, similar to pack carburizing. It is necessary to cover the surface to be borided with a sufficient amount of boriding powder (about 10-20 mm thick). On top of the filled box an inert filler material (e.g., SiC), about 100 mm thick, has to be added. The contents of the box are then covered with a lid. The box should be not greater than 60% of the furnace chamber, and it should be suited to the shape of the objects being borided (e.g., long shafts should be treated in a long round box).

The temperature ranges typically between 840 and 1050°C, although about 900°C is the temperature frequently used. During heating, the temperature range between 700°C and the boriding temperature should be passed rapidly to reduce the formation of FeB (Lu, 1983). When the boriding process is finished, the boxes are removed from the furnace chamber and cooled in air.

Mareels and Wetterich (1985) showed that the reaction takes place as follows:

$$\text{KBF}_4 \xrightarrow{\text{T}\uparrow} \text{KF} + \text{BF}_3$$

$$4\ \text{BF}_3 + 3\ \text{SiC} + \frac{3}{2}\ \text{O}_2 \rightarrow 3\ \text{SiF}_4 + 3\ \text{CO} + 4\ \text{B}$$

$$3\ \text{SiF}_4 + \text{B}_4\text{C} + \frac{3}{2}\ \text{O}_2 \rightarrow 4\ \text{BF}_3 + \text{SiO}_2 + \text{CO} + 2\ \text{Si}$$

$$\rule{9cm}{0.4pt} \tag{2}$$

$$\text{B}_4\text{C} + 3\ \text{SiC} + 3\ \text{O}_2 \xrightarrow[\text{SiF}_4]{\text{BF}_3} 4\ \text{B} + 2\ \text{Si} + \text{SiO}_2 + 4\ \text{CO}$$

Not all experts agree with this reaction sequence. But it appears that, at least at high temperatures, silicon carbide is an active reaction partner.

It can be seen from the above equations that oxygen is necessary for the reaction. The component KF is liquid at the boriding temperature and is responsible for sintering the particles of the boriding powder. Goeuriot et al. (1981) proposed the use of $\text{BF}_3$ instead of $\text{KBF}_4$ as activator and developed a special boriding process called Borudif.

As already mentioned, a monophase iron subboride layer is desired. This cannot be obtained with higher alloyed steels when commercial boriding powder in the delivered state is used. Table 3 shows the influence of the $\text{B}_4\text{C}$ content in the powder mixture on the constitution of the boride layer of different steels.

The influence of the $\text{B}_4\text{C}$ content in the boriding powder on the constitution of the boride layer of steel 42 CrMo 4 (0.42% C, 1% Cr, 0.25% Mo) is shown in Fig. 9. By decreasing the $\text{B}_4\text{C}$ content to 2.5%, we can obtain a FeB-free boride layer. With higher alloyed steels this method is no more successful. In such cases it is recommended that 5 to 15% chromium powder be added to the boride mixture (10 to 15% for high chromium-alloyed steels). Due to the high affinity between boron and chromium, a stable chromium boride ($\text{Cr}_2\text{B}$) will be formed in the powder, and hence the boron availability of the boriding powder is effectively decreased.

A new variant of boriding with solid media is boriding by using a fluidized-bed technique (Fig. 10). The fluidized medium is a special boriding powder. This powder has to be fluidized by an oxygen-free gas (e.g., by a $\text{N}_2/\text{H}_2$ mixture; von Matuschka and Trausner, 1986). Fig. 11 shows a boride layer obtained by treating in a fluidized bed.

When boriding is performed with paste a protective gas or a vacuum furnace is necessary. This procedure is of special interest if partial boriding is desired.

TABLE 3  Influence of the $B_4C$ Content of the Boriding Powder on the FeB Phase of the Boride Layer of Different Steels (borided at 900°C for 5 h, 18/8: austenitic steel)

| steel | $B_4C$, % 2·5 | 5 | 7·5 | 10 |
|---|---|---|---|---|
| C 15 | □ | □ | ◪ | ◨ |
| C 45 | □ | □ | ◪ | ◨ |
| 42 Cr Mo 4 | □ | ◪ | ◨ | ▬ |
| 61 Cr Si V 5 | □ | ◪ | ◨ | ■ |
| C 100 | □ | ◪ | ◨ | ■ |
| 100 Cr 6 | □ | ◨ | ▬ | ■ |
| 145 Cr 6 | ◪ | ▬ | ■ | ■ |
| x 40 Cr 13 | ■ | ■ | ■ | ■ |
| 18/8 | ■ | ■ | ■ | ■ |

□ no FeB

◪ FeB only in corners

◨ FeB individual serrations

▬ FeB no enclosed layer

■ FeB enclosed layer

5% $B_4C$     ▬ 10μm     2,5% $B_4C$

FIGURE 9  Influence of the $B_4C$ content of the boriding powder on the constitution of the boride layer of steel 42 CrMo 4.

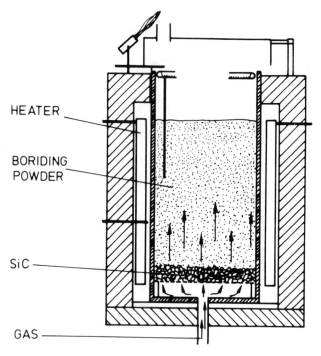

HEATER

BORIDING
POWDER

SiC

GAS

FIGURE 10   Schematic presentation of a fluidized bed for boriding
(from von Matuschka and Trausner, 1986).

FIGURE 11   Boride layer obtained by fluidized bed boriding (alloyed
construction steel).  ×500.

The thickness of the boride layer depends on the temperature, the boriding time, and the composition of the steel (Figs. 12, 13). Recommendations for the thickness of the boride layer of some steels and their boriding and hardening temperatures are given in Table 4.

## Liquid Boriding

Liquid boriding takes place in borax-based salt melts ($Na_2B_4O_7$) with or without electrolysis. The electrolytic liquid boriding, water-free $Na_2B_4O_7$ is used, to which NaCl and $B_2O_3$ are added. The object acts as the cathode; graphite or platinum act as the anode.

Liquid boriding without electrolysis is carried out in a borax-based melt to which $B_4C$ or SiC is added. The boride layer obtained in a borax-based melt to which 30% SiC was added is shown in Fig. 14. A monophase $Fe_2B$ layer has been obtained.

The electrolytic procedure may have some disadvantages, such as nonuniform thickness of the boride layer due to different current density or due to the shadow effect on the anodic side.

## Gas Boriding

In principle, it is also possible to boride by using different gaseous media, such as $BCl_3$ and $B_2H_6$, and several researchers have investigated this, but no industrial process is available. This is largely due to the corrosive and/or toxic behavior of these gaseous media, which lead to difficulties in handling. Hegewaldt et al. (1984) found that gas boriding can be applied to unalloyed steels at tempera-

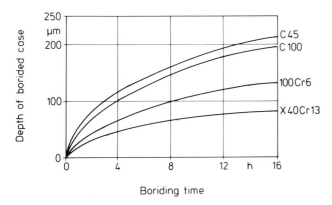

FIGURE 12 Influence of the composition of the steel on the thickness of the boride layer. Boriding temperature, 900°C (commercial boriding powder).

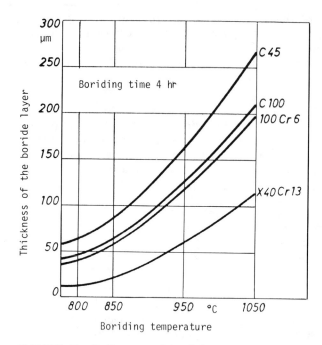

FIGURE 13   Influence of boriding temperature and steel composition
on the thickness of the boride layer (courtesy of Degussa, Hanau).

tures about 650°C. Figure 15 shows a schematic presentation of gas
boriding equipment. It is a pit-type furnace which is vacuum sealed.
All parts in contact with the hot reaction gases are coated. The
boriding gas consists of a $N_2/H_2$ $BCl_3$ mixture. The $BCl_3$ portion
is very small, so a very fine regulation of this gas is necessary.
This small amount of $BCl_3$ is also important for avoiding the formation
of $FeCl_2$ (corrosion).

*Plasma Boriding*

Plasma boriding can also be performed with similar media, as in
the gas boriding process. There have been several investigations
in this field (e.g., Casadesus and Gantois, 1978; Dearnley et al.,
1986), but no industrial process is known to be available for commercial treatment.

*Pretreatment*

Engineering components to be borided should be cleaned before
treatment. Components susceptible to distortion should be stress

TABLE 4 Recommended Boriding and Hardening Temperatures and Thicknesses of the Boride Layers Applied in Cold-Work Technique

| Steel | Boriding temperature (°C) | | Hardening after boriding quenching | | Thickness of the boride layer (μm) | |
|---|---|---|---|---|---|---|
| | Without quenching | With quenching | Temperature (°C) | Medium | Without hardening | With hardening |
| C 45 | 920 | 880 | 840 | Oil/salt bath | 40–200 | 60 |
| 42 CrMo 4 | 920 | 880 | 840 | Oil/salt bath | 20–150 | 60 |
| 90 MnCrV 8 | 880 | 840 | 800 | Oil/salt bath | 40–100 | 40 |
| 105 WCr 6 | 900 | 840 | 820 | Salt bath | 40–100 | 50 |
| X 165 CrMoV 12 | 880 | 1020 | 1020 | Salt bath | 20–80 | 40 |
| X 210 CrW 12 | 880 | 980 | 950 | Salt bath | 20–80 | 40 |

FIGURE 14   Boride layer on steel C45 obtained in a borax-based
melt with 30% SiC. Boriding temperature, 950°C; boriding time, 5 h.

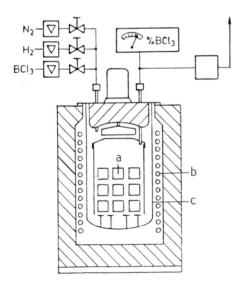

FIGURE 15   Schematic presentation of a gas boriding furnace: (a)
parts to be borided, (b) heater, (c) coated furnace wall (from
Hegewaldt et al., 1984).

relieved. Further, to obtain a good layer quality, the surface should be ground if possible.

Sharp edges and corners should be avoided to avoid the danger of a buildup effect due to diffusion from two sides. Here also the formation of FeB will be favored (Fig. 16). It is obvious that such edges may crack off when loaded.

Generally, borided objects show a positive dimensional change, which means the dimensions of the component are increased. This is due to the differences of the specific volumes between the substrate and the boride layer. These changes can be estimated and thus taken into consideration during manufacturing. The changes are about 10 to 25% of the thickness of the boride layer and depend on the composition of the base material.

### After Treatment

Borided components can be quench-hardened and tempered. The heating to hardening temperature should be carried out under oxygen-free protection gas or in a neutral salt melt (borided layers are oxidized in oxygen-containing media at temperatures above 700°C). The highest hardening temperature should be 1050°C (1080°C) (eutectic, 1174°C). Ammonia should be avoided, since the iron borides transform to iron nitride at 600°C. FeB will be resistant in ammonia-containing atmospheres up to 400°C, and $Fe_2B$ up to 352°C.

Recommended quenching media are oil, salt baths, and air. The hardening data should be suitably chosen to avoid cracking. Perpendicular cracks are mostly due to unsuitable volume changes of the substrate.

Borided components should always be quench-hardened if they have to undergo a certain surface pressure (Hertz's pressure).

Surface roughness will be increased by boriding, especially when there was a fine surface finish before treatment (Fig. 17). If the borided surface is too rough for the application assigned, it can be ground (with special polishing disks), polished, or otherwise surface treated.

### 2.2 Nonferrous Metals

Boriding of nonferrous metals (titanium and its alloys, tantalum, etc.) is usually carried out in amorphous boron. Some of the nonferrous metals (e.g., titanium and its alloys) have to be borided under high-purity argon or under high vacuum. In these cases oxygen has to be removed from the amorphous boron by a suitable annealing treatment (e.g., vacuum degassing) before the boriding process starts.

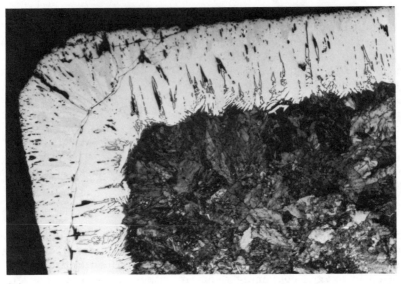

(a)

FIGURE 16   "Cushion-effect and formation of FeB at sharp corners and edges.

Titanium and its alloys are borided preferably between 1000 and 1200°C. The thickness of the compact close $TiB_2$ layer is about 12 μm after 8 h and about 20 μm after 15 h treatment at 1000°C. A thickness of 20 μm will be obtained at 1200°C after 8 h. The boride layer is strongly interlocked with the base metal so that outstanding cohesion of this layer with the base metal is ensured.

Similar conditions apply for boriding tantalum. A single-phase tantalum boride layer is obtained. The thickness of the layer after 8 h at 1000°C is about 45 μm (Fig. 18).

Boriding of nickel-based alloys can be carried out, e.g., at 940°C. On the nickel alloys IN 100 and Nimonic 80, a boride layer 60 μm thick was obtained after 8 h at 940°C (Fig. 19). With tantalum and nickel-based alloys the boride layers show no serrations as with the titanium samples. The layers mostly lie smoothly on the base metal.

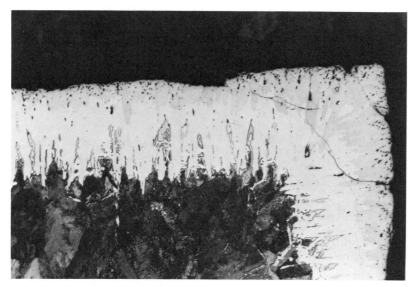

(b)

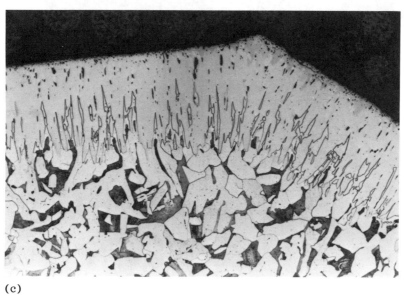

(c)

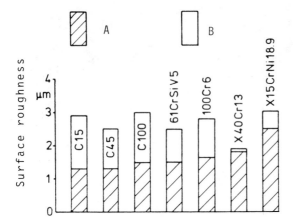

FIGURE 17   Surface roughness of borided steels: (A) roughness
before boriding, (B) roughness after boriding.

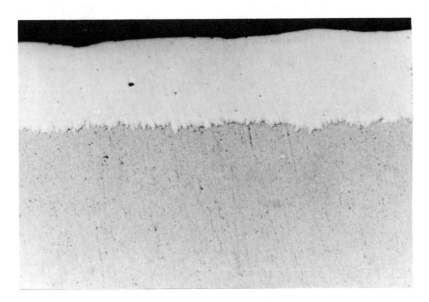

FIGURE 18   Boride layer on tantalum (borided at 1000°C for 8 h).
×500.

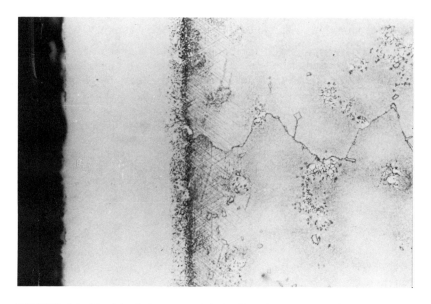

FIGURE 19   Boride layer on nickel alloy IN 100 (borided at 940°C for 8 h). ×500.

## 3   CONTROLLING THE BORIDE FORMATION

The extent to which boriding occurs can be controlled by

Hardness testing
Determining the thickness of the boride layer
Determining the constitution of the layer (e.g., the ratio between
   FeB and $Fe_2B$)

### 3.1   Hardness Test

Hardness is often determined by the microhardness method, either on the surface or in the cross section. The hardness of iron boride on unalloyed steels is about 1500 to 1600 HV 0.2. With high chromium-alloyed steels it may increase up to 1900 to 2000 HV 0.2.

The hardness of the $TiB_2$ layer on the titanium alloy Ti-Al6-V4 is higher than 3000 HV 0.1. In tantalum borides the layers show a hardness of about 3200 HV 0.2. The hardness of the borided IN 100 samples is about 1700 HV 0.2.

## 3.2  Determining the Thickness of the Boride Layer

Generally, determining the layer thickness is achieved by calculating
the values from a micrograph. An example for the metallographic
estimation of the layer thickness is shown in Fig. 20. The distances
between the surface and the interface boride layer-base material
are generally determined at certain specific intervals (e.g., each
2 mm). The thickness d will be calculated by the arithmetic mean.

By using the eddy current method, it is possible to determine
the thickness of the boride layer without destroying it. A schematic
presentation of the equipment is given in Fig. 21. Its main parts
are the sensor S and the microcomputer. The procedures will be
described by the impedance of the sensor

$$Z = \frac{U}{I} \tag{3}$$

where U is induced voltage and I is induction current, which is
a function of the permeability, the specific electrical conductivity,
and the thickness of the layer.

## 3.3  Determining the Constitution of the Boride Layer

As already mentioned, a monophase $Fe_2B$ layer is desired. Often
it is difficult to avoid the formation of FeB in practice. Therefore
the ratio $FeB/Fe_2B$ will be determined, and/or the constitution will
be specified as shown in Fig. 22.

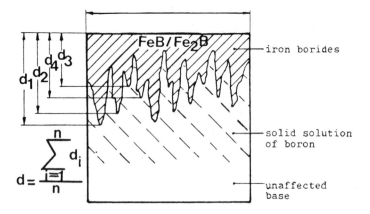

FIGURE 20  Metallographic estimation of the layer thickness d.

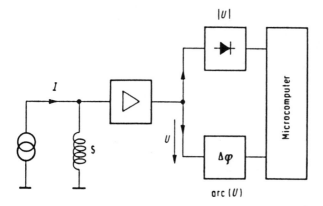

FIGURE 21 Schematic presentation of the equipment for determining the depth of the boride layer (from Ott, 1983).

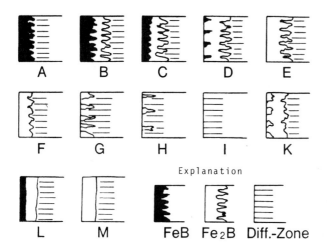

FIGURE 22 Classification of the types of boride layers on steel (from Kunst and Schaaber, 1977).

FIGURE 23   Cracks at the interface FeB/Fe$_2$B.

The phases FeB and Fe$_2$B have different thermal expansion coefficients. FeB is produced under a tensile stress, while Fe$_2$B is formed under a compressive stress. At the interface of the two phases, cracks frequently occur. These cracks are parallel to the surface (Fig. 23) and may lead to spalling under load. Therefore a monophase layer is preferred.

Principally, the phases can be determined by x-ray phase analysis. This is true for all borided alloys.

## 4   PROPERTIES

### 4.1   Wear Resistance

The wear resistance of borided components is often tested by the Faville test (the cylinder is rotating in a similar treated prism), by the pin-and-disk test (adhesion as well as tribooxidation), and by the grinding test (abrasive wear). Examples are presented in Figs. 24 to 26. It is obvious that the wear behavior can be decisively improved by boriding in all cases. It also can be seen that with abrasive wear, the composition of the steel is also of importance. This is a general observation. Sometimes it has been observed that hardened samples behave better than slowly cooled ones. This may

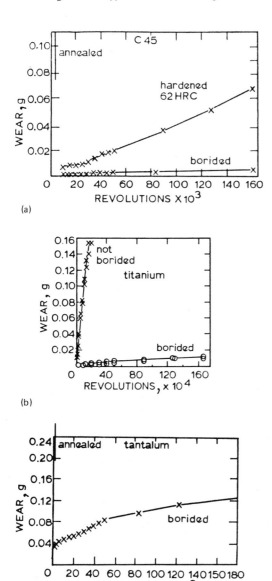

(a)

(b)

(c)

**FIGURE 24** Influence of boriding on the wear resistance (Faville test): (a) steel C45 (borided at 900°C for 3 h), (b) titanium (borided at 1000°C for 24 h), (c) tantalum (borided at 1000°C for 8 h).

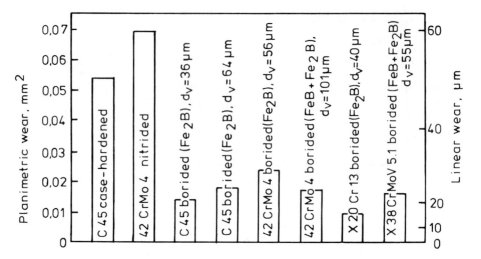

FIGURE 25  Influence of steel composition on wear resistance under tribooxidative wear ($d_V$ = thickness of the boride layer). Test conditions: pin-disk test; relative humidity, 50%; temperature, 23°C; velocity, 0.1 m/s; test way, 1000 m.

be partially due to the structural changes by hardening and/or to the higher hardness of the base material.

The best wear behavior has been observed when wear takes place within one phase. Investigations with higher alloyed steels show that this one phase can also be FeB, but the FeB layer has to be thick enough that wear is limited to this one phase. With a thin, interrupted FeB layer, wear takes place in both phases (Fig. 27). The inhomogeneous microstructure may be responsible for the high wear rate and/or the great scatter band.

## 4.2  Corrosion Behavior

Borided steel components also show a good corrosion resistance in various aggressive media. Good resistance to corrosion has been observed in chloride-containing media, specifically in 20% HCl-containing liquids. The iron boride teeth can be isolated by etching in a hot 18% HCl-containing solution. The base material will be dissolved, and the boride teeth will remain unaffected (Fig. 28).

In humid atmospheres borided components often appear corroded on the surface, but this "corrosion" can easily be wiped off.

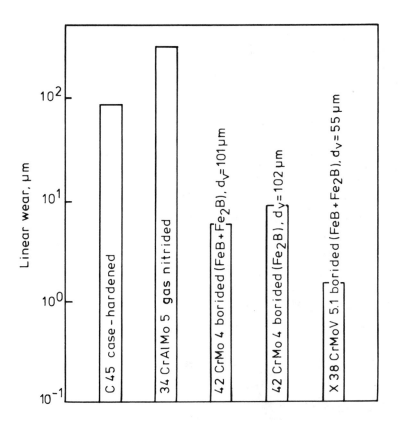

**FIGURE 26** Influence of steel composition on wear resistance under abrasive wear ($d_v$ = thickness of the boride layer). Test conditions: DP-U grinding tester, SiC paper 220, testing time, 6 min.

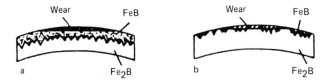

**FIGURE 27** Wear in boride layers: (a) thick, continuous FeB layer; (b) thin, interrupted FeB layer.

FIGURE 28   Scanning micrograph of isolated boride teeth.


4.3   Mechanical Properties

Static bending tests on unnotched circular samples of 42 CrMo 4
of different diameters and a boride layer about 80 μm thick showed
that in both slowly cooled and quench-hardened tempered states
the bending strength was not noticeably impaired. The same applies
for deflection (as index of ductility).

The endurance limit of samples is generally not negatively
affected by boriding. With notched samples sometimes even an
increase was observed.

## 5 MULTICOMPONENT BORIDING: A SPECIAL TREATMENT

Multicomponent boriding is a thermochemical treatment involving the consecutive diffusion of boron and one or more metallic elements. This method was first mentioned by Zemskov and Kaidash (1965).
This treatment will be carried out in two steps:

1. Boriding in the conventional manner, producing a compact layer at least 30 μm thick. The presence of FeB can be tolerated and may even prove useful in some cases, since the borides of the metals diffused into the borided surface in the second step may need more boron than iron.
2. Diffusing metallic elements into the borided surface, which is performed in powder mixtures (e.g., in a mixture of ferro-chromium (as Cr-donor), $NH_4Cl$ (as activator), and $Al_2O_3$ (as diluent)) or in a borax-based melt (Chatterjee-Fischer, 1986). If the powder route is chosen, sintering of particles can be prevented by introducing argon or hydrogen gas into the reaction vessel. The temperature ranges between 850 and 1050°C.

The most interesting coatings are obtained by

Borochromizing—coatings are both wear resistant and highly corrosion resistant
Borovanadizing or borochromvanadizing—coatings are highly wear resistant
Borochromtitanizing—coatings have high resistance to abrasive wear and corrosion
Boroaluminizing—coatings have good corrosion resistance, particularly in humid environments

Figure 29 is a micrograph of a borochromtitanized engineering component made of an alloyed construction steel. The outer gray layer consists of titanium boride; the brighter layer beneath is iron-chromium boride, as can be seen in Fig. 30. The surface hardness of this layer can be extremely high.

Borovanadized and borochromvanadized layers are quite ductile, though their hardness exceeds 3000 HV 0.015. Thus, there is little danger of spalling under typically encountered impact loading conditions.

Wear resistance has been tested by the Faville test and the grinding test. The results obtained by the Faville test are presented in Fig. 31, and those obtained by the grinding test are shown in Fig. 32. The corrosion behavior of these layers in an atmosphere with 100% relative humidity and at 40°C is shown in Fig. 33.

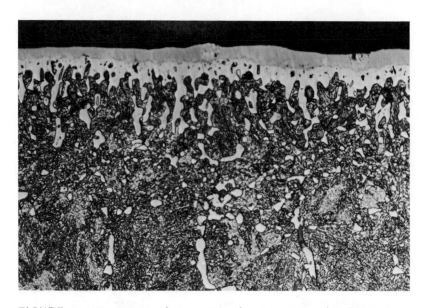

FIGURE 29   Structure of the case of a borochromtitanized alloy construction steel. ×500.

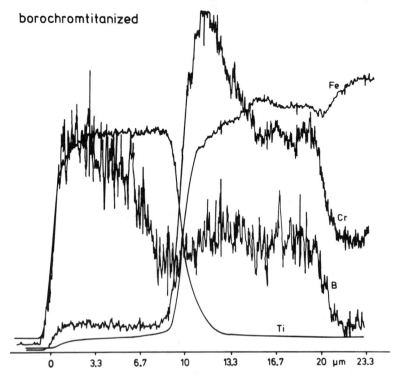

FIGURE 30   Element distribution in the case as determined by electron microprobe analysis.

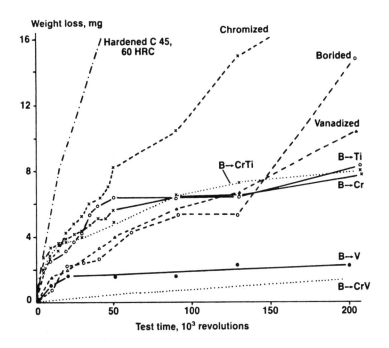

FIGURE 31  Performance of various coatings on C45 steel under metal/metal wear (Faville test).

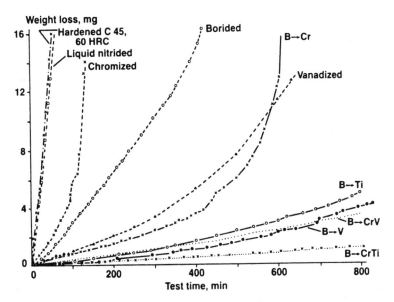

FIGURE 32  Performance of various coatings on C45 steel under conditions of abrasive wear (grinding tester, SiC grinding paper).

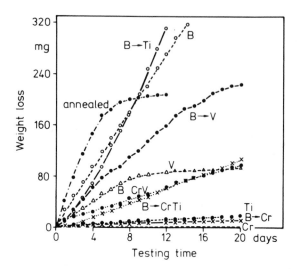

FIGURE 33  Performance of various coatings on C45 steel in an
atmosphere with 100% relative humidity at 40°C.

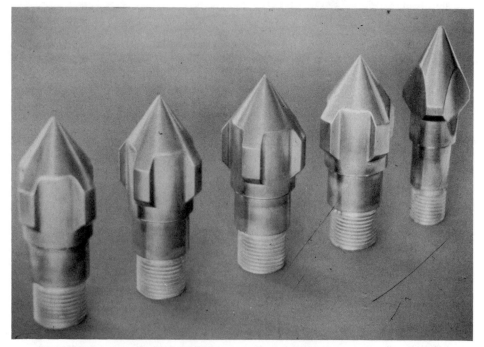

FIGURE 34  Extruder tips, partial borided and quench hardened
(at 1030°C for 1.5 h with pastes under argon, tempered at 300°C)
(courtesy Elektroschmelzverk, Kempten, Federal Republic of Germany).

## 6 APPLICATIONS OF BORIDING

Boriding can be applied to all engineering components and a series of tools which undergo high abrasive wear, such as extruders (Fig. 34), and other components used in transporting or manufacturing plastic materials, different types of dies, components for pumps (Fig. 35), components used in the ceramic industry, combustion parts (Fig. 36), links of conveyer chains, and so on. With great success, screw gears are borided, the only treatment which meets the high requirements of these engineering components (Fig. 37).

(a)

(b)

FIGURE 35  Borided pump components: (a) pump housing (partial borided), (b) impeller (rotor 1) (courtesy Electroschmelzverk, Kempten, Federal Republic of Germany).

FIGURE 36   Combustion parts (pack borided at 900°C for 6 h).

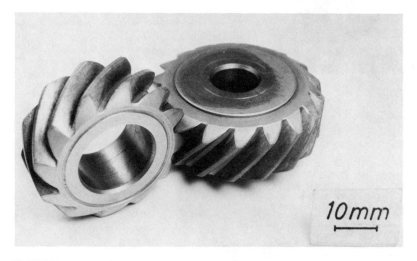

FIGURE 37   Borided screw gears (courtesy Elecktroschmelzverk, Kempten, Federal Republic of Germany).

## 7  DIFFUSION METALLIZING

Special case properties can also be obtained by diffusing metallic elements into the surface of components. Several metallic elements can be diffused into iron and steel such as chromium, vanadium, aluminum, titanium, niobium, and silicon. Because of the substrate and the treatment conditions, either a substitutional solid solution or carbide layers are obtained. When carbide layers are desired, the carbon content of the substrate should be at least 0.45%. Materials with lower carbon content can be carburized before the diffusion metallizing process.

In principle, it is also possible to diffuse simultaneously two metallic elements into the surface of the component (e.g., chromium + vanadium or chromium + titanium).

The treatment can be carried out in donors of all states (gaseous, solid, liquid, and plasma) but there are preferred techniques for the different diffusion processes. At present, the treatment is mostly

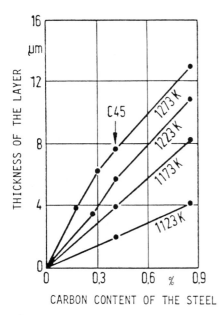

FIGURE 38  Influence of the treating temperature and the carbon content of the substrate on the thickness of the vanadium carbide layer, treating time 4 h (obtained by the TD process) (from Arai, 1979).

performed in powder mixtures and melts. The treating temperature
is usually between 900 and 1100°C. The treating time depends on
the thickness of the layer, the treating conditions (temperature,
composition of the donor, etc.), and the carbon content of the
substrate (Fig. 38).

The growth rate of the layer thickness follows the parabolic
time law

$$x^2 = kt \tag{4}$$

where

x = layer thickness
k = factor (e.g., depending on the temperature)
t = treating time

## 8  EXAMPLES OF DIFFUSION METALLIZING

### 8.1  Chromizing

Chromizing is defined as a thermochemical treatment involving the
enrichment of the surface layer of an object with chromium. The
treatment is preferably carried out in powder mixtures or in borax-
based melts (Arai, 1979). The desired layer is a carbide layer,
usually one of type $Cr_7C_3$ (9% C), but sometimes of type $Cr_{23}C_6$
(5.68% C). The hardness of the $Cr_7C_3$ layer is about 1700 to 2300
HV 0.002 while that of the $Cr_{23}C_6$ type is significantly lower (1100
to 1300 HV 0.02); (Child et al., 1984).

Chromium carbide layers can be obtained on steel as well as
on sintered steel and cast iron with sufficient carbon content.
Examples are shown in Figs. 39 to 41. Figure 39 shows the carbide
layer obtained on an unalloyed steel with 1% C by pack chromizing
(left) and by liquid chromizing (right). Figure 40 shows the cross
section of a pack chromizing cast iron sample, and Fig. 41 gives
an example for a pack chromized sintered steel with 1% C.

When carbon-free iron is chromized, the result is a substitutional
solid solution of low hardness (150 to 250 HV 0.02) but good corrosion resistance. The reaction may be a reduction reaction

$$CrCl_2 + H_2 \rightarrow (Cr) + 2\ HCl \tag{5}$$

or a replacement reaction

$$CrCl_2 + (Fe) \rightarrow (Cr) + FeCl_2 \tag{6}$$

C100

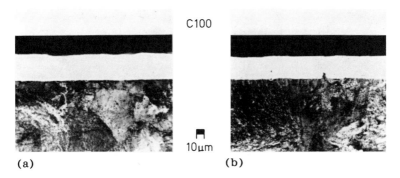

10μm

(a)                                    (b)

FIGURE 39  Chromium carbide layers on an unalloyed steel with
1% C: (a) pack chromizing for 5 h at 1000°C in a powder mixture
consisting of 78% ferrochromium, 20% $Al_2O_3$, and 2% $NH_4Cl$; (b)
liquid chromized for 4 h at 1050°C in a borax-based melt with 30%
ferrochromium.

FIGURE 40  Chromium carbide layer on cast iron. (×500:1; pack
chromized for 5 h at 1000°C.)

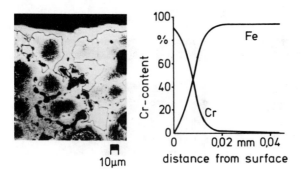

FIGURE 41 Chromium carbide layer and chromium distribution sintered steel with 5% Ni, 1% Si, and 1% C (pack chromized for 2.5 h at 1050°C).

Mazars (1981) has shown that the growth rate can be estimated by the following equation (valid for Cr contents > 12%):

$$x = 14,800 \sqrt{t}e^{(-9350/T)} \tag{7}$$

where

$x$ = layer thickness ($\mu$m)
$t$ = treating time, h
$T$ = treating temperature, K

Because of the great affinity of chromium for oxygen, it is advisable to rinse the powder with argon or hydrogen before and during the process. Otherwise there is the danger that the powder will be baked and the powder particles on the surface of the component will sinter (Chatterjee-Fischer, 1981).

Chromizing is mainly carried out to improve wear resistance and corrosion behavior.

## 8.2  Vanadizing

Vanadizing is defined as a thermochemical treatment involving the enrichment of the surface of an object with vanadium. This treatment is similar to chromizing and will be performed in powder mixtures (ferrovanadium as V donor, $NH_4Cl$ as activator, and $Al_2O_3$ as diluent) or in borax-based melts. The desired layer is a monophase vanadium carbide layer of the type VC (13.5 to 19.8% C). The element distribution obtained by pack vanadizing of a steel with 1.4% C is presented in Fig. 42.

The vanadium concentration in the layer depends on the vanadium content of the powder mixture. It can be decreased by diluting the powder mixture with $Al_2O_3$, for example.

Młynarczak and Kazimierz (1980) have postulated that the reaction can take place as follows:

$$Fe_3C + 3\ VCl_2 \rightarrow VC + 3\ FeCl_2 + 2(V) \tag{8}$$

Investigations on liquid vanadizing have shown that the reaction should take place by forming oxides of the added element (e.g., $V_2O_3$; Child et al., 1984). Therefore it is necessary to revolutionize the melt, first to avoid sinking of the metallic powder to the bottom of the furnace and, second, to provide the melt with sufficient oxygen.

The thickness of the layer depends on the treating time and the temperature. For liquid vanadizing an example is shown in Fig. 43.

Vanadizing is mainly carried out to improve wear behavior. The hardness of the carbide layer on an unalloyed steel with 1% C was determined to be between 3000 and 3300 HV 0.05. Though the hardness is very high, vanadium carbide layers show good ductility.

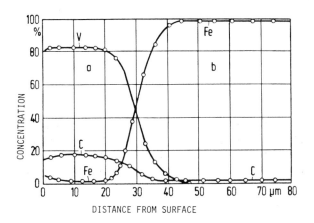

FIGURE 42  Element distribution in the vanadium carbide layer of a specimen made of a steel with 1.4% C, 0.63% Mn, and 0.41% Cr, obtained by pack vanadizing for 7 h at 1000°C in a powder mixture consisting of 98% ferrochromium and 2% $NH_4Cl$ (from Młynarczak et al., 1980).

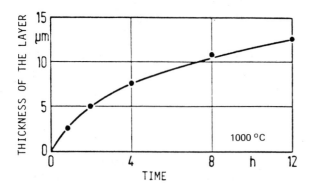

FIGURE 43  Influence of the treating time on the thickness of the vanadium carbide layer, obtained on an unalloyed steel with 0.45% C by the TD process (from Child et al., 1984).

### 8.3  Aluminizing

Aluminizing is a thermochemical treatment involving the enrichment of the surface of an object with aluminum, preferably applied to low-carbon steels. It is carried out in aluminum melts or in aluminum-containing powder mixtures. The treating temperature ranges between 700 and 1100°C. The influence of the treating temperature and the treating time for pack aluminized specimens is shown in Fig. 44.

   Eggeler et al. (1985) propose the following reaction sequence for this process:

$$6 \ HCl + 2 \ Al \rightarrow 2 \ AlCl_3 + 3 \ H_2$$

$$AlCl_3 \rightarrow AlCl + Cl_2$$

$$3 \ AlCl \rightarrow AlCl_3 + 2(Al) \tag{9}$$

The layer consists of a solid solution of $\alpha$-iron and some Al-Fe compounds. The aluminum content at the surface is as high as 14.3%. Aluminizing is used when a need exists to increase the oxidation resistance (up to 800°C) and the corrosion behavior against sulfurous media (e.g., $H_2S$).

### 8.4  Siliconizing

Siliconizing is a thermochemical treatment involving the enrichment of the surface of an object with silicon, preferably applied to low- and medium-carbon steels. The desired layer consists of solid solution of $\alpha$-iron containing no more than 15% Si. The thickness may be up

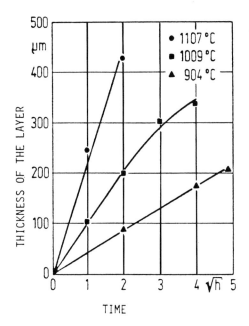

FIGURE 44  Influence of treating time and temperature on the layer thickness of a sample made of steel 13 CrMo 4.4 pack aluminized in 14% Al powder + 84% $Al_2O_3$ + 2% $NH_4Cl$ (from Eggeler et al., 1985).

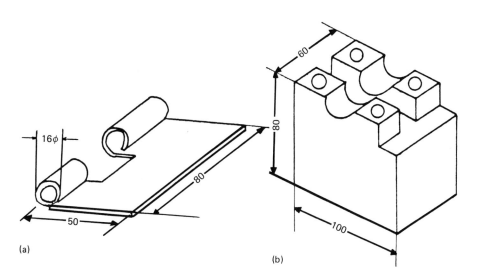

FIGURE 45  Bending die, vanadized by the TD process: (a) sketch of product (mild steel); (b) sketch of the punch (D2 tool steel—1.5% C, 12% Cr, 1% Mo, 0.75% V) (courtesy Toyo Carlorizing Company, Japan).

to 1 mm. When an activator (e.g., $NH_4Cl$) or a high diluent portion is used in the mixture the layer may be porous.

The treating temperature ranges between 900 and 1100°C. Siliconizing is carried out to improve corrosion resistance against several acids.


## 8.5  Application

The pack method is generally restricted to chromizing (e.g., small components for bicycles). With the liquid method vanadizing is usually performed. But the diffusion of niobium and chromium has also been investigated (e.g., Child et al., 1984). It is reported that many tools and some engineering components are treated on an industrial scale (e.g., roller chains, forming rolls, scissors, screws for injection molding (Arai, 1986), dies for drawing (cold-working die steel with 9% Cr), burring punches, piercing punches, blanking dies, etc. Figure 45 shows a bending die. Its lifetime could be multiplied by vanadizing (compared to chromium plating).


## REFERENCES

Arai, T. (1979). *J. Heat Treating, 1*:15.
Arai, T. (1986). *TZ Met. Bearb., 80* (3):27.
Arai, T. and Komatsu, N. (1978). *Metal Stamping, 12* (4):6.
Casadesus, P. and Gantois, M. (1978). *Härterei-Tech. Mitt., 33*:202.
Chatterjee-Fischer, R. (1981). "Thermochemical treatment in powdered media," Proceedings of the Eighth International Conference on Chemical Vapour Deposition, Paris, pp. 508-515.
Chatterjee-Fischer, R. (1986). *Metal Progress, 129*:24.
Chatterjee-Fischer, R. and Schaaber, O. (1976). "Boriding of steel and non-ferrous metals," Proceedings of Heat Treatment '76, Stratford-upon-Avon, Metals Society, London, pp. 27-30.
Child, H. C., Plumb, S. A., and McDermott, J. J. (1984). "Carbide layer formation on steels in fused borax baths," Proceedings of Heat Treatment '84, London, Metals Society, London, part 5.1.
Chou, G. P., Davis, A. L., and Narasinha, M. C. (1977). *Scripta Metallurg., 11*:417.
Dearnley, P. A., Farrell, T., and Bell, T. (1986). *J. Mater. Energy Syst., 10*:126.
Eggeler, G., Theuerkauf, T., and Auer, W. (1985). *Z. Werkstofftech., 16*:359.
Fiedler, H. C. and Hayes, H. J. (1970). *Metallurg. Trans., 1*:1071.
Goeuriot, P., Thevenot, F., Driver, J. H., and Laurent, A. (1981). *Traitement Thermique, 152*:21.

Habig, K.-H. and Chatterjee-Fischer, R. (1981). *Tribology Int.,*: 209.

Hegewaldt, F., Singheiser, L., and Türk, M. (1984). *Härterei-Tech. Mitt.,* 39:7.

Krishtal, M. A. and Grinberg, E. M. *Metals Science and Heat Treatment,* 16:283.

Kunst, H. (1972). *Kunststoffe,* 62:726.

Kunst, H. and Schaaber, O. (1967). *Härterei-Tech. Mitt.,* 22:1.

Lu, Ming-Jiong (1983). *Härterei-Tech. Mitt.,* 38:156.

Lyachova, L. S. and Kuli, A. J. (1972). *Zavodskaja Laboratorija,* 38:981.

Mareels, S. and Wetterich, E. (1985). *Härterei-Tech. Mitt.,* 40:73.

Massalski, Th. B., ed. (1986). *Binary Alloy Phase Diagrams.* ASM, Metals Park, Ohio.

Matuschka, A. von (1977). *Borieren,* Carl Hanser Verlag, Munich.

Matuschka, A. von and Trausner, N. (1986). Method for boriding metal and metal alloys with solid boration agent Ausz. Eur. Patentanmeld. 26.März 1986, Patent No., EPO 175157 (EUR).

Mazars, P. (1981). "The chromizing of mild steel," Proceedings of the Eighth International Conference on Chemical Vapour Deposition, Paris, pp. 516-524.

Młynarczak, A. and Kazimierz, J. (1980). *Neue Hütte,* 25:259.

Ott, A. (1983). *Härterei-Tech. Mitt.,* 38:268.

Zemskov, G. V. and Kaidash, N. G. (1965). *Metal Science and Heat Treatment,* 1:256.

# Index